The Red-winged Blackbird

The Red-winged Blackbird

The Biology of a Strongly Polygynous Songbird

by

Les Beletsky

Illustrated by David Beadle

ACADEMIC PRESS
Harcourt Brace & Company, Publishers
LONDON SAN DIEGO NEW YORK
BOSTON SYDNEY TOKYO TORONTO

ACADEMIC PRESS LIMITED
24–28 Oval Road
LONDON NW1 7DX

U.S. Edition Published by
ACADEMIC PRESS INC.
San Diego, CA 92101

This book is printed on acid free paper

A catalogue record for this book is available from the British Library

ISBN 0-12-0847450

Typeset by Phoenix Photosetting, Chatham, Kent
Printed in Great Britain by the University Press, Cambridge

CONTENTS

PREFACE

The Red-winged Blackbird is one of the most abundant and widespread of North American songbirds, and also one of the most closely studied. Research attention during the past three decades often centered on redwings primarily because of their strongly polygynous mating habits – several females nest simultaneously on a male's breeding territory, which is relatively rare behavior in birds. A coterie of biologists with interests in the ecology and behavior of which birds, which I am a member, spent many years characterizing the breeding biology of the species and testing ideas of how the breeding system works and why it functions as it does. So much information has been collected – so many experiments conducted, so much analysis performed – that at this point we can make the argument that several aspects of avian biology are better understood for redwings than for any other species. Yet, to date, there has been no attempt to synthesize the knowledge we have of redwings. I was pleasantly surprised, therefore, when approached by Academic Press to produce a book with that objective.

The best part of professional participation in science is the continual opportunity to learn. I looked forward to writing this book, my first, because I suspected that it would be a seminal learning experience and, indeed, during its preparation, I significantly increased my store of knowledge. For instance, I learned that writing a book chapter on an important aspect of avian biology takes a good, long while. I started the book during a long holiday weekend early one July with a chapter on redwing communication, with the aim of completing the chapter in only 4 days. That weekend I happily witnessed a parade and a pyrotechnics display but I certainly did not finish the chapter. It turns out that the general formula for making reasonably correct estimates of how long each step takes in writing a book is to make an educated guess and then multiply by about seven.

I also learned of a physical constraint on writing that I had not even considered before taking on the job – the limits of the body. A few months into writing I realized that I might actually suffer physical harm during the book's preparation. With only about half the chapters written, my wrists were feeling twinges of repetitive strain injury, the bane of secretaries and other keyboard denizens, and my eyes were starting to hurt. At that point I thought 'To complete the book, how can I possibly read all the material I have to read, write all the drafts I have to write, and type all the pages I have to type?' Well, the book is done and I survived, reasonably intact. But let me report that it took a good number of hours. On balance, however, I found producing a book a very positive and satisfying enterprise.

Writing a book such as this one requires many decisions – at what level to write, how much data from published sources to include, how much to write on each subject, etc. As I labored on the chapter on communication, for example, I knew I could simply keep scribbling and eventually complete an entire volume on the one subject. I stopped when I realized that the market for a book on redwing communication might be vanishingly small : in fact, I was personally acquainted with the six or seven people who might consider a purchase. As far as the level of writing, I attempted to make the information in the book accessible to all – researcher, student and layperson interested in birds. Many authors of scientific books describe a fundamental decision, or trade-off, in their writing, between subject inclusivity and insight. I came rapidly to understand the dilemma. In general, I opted for greater inclusivity because I wanted the book to be an informative, up-to-date reference for anyone interested in the biology of redwings. For some types of information, I tried to be exhaustive, including in tables, for example, information from all relevant studies.

Another decision concerns when to stop writing and submit the manuscript. Redwing researchers are an active lot. A writer trying to finish a book on the species frustratingly finds a continual onslaught of newly published papers that deal with redwings; it seems sometimes as if every issue of every major ecology and behavior journal includes one. For example, at the time of this writing, several major studies of extra-pair mating in redwings are just coming to fruition. On one hand, one wants very much for the book, at its publication, to be the latest word on its subject matter, but on the other hand, at some point one must cease adding and editing and send the thing off as finished. My hope is that this book, in addition to its other uses, can be consulted for many years to come as the basic reference on redwings.

A few administrative notes are in order. Although it may not always seem the most logical presentation , I arranged chronologically, for consistency's sake, several tables that provide information from many different sources. In various parts of the book I use as area measures both hectares and acres; 1 hectare (10,000 m^2) = 2.47 acres. The terms 'mean' and 'average' are used interchangeably. Unless otherwise noted, means are presented plus or minus one standard deviation.

For taking the time to read and comment on sections of the more meandering prose that characterized earlier drafts of this book, I want to thank especially Gordon Orians, who read the entire manuscript, and Ken Morgan, Martha Dunham, Ken Yasukawa, Eugene Morton, Bill Searcy, Scott Edwards, and Andrew Richford. Richard Dolbeer, Frances James, Scott Lanyon, Bill Searcy, Pat Weatherhead, and Ken Yasukawa responded promptly and generously to my requests for information and I thank them for that; the information they provided made the book more interesting. Many people have helped me during my 20 years studying redwings, and it gives me great pleasure to be able to thank them publicly. Douglas Smith, William Thompson, John Wingfield and Gordon Orians served admirably as mentors, advisers, and colleagues. For their help in the lab and with data analysis I thank Stella Chao, Lynn Erckmann, B J Higgins, Sharon Birks, and Carla Calogero, and for their sophisticated knowledge of and assistance with statistical analysis and computer programming I thank George Gilchrist, Mark Wells and especially John Bishop. Finally, a heartfelt thank you to all the assistants, 30 people or more, who over the years helped me collect field data on redwings that formed the basis of a career. Almost without exception they were very bright, highly-motivated, ridiculously underpaid individuals who worked hard, made excellent suggestions for project improvement and were good company during long field days. Several have gone on to graduate study and scientific careers of their own and a few have become my good friends. I also thank the staff of the Columbia National Wildlife Refuge for granting permission for so many years to Gordon Orians and me to study blackbirds on lands they manage, and for their unstinting assistance during the mini-crises that regularly seem to befall all field biology projects.

Finally, colleagues suggested that a book such as this requires a dedication. I pondered to whom to dedicate the book for all of a minute or two. It was the easiest book-related decision that I made; the choice was clear. This book is dedicated to Red-winged Blackbirds and to the people who study them.

Les Beletsky
Summer 1995

Chapter 1

INTRODUCTION

RED-WINGED BLACKBIRDS – WHY SO MUCH INTEREST?

Come upon a marsh during late spring in almost any locale in North America and you will usually find in residence a group of black-and-red birds mixed with smaller, brown ones. The striking black birds with red feathers at their shoulders frenetically patrolling the boundaries of their territories, and making most of the racket, are male Red-winged Blackbirds; the brown and streaked, more subdued individuals are females. Red-winged Blackbirds, to most people, are small, basic, almost unelaborated birds. Except for the male's bright red wing patches, the species possesses none of the gaudy 'doodads' – wattles, combs, crests, spurs, spots, or bare areas – common to many other species, and they are of a quintessential bird shape. But, although basic and common, Red-winged Blackbirds hold a special place in the ornithological world and also in mine: I am one of a healthy number of scientists who spend major portions of their professional lifetimes studying these remarkable birds. Why we do so and what we have discovered are the subjects of this book.

The Red-winged Blackbird, *Agelaius phoeniceus*, is at once one of the most abundant birds in North America and one of the most studied. Reasons for the latter include the former, but a number of other important factors conspire to stimulate the considerable amounts of research on this animal. The foremost aspect of the species' biology that attracts frequent interest by researchers is its mating system – strongly expressed polygyny. Unlike most birds, which breed monogamously, male redwings attract each year up to 15 or more mates to their exclusive territories. The number of species of birds within which most individuals regularly breed polygynously, and wherein the female makes a greater reproductive investment to her nest than does the male, is comparatively small; most of them, in contrast to the redwing, show only weak polygyny, males having harems of only two or three females. The evolution of nonmonogamous mating systems – how they originate and how they are maintained by natural selection – has been a major interest of behavioral ecologists during the past 25 years. For North American researchers, the redwing has become the species of choice for investigations of avian polygyny.

The ubiquity of the animal – its great abundance, broad geographic range, and close proximity to most North American research institutions – explains part of the redwing's scientific popularity and its exaggerated polygyny explains another. Another major attraction is its relative ease of study in natural situations. Take, for instance, the redwing's preferred habitat. Redwings generally establish their breeding territories in marshes, which offer a number of advantages for investigators, especially when compared to, for example, conducting research on birds in forests. Nests are relatively easy to find in the emergent vegetation of a marsh, greatly facilitating studies of breeding success. (However, penetrating marshes to find nests also poses special risks to researchers. Entering marshes, where water and mud are too deep for boots and the vegetation too dense for canoes, means donning 'chest waders' – rubber boots whose uppers continue to mid-chest level. More than once I have in waders stepped into deep, sucking mud and, because the waders are unwieldy, have come very close to being unable to extricate myself. Struggling, in fact, only got me in deeper. When one is searching for nests alone at a remote marsh, becoming, literally, 'a stick in the mud' is a real possibility.)

The low, uniform stature of marsh plants, which often dominate redwing territories, offers superior bird viewing opportunities. This, combined with the fact that territories are usually fairly small, means that a researcher can often see several redwings out in the open performing their full territorial and reproductive behavioral repertoires from one location. At the Columbia National Wildlife Refuge (CNWR) in

Washington State, where I conduct my own research, small rock cliffs 7 or 8 meters high are adjacent to many of the redwing's breeding marshes; we can sit on these clifftops, look down into the marshes and view essentially entire breeding efforts unfold below us. (The almost unobstructed views of redwing behavior made possible by their marsh dwelling should not be underestimated as a partial cause for the redwing's popularity as a research subject. Ornithologists usually cannot study what they cannot see. I recall that initially I chose as the subject of my graduate research a field investigation of the behavior of Northern Orioles, a decision made against the advice of my more ornithologically seasoned advisor. I did succeed in *studying* orioles for one entire breeding season, but I actually *saw* the animals during that Michigan spring for a total of about 30 minutes. They mostly appeared to me as small, orange streaks flitting between the maples and oaks in the deciduous woodland canopies that they preferred.) Redwings are also relatively easy to catch and to manipulate experimentally, and are hardy birds, faring well in captivity.

Another reason that the redwing continues to draw disproportionate interest from behaviorists is that, within the species, there exists extensive regional variation in behavior. This variability can be exploited to test many kinds of ecological–behavioral hypotheses, several of which will be considered in this book. For example, the factors influencing why harem sizes vary broadly among various North American geographic areas may be investigated and explained ecologically by comparing the typical habitats in which the birds breed in each region (see Chapter 6).

Finally, redwings have been considered major agricultural pests, destroyers of crops such as rice and corn. During the nonbreeding portion of the year, redwings gather with other blackbirds into immense flocks, often millions of birds, that damage roosting areas and attack field crops and livestock feed. Financial losses to farmers have stimulated significant research during the past 35 years into redwing biology as it affects agriculture.

As a result of these extensive research efforts, perhaps more ecological and behavioral studies have been concentrated on this bird than on any other North American species. In a sense, it has become the 'white rat' of some research areas, attaining within avian field studies an almost unique position. In Europe, the 'functional equivalents' of the redwing are the Great Tit and Pied Flycatcher, species that, like the redwing, make frequent and fundamental contributions to the fields of animal behavior and behavioral ecology.

THE PURPOSES AND PLAN OF THE BOOK

A compendium of information on the general biology of redwings, which this book strives to be, is especially timely because, as a consequence of many years of research attention, there are now several aspects of avian biology that are arguably better understood within this species than within any other. These areas of knowledge include aspects of polygynous mating and mate choice, of sexual dimorphism in body size and plumage pattern, of vocal and visual signalling, and of aggression and territoriality. The main purpose of this book, then, is to bring together in one place information on many facets of the biology of Red-winged Blackbirds, and to summarize and review the information, so that it is readily available to all; to those interested in birds generally or in redwings specifically, as well as to active researchers. Although there are other books that feature redwings, there is no other that covers so broad a range of topics, nor one that synthesizes as much information; thus, this book supplements more narrowly-focused publications.

In this book, the reader can expect to find an emphasis on breeding ecology and behavior because these are my own areas of expertise. Specifically, I will stress the redwing's mating system, its marsh breeding habits and how the redwing's penchant for marshes has influenced other aspects of its biology; also stressed will be communication, and territoriality and territory acquisition. I also emphasize the contributions of redwing studies to broader scientific issues.

First, in Chapters 2 and 3, I cover some general redwing biology, addressing such topics as taxonomy, geographic distribution, and morphology. Next, in Chapter 4, I discuss nonbreeding behavior, such as food preferences and foraging behavior, and behavior while in groups. Chapter 5 goes into some detail describing redwing communication. The main body of the book, Chapters 6 through 10, covers breeding biology, broadly-defined, and reproductive success. Chapter 11 delves into the positive and negative relationships between people and redwings, focusing on the redwings' impact on agriculture. In the final chapter, I discuss geographic variation in redwing breeding biology, summarize the contributions of redwings to scientific inquiry, and discuss future directions for research.

THE REDWING PERSONALITY

Many aspects of redwing biology will be described in great detail in subsequent chapters. But much of what follows will be clearer if the reader has an intuitive feel for the redwing's 'personality.' Species often have peculiar suites of behavioral traits (personalities, if you will) that set them apart from even their close relatives, and that come to characterize them for people: crows are raucous and curious; coyotes and foxes, smart and cunning, etc. Knowing of the redwing's spirited and pugnacious personality is important because these traits doubtless contribute to the bird's great ecological success and also facilitate research on the animal.

First, redwings are tenacious birds and often benefit from human manipulations of landscapes. Where populations of other songbird species decline, redwing populations often grow. Redwings are not 'skittish' around humans. They often breed in areas in which they are exposed daily to people's activities and disturbances. They do not terminate their nesting merely because people are near or because they are under close scientific scrutiny, and tolerate field workers periodically peering into their nests and handling their eggs or chicks. When redwings are killed as crop pests in great numbers, sometimes by the hundreds of thousands, populations quickly rebound. Individuals show very high fidelity to their breeding sites from year to year, and even when their territories are temporarily destroyed by fire or floods they typically remain and wait for the vegetation to recover (see, e.g., Chapter 11). The redwing's tenacity is demonstrated in their rivalry with Yellow-headed Blackbirds. Male yellowheads, also polygynous marsh-breeders, claim some of the same marsh areas for their territories as do male redwings. Because the yellowheads are much larger, they usually get the sites they want. But often, if redwing nests are already underway when yellowheads arrive on the scene, the male redwings will repeatedly battle the larger birds in efforts to prevent eviction and loss of the nests. Female redwings, moreover, sometimes will not abandon their nests under threat by male yellowheads, but stay on their eggs or guard their nestlings, often sustaining injury as the yellowheads, more than twice their weight, repeatedly harass and assault them.

One particularly tenacious redwing comes to mind. Trapping one day near a breeding marsh, I caught an unbanded male who could not fly. I put a single blue band on his leg and released him into the tall grass adjacent to an agricultural field about half a kilometer from a road, where I thought he might find food and cover from predators. Unless his flight capability returned quickly, his prospects were poor for survival for any length of time. The next day I sat in a pick-up truck along the road, waiting for birds to enter traps. My eyes caught movement in the rear-view mirror. I watched in amazement as the little fellow with the single blue band *walked* under the fence that bounded the large alfalfa (lucerne) field he had had to cross, down the steep, dirt embankment, across the road, under the fence to the adjacent cattle pasture, and then down to the stream that runs through it. His thirst sated, the redwing promptly reversed direction, walked across the road, up the embankment, and disappeared back into the alfalfa field. Flightless he may have been, but he surely was going to continue with food and drink.

Redwings are active, aggressive birds. A male redwing on a crowded breeding marsh in early spring is almost a blur of motion – constantly patrolling his territory borders, advertising his territory with vocalizations and up-raised wing displays, forcefully evicting male territorial intruders, fighting with neighbors, and chasing females. I often refer to the male redwing lifestyle as 'hyperactive,' which becomes especially evident when their behavioral style is compared with that of even such a closely-related bird as the Yellow-headed Blackbird. Yellowhead males often sit quietly on their territories for long stretches of time, fly silently from perch to perch, and occasionally rattle off their strange and distinctive vocalizations. Redwing males, in contrast, seldom sit still and are rarely quiet for more than a minute or two (see Chapter 5).

Redwings are smart. In our work we catch most of them with funnel traps, which are made of wire netting and into which birds drop through wire 'funnels' to get at seed bait. The funnel area occupies only a small portion of the main body of the trap, so the birds cannot locate the funnel again to exit. Several males had territories near to one of our main trapping locations, and I knew from watching that the banded owners of those territories regularly went to the large trap to gorge on sunflower seeds. However, to my surprise one year, as I inspected our trapping records, I found that we rarely captured these males. Suspicious, the next time I set the trap I watched it from afar with a telescope. Sure enough, some of the males who frequently visited the trap not only knew how to enter, but had also learned how to exit through the funnel.

But redwings do not always need repeated exposure to a stimulus to learn. One of my favorite redwing stories was told to me by Dave Gori, who worked for many years studying the breeding behavior of redwings and other blackbirds. For a crucial experiment, Dave needed to catch rapidly all the male redwings that owned territories around a small lake. After consulting with experts he decided that the way that would give him the best chance of quickly catching the males was a custom-made shotgun that fired a weighted net, bringing down any bird at which it was aimed. The gun cost $1 500, a price so steep that Dave had to convince several other scientists to contribute funds toward its purchase, with the understanding that they, too, might use it. The first day, Dave went to the lake, aimed the shotgun at one of the flying male redwings, and fired. The bird turned in mid-air, easily evading the hurtling net. Other redwings left the lake after hearing the shotgun blast. Dave returned to the lake the next day to try again. He unzipped his duffle bag and removed the shotgun. Immediately, all the redwings on the lake left the area. The shotgun still sits, unused, in Dave's closet.

A HISTORY OF RESEARCH ON RED-WINGED BLACKBIRDS

The redwing has a long history of being singled out among songbirds for detailed studies of breeding behavior. Its curious habits have long attracted special notice. Ninety years ago M. O. Wright (1907) wrote:

> 'When redwings live in colonies it is often difficult to estimate the exact relationship between the members, though it is apparent that the sober brown, striped females outnumber the males; but in places where the birds are uncom-

mon and only one or two male birds can be found, it is easily seen that the household of the male consists of from three to five nests each presided over by a watchful female, and when danger arises this feathered Mormon shows equal anxiety for each nest, and circles screaming about the general location.'

The first detailed scientific report is that of A. A. Allen, who published in 1914 a long description of the bird and its breeding behavior in New York State in which he speculated about the relationship between the species' habitat and behavior. G. H. Linford (1935), two decades later, completed a detailed Master's thesis on the breeding biology of redwings in Utah. Contemporary interest in the redwing as a subject worthy of behavioral study is traceable to John T. Emlen who, as a zoologist at the University of Wisconsin during the 1940s and 1950s, suggested redwings as study subjects to his students.

Robert W. Nero studied redwings under Emlen's tutelage from 1948 to 1953. Nero monitored the breeding behavior of redwings on a 2-acre cattail marsh adjacent to Lake Wingra, near the University of Wisconsin's Madison campus (Fig. 1.1). With his wife's help, he caught and color-banded individuals and observed the birds, beginning, he says, almost entirely by simply closely watching, during his first year of field work, the behavior of a single territorial male (Nero 1984). After

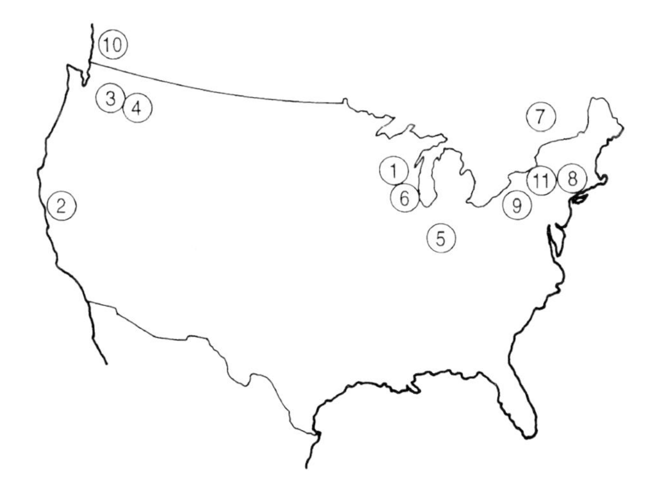

Fig. 1.1. Locations of major field studies of redwing ecology and behavior. (1) Lake Wingra, Madison, Wisconsin; (2) North-central California; (3) Columbia National Wildlife Refuge, Washington; (4) Turnbull National Wildlife Refuge, Washington; (5) Yellowwood Lake, Indiana; (6) Newark Road Prairie, Wisconsin; (7) Queen's University Biological Station, Ontario; (8) Millbrook, New York; (9) Pymatuning Laboratory, Pennsylvania; (10) Delta Marsh, British Columbia; (11) Ithaca, New York.

recording behavior for six breeding seasons, Nero produced a Ph.D. thesis, the bulk of which he published as two descriptive papers, one on mating and nesting activities and one on territoriality (Nero 1956a, b); later, he compiled his observations in book format (Nero 1984). Many of Nero's detailed observations of redwing behavior are still of great use today to ornithologists using contemporary techniques to answer questions of evolutionary biology; the reader will discover references to Nero's work scattered throughout this volume.

The next phase of research on redwing breeding ecology and behavior was stimulated by Gordon H. Orians, one of a small group of biologists usually credited with developing, during the 1960s, the field of behavioral ecology. Orians' work, and that of his students, pushed the redwing to its current position of research prominence; it is probably fair to say that it is their work that is responsible for the wide use of the redwing as an ecological and behavioral study subject.

Gordon Orians, whose undergraduate advisor at the University of Wisconsin was John Emlen, and who was familiar there with Robert Nero and his Lake Wingra research, chose to study redwings for his doctoral work at the University of California at Berkeley. He made a comparative behavioral and ecological study of Red-winged and Tricolored Blackbirds breeding in north-central California. When, in 1960, he moved to Seattle to take a faculty position at the University of Washington, his work with blackbirds continued, as it does to this day. Most of his blackbird research was concentrated at two National Wildlife Refuges, one in extreme eastern Washington (Turnbull NWR), the other in the south-central part of the state (Columbia NWR).

Orians' doctoral work, together with later research carried out in concert with his own graduate students, coalesced into a series of landmark publications on the comparative breeding biology of several blackbird species and the ecological underpinnings of differences in their behavior. Many behaviorists credit Orians' 1961 paper, *The Ecology of Blackbird (Agelaius) Social Systems*, for showing how mating systems might be analyzed from ecological and evolutionary perspectives. Comparative behavior of three species was examined in a mainly descriptive work, published in 1968, *A Comparative Study of the Behavior of Red-winged, Tricolored, and Yellow-headed Blackbirds,* which Orians produced with the assistance of a Berkeley artist, Gene Christman. A major contribution to comparative behavioral ecology was published by Orians in 1980, *Some Adaptations of Marsh-nesting Blackbirds.* This work, based on the author's research during the 1960s and 1970s in Washington and British Columbia, and also on a year's work on Argentinian blackbirds, emphasized using blackbirds to test evolutionary hypotheses of habitat selection, foraging behavior, territoriality, and mate selection. Orians also produced a 1985 book on comparative blackbird biology for a more general audience, *Blackbirds of the Americas*, beautifully illustrated with drawings by Tony Angell. These works, together with a seminal paper on the evolution of nonmonogamous mating systems in birds and mammals (Orians 1969; see Chapter 6), which was based on ideas developed during research on redwings and other songbirds, are perhaps Orians' most important contributions to research on blackbirds.

Several of Orians' postdoctoral and graduate students, among them Mary Willson, Henry Horn, Celia Haig Holm, James Wittenberger, Cynthia Patterson, and Elizabeth Gray, conducted field studies of redwings in addition to other blackbirds, and their work is cited repeatedly throughout this book. William Searcy, in particular, now of the University of Miami, has emerged as one of the foremost avian behavioral ecologists, and uses redwings extensively in his work. Searcy, an Orians Ph.D. student, conducted his graduate research from 1974 to 1976 at Turnbull NWR – a beautiful Ponderosa Pine parkland with scattered lakes, along the margins of which the birds breed in cattails. Searcy then moved to Rockefeller University's Field Research Center in Millbrook, New York, where he met Ken Yasukawa, another postdoctoral student working with redwings, and with whom he started a long-term, highly productive scientific collaboration. In the early 1980s, Searcy moved to the University of Pittsburgh, continuing his redwing studies in marshes and upland areas of that institution's Pymatuning Laboratory in north-western Pennsylvania. Searcy, alone and together with Yasukawa, has published extensively on redwing communication and breeding biology, with special emphasis on the effects of sexual selection on the species' biology. The two researchers recently published a monograph on redwings that concentrates on sexual selection and polygynous breeding (Searcy and Yasukawa 1995).

Ken Yasukawa, along with a number of other students whose graduate research centered on redwings, emerged from Indiana University in Bloomington. There, Val Nolan Jr started banding redwings in 1969 to give students in his ornithology class experience reading bird bands and observing the behavior of individual birds. Nolan and his students, among them Cynthia Patterson, Torgeir Johnsen, and Yasukawa, studied redwings that bred in marshes on Yellowwood Lake, a 54-hectare lake in Yellowwood State Forest, Indiana. The lake is surrounded by mixed deciduous forest and also by some picnic areas and campgrounds. Redwings were studied mostly at the north end of the lake, in emergent vegetation (cattail, bureed and bulrush) that usually supported about 25 male territories. After finishing his doctoral work, Yasukawa was a postdoctoral researcher at Rockefeller University before accepting a position on the faculty of Beloit College in Wisconsin. Since 1983, he and his students have produced a steady stream of publications on redwing communication and breeding behavior. His main study area in Wisconsin is known as Newark Road Prairie, 13 hectares of gently sloping prairie, sedge meadow, and woodland.

Two more researchers who, together with their students, have contributed many studies of redwing ecology and behavior during the past 20 years are Raleigh Robertson and Patrick Weatherhead. Robertson conducted his doctoral work at

Yale University on redwing breeding ecology during the late 1960s and continues his research with redwings at Queen's University in Ontario. Weatherhead, Robertson's former student, studies redwing breeding behavior, ecology, and territoriality. He is presently on the faculty of Carleton University in Ottawa, where he and his students – among them Christopher Eckert, Katherine Muma, Karen Metz, and David Shutler – continue as highly productive redwing researchers. Weatherhead's work with redwings is carried out mainly in and around the Queen's University Biological Station, near Kingston, south-eastern Ontario; he has studied territories in a woodland area, in which there are lakeshore cattail marshes bordered by deciduous forest, and also small cattail marshes that occur in ditches alongside highways. Weatherhead, in addition to his interests in redwing breeding biology, has published extensively on nonbreeding behavior and ecology – how redwings behave when in flocks during nonbreeding months. Sievert Rohwer, a biologist at the University of Washington, together with his students Paul Ewald, Scott Freeman, Nancy Langston, and Dave Gori, has contributed many highly original field studies of redwing biology. Jaroslav Picman, now of the University of Ottawa, has also over many years contributed studies of redwing breeding behavior, chiefly using a population in south-western British Columbia. Finally, David Westneat, of the University of Kentucky, making use of the latest molecular techniques, has for several years studied the breeding biology and genetic relatedness of redwings in Ithaca, New York. Many other biologists have contributed to the wealth of knowledge about redwings, and their work is cited liberally through this book.

During the last 30 years, another body of information on redwings has accumulated on applied problems, primarily on the conflict between redwings and the farmers whose crops they damage during their nonbreeding months. These problems are most severe in the midwestern and southern USA, and it is in these regions where most of the research has taken place. Brooke Meanley, a US Fish and Wildlife Service biologist, studied the relationship between blackbirds and the southern rice crop from the late 1940s to the late 1960s. Richard Dolbeer, a US Department of Agriculture biologist, has since 1973 investigated the effects of redwings and other blackbirds on Midwest crops. He and his colleagues, in particular, have researched various blackbird control techniques, but they also study redwing breeding biology, dispersal, and demography as they relate to crop damage.

As for myself, I have worked with redwings for all of my professional life. I began about 20 years ago, as an undergraduate biology student in New York offering his untrained services as research assistant to his instructor in a course in ethology (the study of animal behavior from biological – evolutionary and comparative – perspectives). The teacher happened to be Douglas G. Smith, a field biologist interested in animal communication who used redwings as study subjects. Smith's chief claim to redwing fame was that he had formulated during his graduate research the simple yet clever experiment of testing for the function of the male redwing's bright red shoulder epaulets by dyeing them black, which caused the birds all sorts of problems (see Chapter 5). My years obtaining graduate degrees were filled with early mornings in New York, and later in Michigan, tramping around cattail marshes, trying to discover, by observation and experiment, the functions of redwing vocalizations. I then became a postdoctoral associate of Gordon Orians and together, in the desert of eastern Washington State, we conducted a long-term study of redwing breeding ecology and reproductive success. My interests during the study came to focus on several specific aspects of avian biology: how birds acquire and maintain breeding territories; how breeding decisions are made; and how the endocrine system affects breeding behavior and territoriality. The study occupied most of my time for about 10 years. The main results of that study are reported in a book coauthored with Gordon Orians, *Red-winged Blackbirds: Decision-making and Reproductive Success* (Beletsky and Orians 1996). Some information gleaned during that long-term study is either discussed or reported for the first time in the current volume and, therefore, below, I provide for the reader information on the study site, the redwing population inhabiting it and our research methods.

THE COLUMBIA NATIONAL WILDLIFE REFUGE (CNWR) STUDY

The Columbia National Wildlife Refuge (hereafter, CNWR) is 23 000 acres of basins, buttes, sagebrush, and rolling desert grassland in the south-central part of Washington State (Fig. 1.1). Its fairly unique terrain, 'channelled scabland,' as it is called, is the result of Pleistocene floods that eroded ancient lava flows, scratching out of the basalt steep depressions and twisting channels. Although in a desert area, CNWR is heavily spotted with lakes and marshes on which blackbirds breed (Fig. 1.2). Most of these wetlands are due to Columbia River irrigation canals that criss-cross the region; seepage from the canals, as well as from crop irrigation in surrounding areas, has led to higher water tables and hence to lakes and ponds filling many low-lying areas. Gordon Orians began field studies of blackbird breeding at CNWR during the early 1960s. In 1977, he initiated there a study of lifetime reproductive success of redwings; I joined him in 1983. We planned to mark individual breeders and monitor their breeding success until we knew how many offspring males and females produced over typical lifespans. Along the way, we would test various hypotheses concerning factors that contribute to variation among individuals in breeding success. Such projects take many years to complete because individuals need to be followed during their entire breeding lives and also a large enough sample of individual lifetimes needs to be collected for statistical analyses. Monitoring breeding success continued for 16 years, through 1992.

Our information on territory ownership and breeding success was for the most part collected on a core study area that included eight separate marshes located on five lakes. The lakes, in a single drainage system, were close to each other, with the northern-most about 3 km from the southern-most. The marshes, dominated by

(a)

(b)

Fig. 1.2a and b. Photographs of the Columbia National Wildlife Refuge (CNWR), site of a long-term study of redwing ecology and behavior.

cattail or cattail/bulrush mix, supported 70 to 80 male redwing territories most years. For analysis, we grouped marshes into two types based on shape. There were four 'strips,' in which emergent vegetation bordered lakes in patches only a few meters wide; hence, all territories in strips were long and thin and all bordered both the shore and the open water of the lake. Three others were 'pocket' marshes, irregularly-shaped expanses of emergent vegetation in which territories took many shapes, did not necessarily border shore or open water, and had up to five adjacent neighboring territories. The eighth marsh was part strip, part pocket. Marsh shape was important because it influenced nest predation rates as well as many redwing behaviors.

We caught and colorbanded all territorial males in the core study area and many other males – some with territories elsewhere and others who did not own territories ('floaters') – by operating large traps some distance from the study marshes. Individually unique colorband combinations on their legs permitted us to identify individual birds throughout their lives. Most females were caught in smaller traps placed near their individual marshes. We recorded, weighed, and if necessary, banded all individuals that we trapped; some were caught only once, some once a year, and some repeatedly. Each year we found and monitored all redwing nests on the eight core marshes, and we tried to determine the female owner of each one (see Chapter 6). Starting about 1 April and continuing through June, marshes were searched for new nests every 6 days; nests were numbered and marked with small pieces of yellow and orange tape placed in nearby vegetation. Nests were checked for progress every 3 days. Nestlings were banded with single US Fish and Wildlife Service bands when they were 6–8 days old. We determined if nests failed or successfully fledged young.

We also devoted considerable amounts of time investigating territory acquisition and monitoring territory ownership of males. We wanted to determine how long individuals were floaters before they obtained territories, how and where they did so, and how their territories changed over time. We monitored territory ownership, shape, and size and how these changed within and between years by making territory maps every 2 weeks during breeding seasons (early March through late June). A map was produced by observing the space usage on a marsh by particular males, i.e. defining the areas that they defended from other males. These maps permitted us to assign putative fathers to the nests that fell within each male territory. Because we were interested in male movements, both of floaters and of territory owners who changed territory locations, we also searched for banded territory owners on all other redwing breeding habitat that surrounded our core marshes. We searched in all directions from the center of the core area, to a distance of 5 to 6 km; at the farthest distances, almost all territory owners were unbanded, indicating that few males moved such distances from our core-area banding locations to look for territories.

The CNWR study has made substantial contributions to understanding the biology of redwings and, in a larger sense, some broader ornithological issues. I include our CNWR results in many parts of this book, chiefly in the sections that deal with reproductive biology. First, however, we must begin with some basics, in Chapter 2, by exploring the taxonomic identity of redwings and where they are found.

Chapter 2

TAXONOMY, DISTRIBUTION, AND MOVEMENTS

INTRODUCTION

This book concentrates on the behavior of Red-winged Blackbirds. Prior to examining an animal's behavior, however, it is always prudent to know something of the species' 'background' – its basic biology. General biology is especially important because aspects of a species' evolutionary development and relationships, or its morphology, for example, may influence or constrain its behavior. In other words, understanding redwing behavior may be facilitated by knowing the position of redwings in the taxonomy of birds, how redwings are classified, who their close relatives are, where they are found and at what times of year, and why they take the forms – size and coloring – that they do. In this chapter and the next, I will discuss various facets of redwing biology and natural history that are of inherent interest but also may be of value in later chapters, for deciphering behavior.

TAXONOMY

The Red-winged Blackbird is one member of a extraordinary assemblage of songbirds, the New World Blackbirds. The group experienced a spectacular, relatively recent radiation, or evolutionary division, into a large array of species (94 at last count) that exploit diverse habitats and prey, and whose ranges essentially span the Americas. More so than many other groups, the New World Blackbirds are highly variable in size and coloration, and include such exquisitely marked birds as the orioles and oropendolas. They also exhibit almost every known type of avian mating system. Thus, in several ways, these blackbirds represent a 'microcosm of songbird evolution' (Freeman 1990), making them an ideal group for evolutionary biologists to use as subjects for comparative studies of ecology and behavior.

Blackbirds gape, which is a particular way of searching for food: they place their bills in crevices, under rocks or other objects, between plant parts, and into floating debris, forcibly open them, and thereby expose prey within or underneath; this prey is unavailable to most other birds. This behavior and the associated anatomical adaptations (elongation of the bill and the skull and associated muscle changes) are the definitive distinguishing characteristics of the group. The radiation of blackbirds into so many different habitats and the ability to exploit the variety of prey types that they do may have depended almost entirely on this feeding adaptation (Beecher 1951, Orians 1985).

The taxonomy of the New World Blackbirds has a somewhat checkered history. Holding until recently family status as the Icteridae, new classifications have 'dissolved' the family and placed its members in other groups. (However, during my

Other 'icterines' – close redwing relatives of North America: (clockwise from upper left) Northern Oriole, Bobolink, Yellow-headed Blackbird, Common Grackle; (center) Western Meadowlark.

scientific training blackbirds were 'icterids,' so I always think of redwings and their cousins as such.) Early taxonomic work (Ridgway 1902) characterized icterids as having nine primary feathers (the long, outer-most feathers of the wing, essential for flight) and variable but typically long, pointed, often conical, bills. The conical bills and the positions of the eyes allows icterids to sight directly to the point of the bill and to see prey in the small areas exposed by gaping (Lorenz 1949, Orians 1985). Friedmann (1929) suggested that the thick-billed Brown-headed Cowbird (*Molothrus ater*) is the most primitive type of icterid and that the group originated from an ancestral species that was finch-like. After conducting a comprehensive study of the musculature and skeleton of blackbird jaws, Beecher (1950, 1951) concluded that the cowbird's (*Molothrus*) morphology was indeed primitive and that three major icterid lines probably arose from an ancestral cowbird species: the Agelaiine line, containing the orioles, meadowlarks, Bobolink, and many of the blackbirds, including *Agelaius*; the Quiscaline line, containing the grackles and some of the blackbirds; and the Cassicine line, containing the caciques and oropendolas. Raikow (1978) studied the appendicular musculature of the New World songbirds, including that of several blackbirds. His work also pointed to three distinct blackbird groups, but they differed somewhat from Beecher's: Icterines (orioles, caciques, oropendolas), meadowlarks, and the remainder (Bobolink, Brown-headed Cowbird, grackles, blackbirds). Raikow's work did not suggest a cowbird-like ancestor for the blackbirds. The first molecular investigation of blackbird phylogeny, which used comparative protein electrophoresis of allozyme (genetically controlled variants of an enzyme) frequencies to examine relationships of seven North American blackbirds, generally supported Beecher's earlier, morphological classification, including evidence for a cowbird-like ancestor for the blackbirds (Smith and Zimmerman 1976).

Freeman (1990) constructed a blackbird phylogeny based on the genetic distances of the nucleotide sequences of mitochondrial DNA from 47 species. His resulting classification had, like the previous efforts, three main branches: an oriole group, the caciques and their allies, and a third group containing the grackles, blackbirds, and cowbirds. There was also an indication that some of the North American blackbirds – Red-winged and Yellow-headed Blackbirds, Bobolink – constituted a fourth line arising from an ancestor that may have evolved independently of the South American blackbirds. The mitochondrial molecular clock, as calculated by Freeman, suggested that most icterine genera originated 5 to 7 million years ago, in the Pliocene; so some sort of Agelaiine species may have existed for millions of years. His data, however, could not be used to determine the timing of the separation of the lineages leading to today's species.

The most recent 'official' classification (AOU Checklist of North American Birds, 1983) moves the North American blackbirds to the family Emberizidae (with warblers, sparrows, buntings, and tanagers), relegating them to subfamily status, the

Icterinae. Sibley *et al.* (1988) and Sibley and Ahlquist (1990), however, with what could perhaps be the last word on the subject for some time, used their comprehensive study of DNA–DNA hybridization of the world's living birds to lower the blackbirds to Tribe Icterinii, subfamily Emberizinae, placed within the family Fringillidae. The Fringillids, in this view, have undergone relatively recent, rapid speciation, resulting in a huge and varied assemblage that includes chaffinches, goldfinches, crossbills, honeycreepers, buntings, wood warblers, tanagers, cardinals, and the New World Blackbirds. Thus, a number of differing taxonomies exist for the blackbirds. They are similar, at least, in that all agree that the group is monophyletic, i.e. descended from a single ancestor. Whether they are classed as Emberizids, Fringillids, or Icterids is probably irrelevant to the purposes of this book: the blackbirds remain an ecologically and morphologically diverse group that often lends itself to relatively easy study, and the Red-winged Blackbird is the prime example.

Agelaius, the genus in which the Red-winged Blackbird is placed, contains 8 other species, only one of which occurs in North America. That species is the Tricolored Blackbird, which breeds only in California and Oregon. The other *Agelaius* species are widely distributed in South America and the Caribbean. A few of them, along with redwings, are polygynous marsh-nesters, e.g., Yellow-hooded Blackbird, Chestnut-capped Blackbird. Freeman's (1990) work suggested that the Agelaiine blackbird lineage, which contains *Agelaius*, is not monophyletic, as previously considered, but that North and South American groups developed independently. A recent comparison of mitochondrial DNA from all *Agelaius* species (using the cytochrome-*b* gene) strongly suggests that the group is actually polyphyletic, i.e. the species it includes descended from more than one ancestor (Lanyon 1994). Because the taxonomic standard is that species grouped within a genus be monophyletic, these investigations of molecular similarities and differences indicate that eventually *Agelaius* may be split into monophyletic groups or else its species placed into other genera (Lanyon 1994).

The redwing's scientific name refers to both behavior and form: *Agelaius*, from the Greek *agelaios*, 'flocking', a strong redwing trait, and *phoeniceus*, from the Greek *phoinikeos*, 'red', the color introduced to the Greeks by the Phoenicians.

DISTRIBUTION

The Red-winged Blackbird breeds over much of North America, from east-central Alaska and the Yukon, across a wide swath of subarctic Canada to Newfoundland, over the entire continental USA, across most of Mexico, south to northern Costa Rica, east to the north-western islands of the Bahamas and (although a bird of capitalist outlook) Cuba (Fig. 2.1). Over this range the species goes by various local names, most inspired by the male's military-style epaulets: in Mexico, Sargento, Tordo charretero (Epauletted Blackbird); in Cuba, Mayito de la Ciénaga; in Belize, Soldierbird; in Guatemala, Tordo Capitan (Captain's Blackbird), Tordo Alirrojo (Red-winged Blackbird); in Costa Rica, Sargento. Over most of this area it is the most abundant marsh-breeding passerine (Orians 1980); where there are marshes, there are usually breeding redwings. Areas lacking water and dense emergent vegetation in which to build nests, such as deserts, high mountains, dense forests, and arctic regions, have no breeding redwings. Since Europeans arrived in North America redwings, particularly those in the eastern USA and Canada and in California, have added upland sites such as fields and forest edges to their breeding habitat, as long

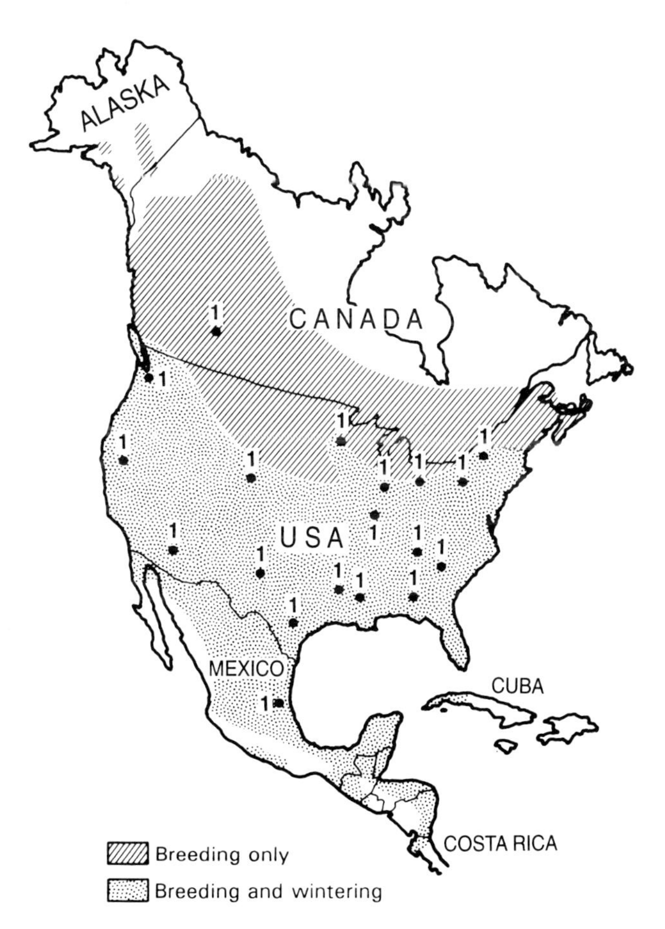

Fig. 2.1. Map of North and Central America showing the breeding and winter ranges of the redwing. Also shown are the 19 sampling sites used for a study of genetic variation among redwing subspecies (Ball et al. *1988; see text).*

as there is water available close by. People's alteration of natural habitats over the past 300 years, primarily the clearing of forests for crop agriculture, has dramatically increased redwing breeding habitat and propelled redwings to their current standing as one of the most abundant birds on the continent (see Chapter 11). In the redwing's more northerly reaches, particularly Alaska and Canada, they are migratory, wintering in the USA mainland, but many populations, especially those in lower-latitude temperate and tropical areas, are resident all year round.

Redwings vary considerably across North America in size, shape, and even plumage (Power 1970, James 1983). Consequently, among breeding populations north of the Rio Grande, 14 subspecies, or 'races,' based on phenotypic differences, have been recognized (AOU 1957, Bent 1958; Fig. 2.2, Table 2.1). Up to eight additional subspecies, according to taxonomies based on female plumage color, occur south of the Rio Grande (Fig. 2.3, Table 2.1). The smallest redwings in the USA are from south-eastern Florida. They gradually increase in size northward and westward (James 1983, James et al. 1984). Three independent studies, each of which utilized a different method, indicate that all redwings are genetically quite similar, which suggests that a reduction in the number of defined subspecies may be warranted. Redwing karyotypes, i.e., the number and morphology of chromo-

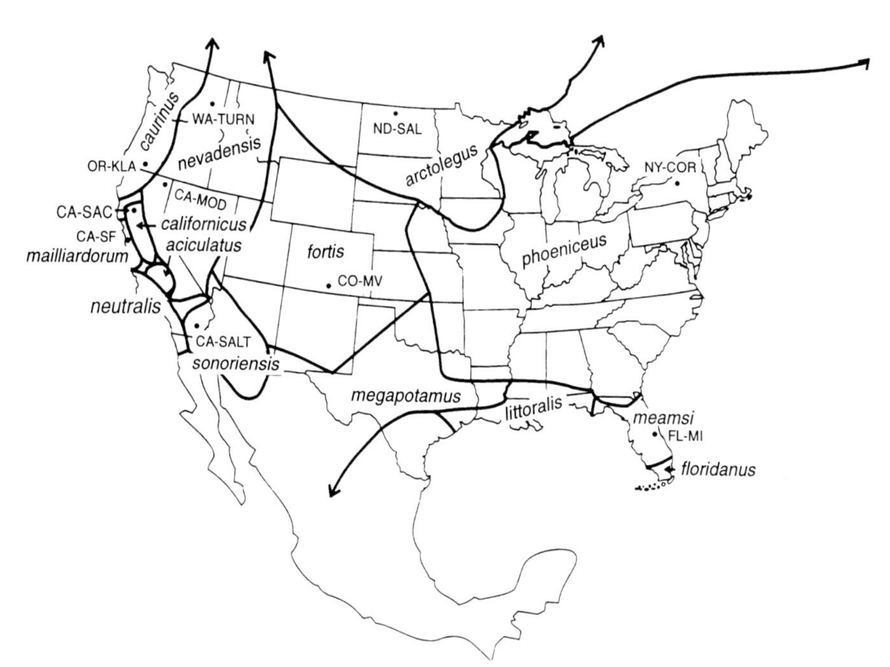

Fig. 2.2. Map showing the approximate distribution of redwing subspecies in the continental USA and southern Canada (from Gavin et al. *1991).*

Table 2.1. Names and breeding ranges of 22 Red- winged Blackbird subspecies. The 14 that breed in USA and Canada were listed by the American Ornithologists Union (AOU) checklist of North American Birds (1957).

Subspecies	Range
phoeniceus	Eastern and central USA, north to southern Ontartio and Quebec, south to northern Mississippi and northern Alabama
mearnsi	Northern and central Florida
floridanus	Southern Florida
littoralis	Southeastern Texas, southern Louisiana, southern Mississippi, southern Alabama, and northwestern Florida
megapotamus	Central Texas south to southeastern Coahuila and northern Veracruz
arctolegus	East-central, south-central, southeastern Alaska, southeastern Yukon, northwestern Saskatchewan, north-central Manitoba, northeastern Ontario, south to central British Columbia, southwestern Alberta, eastern Montana, southern South Dakota, and northern Iowa
fortis	Western Nebraska and western Kansas, Wyoming, Colorado, central and eastern Utah, south to central Arizona and New Mexico, northwestern Texas
nevadensis	South-central British Columbia, eastern Washington and eastern Oregon, Idaho, Nevada, eastern California
caurinus	Southwestern British Columbia to northwestern California
mailliardorum	Central coastal California
californicus	Central Valley of central California
aciculatus	South-central California
neutralis	Southwestern California and northwestern Baja California
sonoriensis	Southeastern California, southwestern Arizona, south to Baja California and northern Sonora
gubernator	Central Mexican Plateau
nelsoni	Morelos, Pueblo, Guerreo, western Oaxaca
nayaritensis	Coastal Nayarit to El Salvador, Highlands of Chiapas and Guatemala
pallidulus	Northern Yucatan Peninsula
richmondi	Central Veracruz to southern Yucatan Peninsula, south to southeastern Nicaragua and adjacent Costa Rica
arthuralleni	Northern Guatemala
grinelli	Nicaragua, El Salvador, Costa Rica
assimilis	Western Cuba and Isle of Pines

somes, are essentially invariant among several subspecies (the sizes and shapes of the seven largest redwing chromosomes were compared in one study using samples from Colorado, Minnesota, and Florida; Cox and James 1984). Ball *et al.* (1988) conducted a survey of variation in the nucleotide sequences in the mitochondrial DNA of redwings collected from 19 sites throughout the USA and in central Mexico and central Canada, including 11 defined subspecies (Fig. 2.1). They found only small genetic distances among samples, suggesting only very minor population differentiation.

Gavin *et al.* (1991) electrophoretically examined variation in allozymes from several gene loci in 10 USA breeding populations that included nine subspecies. They found little genetic variation for seven populations, stretching from New York to Florida to Oregon. There was substantial variation, however, among

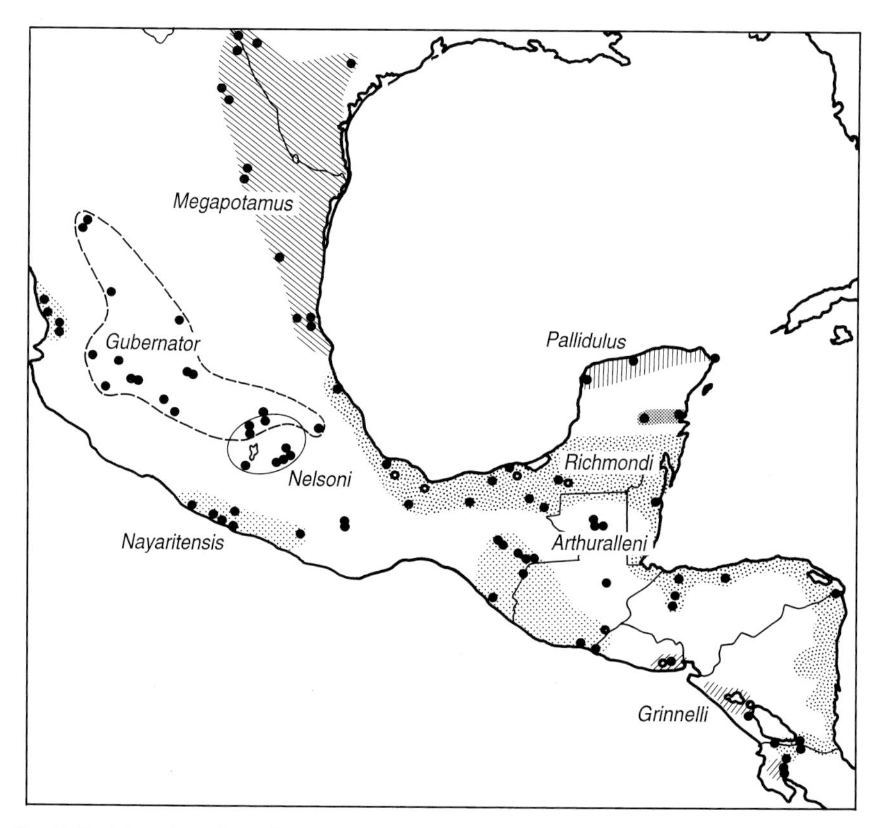

Fig. 2.3. Map showing the approximate distribution of redwing subspecies in Mexico and Central America. (Reproduced with permission from Dickerman 1974.)

some of the California populations. California is a large state that encompasses several habitat types and contains full or partial ranges of seven redwing subspecies (Fig. 2.2.). A population in the south-central portion of the state (Salton Sea NWR) was genetically distinct, as were two geographically close populations in the central part of the state at San Francisco (sampled at San Francisco Bay NWR) and Sacramento (sampled at Sacramento NWR). California redwings are less migratory than redwings at the other sites sampled. Gavin *et al.* conclude that there is much genetic mixing in migratory populations, which would explain the low degree of genetic variation among far-flung populations discovered by Cox and James (1984) and Ball *et al.* (1988). Resident populations, such as some of those in California, probably experience lower degrees of mixing and hence, greater genetic isolation. Thus, although historically redwings have been separated into many subspecies based on lengths of wings and tails, bill size, and coloration, this morphological variation is not accompanied by large genetic differences. Indeed, many ornithologists now believe that it is not biologically justifiable to assign subspecies to sections of the ranges of continuously–distributed species, such as the redwing. The latest edition of the American Ornithologists' Union *Checklist of North American Birds* (1983), consistent with this philosophy, does not list redwing subspecies.

The small amount of geographic variation in redwing genetics raises the possibility that environmental differences during development might account for the geographic variability in their morphology. Frances James (1983) investigated the possibility with clutch translocation experiments. She noted that both nestlings and adults in northern Florida had higher bill depth to tarsus length ratios and lower bill length to tarsus length ratios than nestlings from southern Florida. Nestlings and adults in Minnesota had higher ratios of wing length to tarsus length than individuals from Colorado. During the last half of incubation, James transported three-egg clutches from southern Florida (Everglades) to northern Florida (Tallahassee), others from northern Florida back to southern Florida, and some from Colorado to Minnesota. In all cases she put the transported clutches into the nests of 'foster' females after first removing the foster females' own eggs. The foster mothers accepted and finished incubating the transported eggs. Control clutches were removed from nests, held 2 days in an incubator (as had been all the transported clutches) and placed in other local nests. The result was that young from transported clutches developed the morphological characteristics of the foster population, not the characteristics of the population in which they were born. Interestingly, the characteristics that changed were those that differed most between nestlings from the different regions. Thus, in the 'reciprocal' translocation in Florida, the ratio of bill length to tarsus length, the attribute that best discriminated the two adult populations, shifted most. In the translocation from Colorado to Minnesota, the characteristic that changed most in nestlings, the ratio of wing to tarsus length, also best discriminated the two populations. Thus, not only is redwing morphology geographically variable and maintained by differential development, but the traits that vary most are also geographically variable (James 1983).

Breeding behavior of redwings also varies regionally. For example, average harem sizes (and thus, degree of polygyny), the sizes of territories that males defend, whether males feed their nestlings, and whether females solicit extra-pair copulations (EPCs), all differ markedly between eastern and western USA redwings. Why is there so much variation among different populations? Much of the geographic variation in size might be explained by Bergmann's Rule, that warm-blooded animals in the more northerly parts of their ranges are larger than in southern parts owing to thermoregulatory advantage – large animals have less surface area relative to body volume than small animals, and are thus more efficient heat retainers. Much of the behavioral variation, say between eastern and western USA populations, arises from differences in habitat productivity, which will be discussed in Chapters 4 and 12.

MIGRATION AND WINTER RANGE

After the breeding season is finished redwings undergo an autumnal molt, gather in flocks and, in some areas, migrate. Flocks are usually segregated by sex: females and the current year's young separate from all-male groups. Apparently, the sexes prefer not to associate during their 'time off' from breeding. It is likely that there are advantages in associating with members of one's own sex; it should be useful to be near other individuals seeking precisely the same food. By mid-October to November, migration of eastern redwings is at its height, with flocks of from tens to tens of thousands of individuals traveling diurnally and roosting overnight in marshes. Winter ranges for migratory populations cover most of the mainland USA and even parts of southern Canada. Huge concentrations of redwings, funneling down from the eastern and middle parts of the continent, congregate annually in the south-eastern USA, where they cause extensive crop damage (see Chapter 11). Redwings of the southern USA, the Pacific Coast, Mexico, and Central America are nonmigratory and generally show only minor movement (Dickerman 1974).

Most general information on redwing migration has been obtained by marking individuals at breeding sites and attempting to recover them in wintering areas. Many thousands of birds must be banded in breeding areas to recover even a few at wintering sites. Richard Dolbeer (1982) collated and analyzed all USFWS records for redwings recovered from 1924 to 1979. Figure 2.4 shows the average distances between banding and recovery sites for adult redwings (of both sexes) banded during the breeding period in various regions and then recovered in January of any subsequent year. January is the month that, on average, individuals were recovered at the greatest distances from their breeding sites, suggesting that the farthest extent of migration is reached in that month. The data suggest that individuals from various USA populations migrate about 600–1000 km. Females moved significantly farther during migrations than did males – from 200 to 300 km farther from each of the four regions in Fig. 2.4. Morphological measures by James *et al.* (1984) support the idea: mean winter-size measurements from Florida for males are very similar to those made on males captured during the breeding season, whereas mean female-size measures (e.g., wing length and bill depth) are significantly larger in females captured during winter, suggesting large influxes to Florida of more northerly-breeding females, but not of males. Ketterson and Nolan (1976) suggest that sex differences in winter ranges, particularly when females migrate farther than males, may be because there is an advantage for males to winter

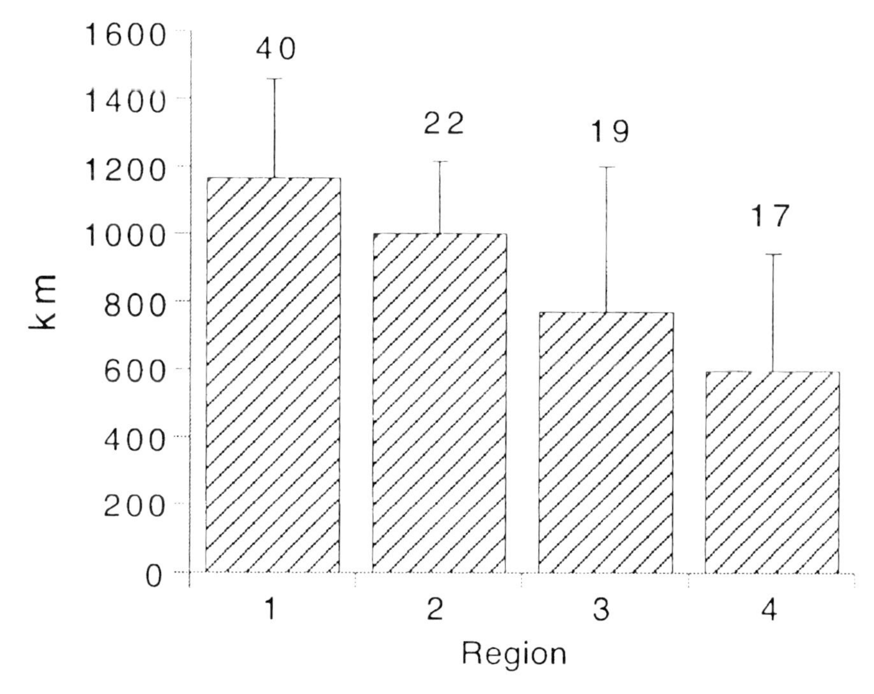

Fig. 2.4. Mean (± SD) distances between breeding-area banding sites and recovery sites during January in wintering areas for redwing populations from four regions of the USA. Region 1, Great Lakes (Minnesota, Wisconsin, Michigan, S. Ontario); 2, Midwest (Iowa, Missouri, Illinois, Indiana, Ohio, Kentucky); 3, Northeast (New York, Connecticut, Massachusetts, Rhode Island, Vermont, New Hampshire, Maine); 4, Middle Atlantic (Pennsylvania, West Virginia, Virginia, Maryland, Delaware). Values above error bars are sample sizes of recovered birds.

closer to their breeding grounds to foster earlier arrival in the spring. Also, the smaller females presumably are less able to tolerate the lower temperatures and intermittent food availability resulting from snow cover in more northerly areas.

Long-term banding and recovery efforts by Dolbeer (1978, 1982) have also shown that individuals often may be faithful to particular roosting areas during a winter, but that they are not necessarily faithful to the same wintering areas between years. For example, in the south-central USA (Arkansas, Colorado, Kansas, Oklahoma, Texas), 13 redwings were recovered at an average distance of 231 km from the locations where they were banded during a previous winter (Fig. 2.5). The average distance between winter (Jan–Feb.) banding sites in the southern USA and subsequent winter recovery sites for 46 males was 57 ± 145 km, and for 10 females, 52 ± 114 km (Dolbeer 1982). These studies are necessarily imprecise because it is impossible to know if a recovery site was an individual's final winter destination or a stop-over point. (In contrast, redwings tend to be highly faithful between years to breeding sites; Chapters 7 and 8.) Dolbeer's analyses also

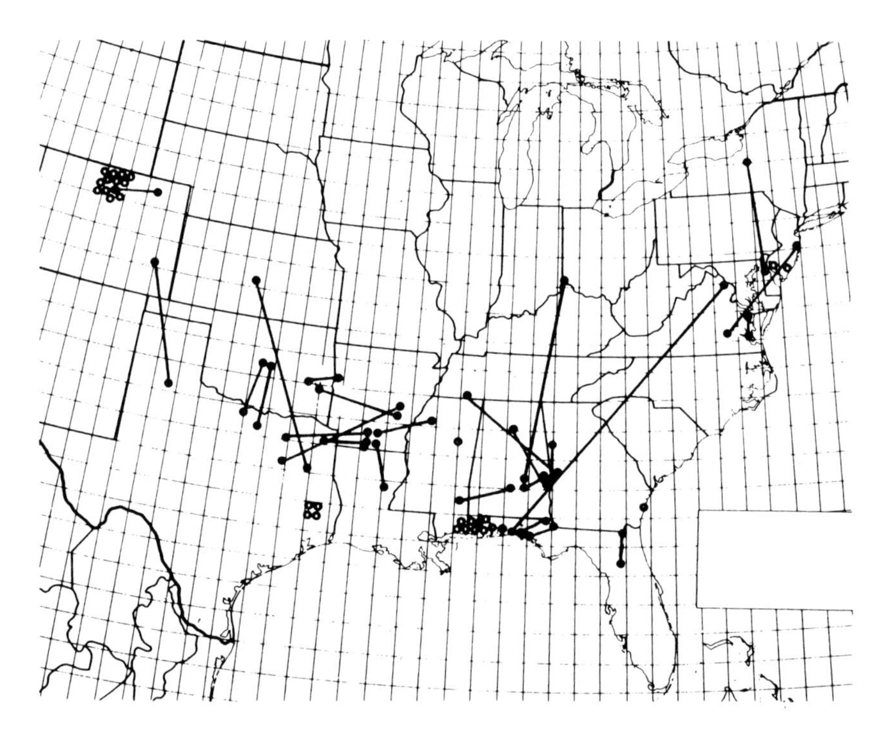

Fig. 2.5. Banding and recovery sites for individual redwings banded and recovered in winter roosts during different years. Lines connect banding and recovery sites for single birds when the sites were greater than 50 m apart (from Dolbeer 1978).

suggest that there is a high degree of intermingling of various breeding populations in their wintering areas, because individuals banded during their breeding or immediate postbreeding periods in many different regions can be recovered in the same wintering area. Furthermore, not all individuals from a single breeding region winter in the same area. For example, birds banded during the breeding season in north-central Ohio were recovered wintering from coastal North Carolina to Alabama (Dolbeer 1978). Mixing of populations in wintering areas provides a basis for the lack of genetic differentiation among distant redwing populations found by Gavin *et al.* (1991) and others, provided that at least a small percentage of individuals switch breeding populations, either by joining the local breeding population in their wintering area, or perhaps by following flock-mates who move northwards to their own breeding regions. As we know from the field of population genetics, even small numbers of breeders switching populations can strongly affect gene flow (Moore and Dolbeer 1989, Gavin *et al.* 1991).

Why do redwings have the migratory patterns they do? One generally accepted explanation of avian migrations is that they occur because food supplies change with season and whereas food might be abundant locally during breeding, it often becomes scarce in winter (Lack 1954, Baker 1978). Birds migrate then not because they cannot tolerate low temperature or snow (several species winter in the high arctic!), but because they cannot locate sufficient food to survive. Redwings are very sensitive to winter conditions because they feed on the ground and therefore snow cover is a particular threat. Thus, it is not surprising that migration patterns change when patterns of winter food supplies are altered. An excellent example is the CNWR redwing population in eastern Washington. Currently male redwings winter in the area, but females and perhaps subadult males migrate. This was not always the case, however. Most inland populations of redwings at the latitude of Washington are entirely migratory. But during the past 50 years, large parts of eastern Washington have been irrigated with Columbia River water, transforming the region into a vast agricultural area, which, in turn, has altered redwing habits. Whereas 75 years ago redwings in eastern Washington could not have survived a winter for lack of food, and so were all migratory, there are now abundant crop fields and livestock feed lots in which to forage. Intriguingly, only males take advantage of these resources to forgo migration. It may be that males are under social pressure to remain near and occasionally visit their territories, the better, presumably, to retain them for the next year's breeding; females are under no such pressure.

NATAL DISPERSAL

Most animals, including birds, disperse, or move to new areas, during one or more of their life stages. The distances involved vary extensively among species. Birds usually undergo 'natal' dispersal – movement from birth sites to first breeding sites – and some continue to move in 'breeding' dispersals – moving between successive breeding attempts. Redwing breeding dispersal is discussed in Chapters 7 and 8. Do redwings breed where they are born or do they disperse long distances? General banding recoveries suggest that almost all redwings breed within 50 km of where they hatched (Dolbeer 1978, Moore and Dolbeer 1989) and most much less than that. At CNWR, between 1977 and 1988 we banded approximately 4400 nestlings that fledged. Through 1990, 143 of these locally-fledged individuals returned as males with breeding territories within the study area (about 6.5% of the estimated 2200 male fledges) and 95 returned as breeding females (about 4.3% of the estimated 2200 female fledges). For 127 of the 143 males we knew exact birth and breeding locations. These males traversed an average of 1420 ± 1350 m from hatch sites to their breeding territories, with a range of 20 to 7200 m (Beletsky and Orians 1993). The 64 females for which we knew exact birth and breeding locations moved 1052 ± 802 m from hatch site to the areas in which, to the best of our knowledge, they built their first nests (range 20–2800 m).

Our dispersal information for males is better than that for females because we searched specifically for banded male territory owners throughout the study area, whereas most of our information on females was gathered in the restricted area of our core breeding marshes. Females redwings, on average, probably disperse after birth farther than do males, as is common in many songbirds (Greenwood and Harvey 1982). Indeed, the fact that we recovered 50% more locally fledged males than females as adults supports the idea that the CNWR females, on average, dispersed farther than did the males. The average dispersal distance we identified for males, 1420 m, is an underestimate of the true value because, although we

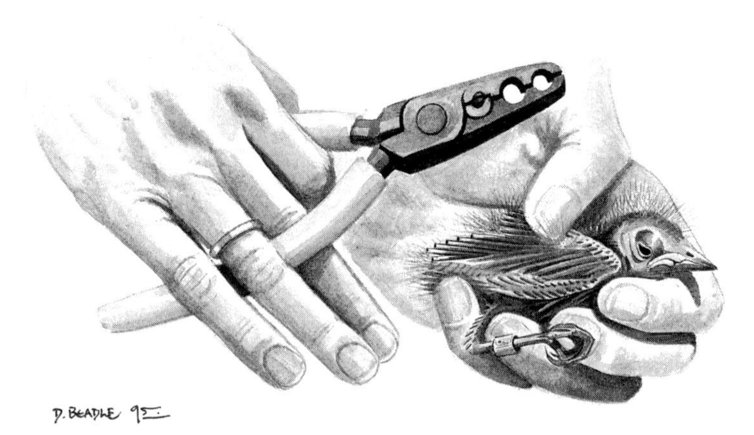

Placing a band on a 7-day old nestling. During the CNWR project, thousands of nestlings were banded, but relatively few returned to the study site in subsequent years or were observed elsewhere.

searched for banded males at all possible territory sites within 5 to 6 km of the center of the study area, and although almost all territory owners at the farthest extent of our censuses were unbanded, a few of our locally fledged males could have moved beyond our search radius. Our information suggests that natal dispersal for most males in the area is limited to 5 or 6 km. Another way of considering this is that the data indicate that all territory owners on our marshes were born either on those marshes or on others not more than 6 km distant. The evidence includes the geographic distribution of territories of the 127 dispersers that we located, our certain knowledge that about 20% of territory owners on the core marshes each year are born on the core, and that there are more than sufficient breeding territories surrounding the core marshes, within 6 km, to supply the remainder of our territory owners (Beletsky and Orians 1993).

I began an overview of the general biology of redwings by examining their taxonomic identity, where they are found, and their movement patterns. The reader can begin to appreciate how such basic knowledge can be of value during investigations of behavior. For example, information on natal dispersal of males will be significant when we consider how males obtain territories (Chapter 10); if males acquire breeding territories near where they are born, they clearly have good, long-term familiarity with those sites, which facilitates establishment of territories. Another example touched on was that different migration patterns of male and female redwings may in some cases be related to breeding season behavioral differences. To breed, males must maintain territories and thus, if it is possible, they may forgo migration; females do not defend territories and so are not similarly constrained. In the next chapter I will continue the overview of the general biology of redwings by examining their morphology – size, weight, and plumage – and development.

Chapter 3

MORPHOLOGY AND DEVELOPMENT

INTRODUCTION

Someone unfamiliar with redwings coming upon an active breeding marsh would find what at first appear to be two songbird species, one black and relatively large, one mostly brown and smaller. They are, of course, the two redwing sexes. The size differential between the sexes is striking; its magnitude – males are up to 20% larger than females, depending on the structures compared, and 1.5 times as heavy – is uncommon among songbirds. A host of selection pressures, some associated with polygynous mating, may have acted on each sex to produce such extreme variation. In this chapter I provide basic descriptions of male and female morphologies and explore the selection pressures that may have driven the size dimorphisms and dichromatisms. In particular, I will seek to answer the questions (1) why do redwings look as they do, (2) why are plumages and sizes of males and females different, and (3) how are plumage, molt, and breeding ecology related? Redwing morphology is also strongly integrated into the animal's visual displays, which will be touched on here but covered in detail in Chapter 5.

GENERAL MORPHOLOGY

Size and weight

To the ornithologically uninitiated I often describe redwings as small, black, robin-sized birds (American Robins, that is), although robins are actually slightly larger than redwings. Of course, in so doing, I refer only to the larger sex, but then so does the English name of the bird. Of the females, I usually say that they are smaller, brown, and streaked, 'like sparrows.' Males range in length from about 20–25 cm from bill tip to end of tail, with a continental average of perhaps 23 cm. The length of the wing is usually considered to be a good indicator of size because, unlike body mass, it does not change within a breeding season or a year, but is usually highly positively correlated with mass (Connell *et al.* 1960, Searcy 1977). (An adult redwing's wing length may increase slightly from year to year, almost certainly due to changes in feather and not bone lengths; Searcy 1979a). Mean wing length of adult males varies regionally, for example, from about 122 mm in New York and Kansas populations, to 131 mm in Washington State. Table 3.1 provides some average wing and tarsus measurements of redwings from various regions. Adult size has been found generally to increase with increasing latitude, for example, from Northern Florida (median wing length of 114 mm), to Arkansas (122 mm), to Kentucky (124 mm; James *et al.* 1984). According to Bergmann's Rule, which states that warm-blooded animals inhabiting colder climates are larger than those in warmer areas, redwings further north should be even larger. Subadult males, 1 year old, are usually slightly smaller than adult males, with average wing lengths of, for example, 119 mm in New York and 124 mm in Washington State populations (Table 3.1). Mean physical measurements of females within populations generally are about 20% smaller than males' (Table 3.1). By wing length measure, yearling females in one study were slightly but significantly smaller than adults (102 mm versus 105 mm, respectively; Langston *et al.* 1990).

Within populations, mean male and female masses usually differ markedly.

Table 3.1. Average size measurements of redwings in various regions. SD of means (or SE, if indicated) are given if provided by authors; samples sizes are in parentheses.

Location	Male subadult	Male adult	Female adult	Male:female ratio	Reference
Wing[1]					
Washington			105.0 ± 0.2[3] (152)		Langston *et al.* 1990
Washington		127.9 (17) 131.6 (21)			Searcy (1979c)
Washington		130.0 (46)	108.0 (9)	1.20	Searcy and Yasukawa (1981)
Washington	124	131	107		Searcy and Yasukawa (1995)
New York	118.9 (16)	122.4 (12)			Searcy (1979d)
Texas	(live) 122.2 (60) (dead) 122.0 (154)	127.9 (24) 116.7 (279)	107.8 (15) 104.5 (101)	1.19 1.17	Johnson *et al.* (1980)
Texas	113.7 (54)	118.0 (7)			Selander and Giller (1960)
Northern Florida		116 (32)	95 (18)	1.24	James *et al.* (1984)
Georgia	113 ± 3 (9)	119 ± 2 (33)	96 ± 2 (33)	1.24	Dunson (1965)
Indiana		125.9 (37)	102.1 (34)		Searcy and Yasukawa (1981)
Ontario		Upland 122.3 ± 0.9[3] (14) Marsh 119.5 ± 0.9[3] (14)			Eckert and Weatherhead (1987b)
Ontario	117.4 ± 3.8 (13)	121.3 ± 3.1 (141)			Shutler and Weatherhead (1991a)
Alberta		126.6 (7)	103.4 (8)	1.22	Power (1970)
Wisconsin		123.1 (40)	101.2 (5)	1.22	Power (1970)

Table 3.1 continued

Location	Male subadult	Male adult	Female adult	Male:female ratio	Reference
Kansas		121.9 (29)	99.4 (25)	1.22	Power (1970)
Colorado		129.4 (16)	106.1 (17)	1.22	Power (1970)
Cuba		108.8 ± 0.4[3] (29)	94.4 ± 0.6[3] (37)	1.15	Whittingham *et al.* (1992)
Tarsus[2]					
Washington			30.4 ± 0.1[3] (152)		Langston *et al.* (1990)
Ontario	25.7 ± 0.7 (13)	25.5 ± 1.4 (114)			Shutler and Weatherhead (1991a)
Texas	(live) 29.6 (60) (dead) 29.3 (154)	31.0 (24) 29.2 (279)	28.0 (15) 25.5 (101)	1.11 1.14	Johnson *et al.* (1980)
Cuba		28.4 ± 0.2[3] (29)	25.6 ± 0.1[3] (37)	1.11	Whittingham *et al.* (1992)

[1]Wing length is usually measured as the unflattened length, or 'chord,' from the carpal joint at the bend of the wing to the tip of the longest primary.
[2]Tarsus is measured as the length between the intertarsal joint and the start of the toes.
[3]SE.

Some average masses for males and females, and the months they were taken, are given in Table 3.2. At Columbia National Wildlife Refuge (CNWR), adult males ranged from 60 to 88 g, with a mean of about 70–75 g, depending on month during breeding (mean mass during the breeding season for about 2500 individuals captured and weighed, in total, more than 13 000 times during the 16-year study, the 'seasonal mean', was 74.1 ± 3.8 g). Subadult males often weigh, on average, about 5% less than adult males (Table 3.2; seasonal mean = 71.2 ± 3.8), although, within the same population there is overlap in adult and subadult weights. Most females weigh 40–55 g (seasonal mean at CNWR = 47.4 ± 3.4 g, range 36–65 g).

Males generally lose mass during the breeding season, probably because they devote increasing amounts of time to advertising, sexual activities, and territory defense and less to foraging. For example, Searcy (1979b) found that adult territorial males in Washington lost, on average, 0.09 g/day during the breeding season of 1974 and 0.11 g/day in 1975, but that there were no significant mass changes in 1976. In central California male masses were significantly higher in late winter than at the termination of the breeding season, but masses increased again by late summer (Payne 1965). At CNWR, the average territorial male lost 3.7 ± 3.6 g from March to June, n = 280 males. Mass also changes on a daily basis; changes in body mass of 3% to 4% in a day or two are not unusual for birds of the redwing's size

Table 3.2. Average masses of redwings in various regions. SD of means (or SE, if indicated) are given if provided by authors; samples sizes are in parentheses.

Location and month	Male subadult	Male adult	Female adult	Male:female ratio	Reference
Washington April/May				48.5 ± 0.3[1] (148)	Langston *et al.* (1990)
Washington March/April		75 (43)			Beletsky and Orians (1987b)
Washington[2]					
February	72.0 ± 3.6 (54)	75.1 ± 3.4 (1094)	47.5 ± 2.0 (34)	1.58	Beletsky and Orians, unpublished data
March	71.1 ± 3.7 (1232)	74.8 ± 3.6 (3770)	46.9 ± 2.5 (963)	1.59	
April	72.1 ± 3.7 (1163)	74.1 ± 3.4 (2298)	48.3 ± 3.2 (2758)	1.53	
May	70.3 ± 3.5 (347)	71.9 ± 3.5 (617)	47.4 ± 3.7 (2732)	1.52	
June	68.8 ± 3.2 (200)	69.6 ± 3.4 (402)	45.4 ± 3.0 (809)	1.53	
July	70.0 ± 3.4 (11)	69.4 ± 2.5 (31)	44.2 ± 2.9 (74)	1.58	
New York October	71.5 (6)	71.2 (12)			Searcy (1979d)
California					
January		70 (14)	46 (6)	1.52	Payne (1965)
June		63 (17)	43 (15)	1.47	Payne (1965)
Georgia June–August	61.2 ± 3.7 (9)	60.7 ± 2.4 (16)	38.8 ± 2.5 (33)	1.56	Dunson (1965)
Ontario April–July	64.7 ± 5.1 (13)	68.5 ± 3.5 (114)			Shutler and Weatherhead (1991a)
Cuba		50.6 ± 0.7[1] (10)	39.4 ± 0.7[1] (11)	1.28	Olson (1985)

[1]SE.
[2]Many individuals during the Columbia National Wildlife Refuge (CNWR) study were weighed multiple times.

(Owen 1954, Kacelnik 1979). We found mass changes in individual males at CNWR of up to 4% in 48 h (Beletsky and Orians 1987a). Females actually tend to gain mass during the breeding season, as they gear up for egg formation, laying, and incubation, but then lose mass later (Table 3.2).

Plumage

Adult males are, overall, a lustrous black, often with some feathers of the head, back, greater wing coverts and tertiaries 'edged with buff and ferruginous brown'

(Dwight 1900). Black is the basic body color of many icterines, and some presently included in the genus *Agelaius* are completely black (Unicolored and Pale-eyed Blackbirds). An all- or mostly-black body may have thermoregulatory advantages in certain climates (Orians 1985) and black feathers may be more wear resistant owing to a higher melanin content (Lucas and Stettenheim 1972), but there is no doubt that a pure-black body has at least one major disadvantage: all-black redwings are plainly highly conspicuous against their mostly brown and green breeding habitats. The buff/brown feather tips on the back and head disappear as feathers wear during the course of a year. This brown coloring may render males

Female plumage varies considerably in brightness over the redwing's range. Western females (for example, the **Agelaius phoeniceus californicus** ***individual on the left) are usually a good deal darker than eastern females (the*** **A. p. phoeniceus** ***female on the right).***

more cryptic and inconspicuous to predators during the autumn and winter (Orians 1985), when their pure black plumage is not essential, for example, for signalling in breeding displays. The shoulder epaulets (lesser coverts of the top portion of the wing, or upper secondary wing coverts) are a bright scarlet–vermilion, and edged by buff/yellow middle coverts (Fig. 3.1). Male redwings that breed in central California sympatrically with Tricolored Blackbirds, which closely resemble redwings except that they have white instead of buff/yellow edging the red lesser coverts, have all-red epaulets; the buff/yellow stripe was apparently lost through the evolutionary process of character divergence. These California redwings were once thought to comprise a separate species, the Bicolored Redwing.

Subadult males (1-year-olds during the breeding season, in the plumage they attained in their first autumn molt) have a variable plumage, ranging from quite female-like to approaching that of an adult male, and all gradations between (Fig. 3.2). Subadult epaulets are usually a dull-to-bright orange instead of the bright scarlet of adults, and each epaulet feather has subterminal bars or spots of black (Dwight 1900). There is some overlap between adults and subadults in both

Fig. 3.1. Extended wings of adult male and mature female redwings, showing relative size and plumage pattern.

epaulet coloration and the amount of brown edging on feathers of the head and back (Selander and Giller 1960, personal observations), making it difficult to guess the age of a few individuals, even in the hand. The source of the variation in subadult male plumage is evidently tied to birthdate: males born earlier during breeding seasons have more time to develop prior to their first autumn molt and so may take on a more adult-like plumage (Selander 1965, 1972, Orians 1985), which presumably is beneficial for their approaching first winter or first breeding season. While a female-like plumage pattern renders a male subadult less conspicuous both to predators and to territory-owning male redwings, thus facilitating both survival and territorial trespassing, it also may reduce the likelihood of its possessor acquiring a breeding territory.

Adult females are, by comparsion, relatively dull and cryptic. They are various shades of brown and tan, and black and white, streaked on the back and chest (Fig. 3.3). However, they do have small, bright patches of pink, orange or yellow, of various sizes on the head, chin, throat, and breast. In fact, these small color patches, in conjunction with variable streaking on the back and on other parts, make each female individually visually very distinctive (Orians 1985). This is in clear contrast to males, which most often are morphologically indistinguishable – they all look alike. Females also have red or reddish epaulet feathers, but the extent and brightness of the red area is highly variable among individuals, and even within

Fig. 3.2. Four specimens of subadult male redwings, arranged from most female-like to most adult male-like, between an adult female and an adult male (from Selander and Giller 1960).

Fig. 3.3. A female redwing near her nest.

individuals between years. Female epaulets even at their brightest extreme, nonetheless never approach the brightness nor the size of the males' (Fig. 3.1).

Yearling females, having undergone their first autumn molt, resemble adult females in overall plumage pattern, but they are less variable (Selander and Giller 1960). They are in general duller than adult females, including the amount of pink or orange on the head, chin, and throat (Nero 1954). Unlike 1-year-old males, who are usually readily identifiable by their age-specific body plumage, 1-year-old females cannot be easily recognized at a distance. First-year birds can often be aged by the degree of skull ossification, i.e. determining whether the frontal and parietal bones are fully or partially developed; the latter condition, which may be ascertained by visual inspection of live birds, identifies an individual as a first-year bird. When females are aged in this manner and divided into yearlings and 'matures', yearlings usually, but not always, have no reddish hue in their epaulets. For example, Selander and Giller (1960) found that 20 out of 47 first-year females from Texas had no red in their epaulets and 16 more had epaulets 'faintly tinged' with yellow, orange, or red; 11 had orange or red areas covering half to all of the epaulet area. All 12 mature females that they scored had bright (orange/red) epaulets. In a California study, all 48 first-year females aged by ossification had either no red or slightly red epaulets, whereas 88 of 97 mature females had large amounts of red (Payne 1969). Inspection of females definitely known to be yearlings provides support for the relationship between age and epaulet brightness. Miskimen (1980) raised 24 females obtained as fledglings and observed molts: 84% acquired rust/brown epaulets during the first late-summer molt and 16% acquired some orange feathers; all acquired bright epaulets after the second late-summer molt. Our findings agree nicely with Miskimen's work; we found at CNWR that 80% of 29 yearlings that we banded as nestlings had no red in their epaulets, whereas all 28 females also banded as nestlings but first trapped and scored when they were older than 1 year had red epaulets.

SEX DIFFERENCES IN MORPHOLOGY

Size

Why do male redwings exceed their mates by 10% in tarsus length, 20% in wing length, and 50% in mass? Although this question resembles a child's simple query – 'Daddy, why is the boy bird bigger than the girl bird?' – the answer, unfortunately, is not so simple. Many correlational and experimental investigations have been devoted to gathering the information required to find an adequate answer. Selander (1958, 1965) first noted that polygamous blackbird species were more intersexually dimorphic than monogamous ones. That pattern was confirmed in a comparative analysis of 62 blackbird species by Orians (1985); it was further supported by recent work by Webster (1991), who, after controlling for various possible confounding factors, found among 35 icterines a positive relationship between degree of polygyny typical of each species (as indicated by mean harem sizes) and sexual dimorphism in size. The mating system/dimorphism pattern also has been identified in other groups; for example, Payne (1984) confirmed it generally across families in lekking birds.

How are such patterns explained? The most widely-accepted, general explanation of sexual dimorphism begins, as is usual in evolutionary biology, with Darwin (1859, 1871), who pointed out that traits that enhance mating success can become exaggerated, particularly if they are associated with intersexual display or intrasexual fighting; in other words, some traits are exaggerated through the action of sexual selection. Because sexual selection depends on variation in mating success, i.e. that individuals with more exaggerated attributes produce relatively more offspring, it follows that sexual selection is most powerful in species in which individuals regularly have more than one mate. Thus, we expect greater degrees of sexual dimorphism in polygynous and promiscuous systems than in monogamous ones. Redwings are strongly polygynous, and therefore strong sexual selection is expected to produce exaggerated characters, size among them, in males.

The argument for size dimorphism continues: because males compete for females, in the redwing's case by fighting among themselves for territories, and provided that larger males are better fighters, males should evolve to become larger. Females, under no such pressure, should remain at some ancestral size or respond to other selection pressures than might lead to larger or smaller size. If larger male size in redwings evolved because it conferred an intrasexual advantage for mating success, then we ought to be able to detect that advantage in the field by relating male size and breeding success (or its correlates). However, traits that evolved through sexual selection are not necessarily under the same selection pressures today that led to their exaggeration (Grafen 1988, Searcy and Yasukawa 1995). Therefore, a contemporary finding that a trait does not appear to be subject to sexual selection does not preclude the possibility of such selection in the past. While a valid concern, especially in the case of the redwing, a species whose habitat has changed so radically during its last 200 generations, this should not prevent us from trying to identify currently operating selection pressures.

Searcy and Yasukawa (1995), in their comprehensive review of sexual selection and polygyny in redwings, compiled the results of 15 field studies that bear on the issue of a potential fitness benefit for larger males. For example, Searcy's (1977) doctoral work in Washington included an analysis of territorial male size (wing and tarsus lengths, mass) versus pairing date and harem size, the predictions being that larger males should be chosen earlier and more often as mates by females (the predictions were supported in only one of three years). Yasukawa (1981c) compared male size (wing length) and harem size in Indiana and found a significant positive correlation. In a study in Ontario, upland breeding males were behaviorally dominant to and larger (by wing length) than marsh-breeders, although marsh territory males, on average, had larger harems (Eckert and Weatherhead 1987b). However, overall, only three of the 15 studies found significant evidence that larger size conferred a mating success advantage. Searcy and Yasukawa conclude that the size dimorphism in redwings evolved owing to sexual selection favoring larger body size in males, but if such selection is currently operating, it is weak. (Even if there is a lack of evidence in redwings, there is some for other icterine species, e.g. the heaviest Montezuma's Oropendolas are dominant and achieve most of the matings in a breeding colony; Webster 1994.)

Provided that larger size conferred reproductive advantages on males, then, generation by generation, they should have grown larger. Such 'directional' selection, however, inevitably is stopped by counter or balancing selection. Selander (1965) proposed two kinds of counter selection, for smaller size, that would act to keep male blackbirds from runaway growth. One type concerns energetics: a larger size requires greater quantities of food, making larger males more susceptible to winter starvation or, at least, making it more difficult for them to maintain good winter condition; males in poor condition at the onset of the breeding season would be poor fighters and, hence, less able to establish and defend breeding territories. The other is that larger individuals, being more conspicuous and perhaps poorer flyers, should be more vulnerable to predation. Using these arguments, Selander predicted a negative relationship between male size and between-year survival. Searcy and Yasukawa (1995) compiled seven applicable field studies. Again, the data provide both positive and negative evidence. For example, survival was measured in one study by looking at return rates of territorial males in Washington; in 2 years returning males were, on average, smaller than nonreturning males, but in one year returning males were larger (Searcy 1979c). Males measured at an autumn roost in Quebec that survived from one year to the next did not differ significantly in size from males that presumably did not survive (Weatherhead *et al.* 1987). Furthermore, there is little evidence that larger male nestlings or fledglings have lower survival rates than smaller ones (Searcy and Yasukawa 1995), although a few studies have found differential nestling survival of males and females, with the smaller females having the advantage (Haigh 1968, Blank and Nolan 1983). Searcy and Yasukawa (1995) conclude that large males today do not pay a 'survival' cost for their size and, in fact, there is a trend, apparent among studies, for large size to provide a slight survival advantage. Recently, however, N. Langston and S. Rohwer (personal communication) demonstrated a significant effect of energetics on male redwing survival that must be related to male size. When these researchers gave supplemental food only to redwing females so that they began breeding earlier than usual, their mates were compelled to display and carry out other breeding activities for longer than usual periods. The consequence for these presumably energy-stressed males was a lowered probability of survival to the next spring.

Although sexual selection has received the lion's share of research attention, there are, of course, other factors that might influence redwing size. For example, Selander (1966) proposed that in some groups niche separation is a potential cause of sexual divergence in form or size. Different sizes could be selected, for example, because there are limited food resources and different foods would be utilized if one sex were larger than the other. This hypothesis appears unlikely to be the primary cause of sexual dimorphism in strongly polygynous species (Searcy 1979c) with powerful sexual selection pressures, and particularly for the redwing, in which the sexes feed in the same manner on much the same items. Furthermore, most considerations of the evolutionary development of size differences in blackbirds assume that is the males who became larger, rather than the females smaller. Langston *et al.* (1990) considered several possible influences on the evolution of body size in female redwings. They found that larger females had advantages in social competition and in breeding earlier. The latter would favor larger size because earlier breeders are usually more successful. However, smaller females had proportionally more fat than large females, which also had a positive effect on reproductive success; thus, there is selection that favors smaller females. Female redwings, therefore, may have at least two opposing selection pressures acting on them, as do males, to fix body size around its current 'optimum'.

Color

Bright colors of birds and their often striking plumage patterns are, to most people, one of their most appealing traits. Plumage colors and patterning probably develop for a variety of reasons, and are used in a number of ways. For example, a bird's color pattern can render it very cryptic against its usual background, reducing its probability of being detected by a predator. Colors and patterns are also used in communication, in a multitude of ways. Some birds' plumages may include bright patches as warning signals, denoting to predators their distastefulness or even that their tissues contain poisons. Colors and patterns may aid birds in species, sex, or individual identification and, in some species, in signalling an individual's social status. Some birds may even choose their mates based on aspects of coloration. Several investigations, notably with monogamous breeders, have shown that female birds may, at least partly, choose their mates on the basis of feather, bill or leg color (Burley 1986, Møller

1994). Selecting mates by color preferences would be advantageous to females only when male plumage correlates with male quality, that is, with such traits as genetic quality (ability to produce high-quality, attractive sons), parental quality (likelihood of being a good parent, e.g. by feeding and/or protecting nestlings), or current health (e.g., parasite load). Female House Finches, for example, prefer to mate with males with brighter, more colorful plumage, and bright plumage in males correlates well with quality measures (Hill 1991). In the redwing, the almost complete lack of overlap in body coloration between males and females argues for a difference in selection pressures on the sexes. The problem is identifying the pressures.

It is doubtful that a simple, all-inclusive explanation of sexual dichromatism, such as Hamilton's (1961) idea that it would promote rapid pair formation by enhancing sex recognition, will be applicable in the redwing's case, in which the entire plumage differs. First, consider the black and red adult males. The black body color, in addition to any thermoregulatory benefits, may be used in communication. Some evidence is available that male redwings are attracted to black-bodied birds, and so may use their main body color as a flocking signal (see Chapter 5). The black and red body may also serve to enhance displays that make males seem bigger than they really are. A common pattern in birds is that when males are larger than females, the male body is used in signalling, especially for displays in which feathers are erected to increase apparent size (Selander 1972). The idea is one of deceit in advertising – smaller birds exaggerating their size to scare other males or impress females. Male redwings often erect their head feathers when held in the hand, presumably an aggressive, submissive, or fear signal, and one that may be enhanced by black coloration. Another example is the most common territorial display of male redwings, the Song Spread, in which the tail feathers are spread and wings held out from the body, significantly inflating the apparent size of the bird; the red feathers, contrasting with the black, may enhance the effect. There is also abundant experimental and observational evidence that males' bright red epaulets have major roles in signalling intrasexual aggression, and perhaps in intersexual signalling during courtship (see Chapter 5).

Whether female redwings use morphological features of males, such as size or brightness of epaulets, to choose their mates is a heavily researched and debated question. To date, there is little evidence that male color or, for that matter size, strongly affects female redwing mate choice. If females were to choose male redwings by their plumage, it is reasonable to assume that the epaulets, which vary subtly among males in size and hue, might have a role in signalling, rather than the more invariable black body plumage. Searcy (1979b) found no significant correlations over a 3-year period between harem sizes of territory males and the length of the red or yellow portions of their epaulets. In only one of the 3 years was there a significant correlation between harem size and male wing length, as would be expected if females chose males primarily on size. Thus, there was little evidence that females in that Washington State population were choosing males by their body or epaulet size. Eckert and Weatherhead (1987a) determined whether several morphological features of territorial males correlated with their parental efforts – nest defense and feeding offspring. They identified a significant correlation between epaulet size and vigor of nest defense, which would suggest that females could predict which males would be good defenders simply by inspecting their epaulets; however, males with larger epaulets also tended to feed offspring less often, potentially negating any beneficial effect of choosing males by this method. Recent correlational studies of male redwing parasite load – both ectoparasites (lice and mites) and endoparasites (including blood protozoa, nematodes, and trematodes) – and morphological features in Ontario have revealed no significant differences in wing length, total epaulet area, or epaulet length, between parasitized and unparasitized birds, indicating that females cannot judge whether males carry heavy loads of parasites by assessing their overall size or epaulets (Weatherhead *et al.* 1993). One constraint on directional selection for larger red epaulets is that, if too large, they would no longer be concealable under black wing feathers, with profound consequences for signalling.

Male redwing coloration has important roles in male–male and perhaps male–female communication (this is explored in detail in Chapter 5). What of female coloration? Most studies of redwing color include an underlying assumption that a hypothetical ancestral blackbird species was monomorphic and dull-plumaged, and therefore that current redwing morphology is derived (i.e. a relatively new trait that was not present in ancestors) especially through the action of sexual selection on males. Could the process have happened in the direction opposite to the one assumed? Could the ancestors have been brightly colored but subjected to selection that favored female dullness?

Females are cryptically colored, and are much more difficult to detect visually against a variety of natúral backdrops than are males. A recent comparative analysis of icterines based on mitochondrial DNA sequences, skeletal characteristics and allozyme frequencies showed that female coloration is more labile evolutionarily than male coloration, and that most of the plumage changes that led to the high degrees of sexual dichromatism in the icterines most likely resulted from changes in female and not male brightness (Irwin 1994). Thus, if it is female rather than male plumage that drives the evolution of dichromatism in the group, adaptive explanations for female redwing coloration, other than crypticity, need to be considered. Irwin (1994) proposed three possibilities: that bright female plumage has an aggressive signalling function, that males choose females at least partly on plumage characteristics, or that there is selection to be dull either as an antipredation device or to reduce aggression from other females within harems. Female redwing plumage appears to have little if any effect on female–female aggressive interactions, athough female epaulets are exposed during these interactions (see Chapter 5). No-one has yet shown a social or reproductive advantage for redwing females with brighter plumage, either in their interactions with males or with other females (Muma and Weatherhead 1989). Because male redwings apparently do not choose females, the only one of Irwin's possible explanations applicable in the redwing's case is that females may be under selection pressure to be dull. This may prove to be correct, in which case the best explanation for the females' reddish epaulets may be that they are neutral traits, present only because of their genetic correlation with male morphology (Lande 1980, Muma and Weatherhead 1989; see Chapter 5).

The Cuban Exception

Redwing morphology certainly varies somewhat among regions over its range. Average size differs slightly from region to region, as does sometimes coloration – females in some California populations, for instance, are quite dark – but overall, the strong male–female differences are preserved. Therefore, the existence of an island population that differs strikingly from mainland redwings in plumage and other attributes is curious. In this population, currently classified as *A. p. assimilis*, which is confined to the western part of Cuba (Bond 1961, Garrido 1970), the sexes are more similar in size than mainland populations and both sexes are black. Further, both Cuban sexes are smaller and lighter than most of their mainland USA/Canada counterparts (Tables 3.1, 3.2). Male weight exceeds that of females by only about 25% (half the typical mainland value of 50%). Males are all black and have red epaulets with yellow borders; females are 'uniformly coal-black,' lacking the reddish epaulets of mainland females (Whittingham *et al.* 1992). In contrast to mainland redwings, the sexes in Cuba are difficult to distinguish, visually or vocally. They give the same type of song (like the 'conc-a-ree' song of mainland males; see Chapter 5), and the males do not expose their epaulets when they sing (they do so, however, during flight).

This morphological similarity of the sexes is associated with, for the species, an unusual mating system: Cuban redwings show evidence of monogamy (vocal duetting between the sexes being one indication; Morton 1983, Whittingham *et al.* 1992). The current best guess is that, because the Cuban redwings breed in the same types of habitats as mainland birds, the unusual morphology may have developed because of a behavioral convergence of sex roles. Both sexes may defend the territory all year, against individuals of both sexes (Whittingham *et al.* 1992). If so, then it is likely that the group is derived from mainland populations, perhaps from the Mearns redwing of Florida, which also in some areas tends toward monogamy (E. S. Morton, personal communication). The Cuban exception clearly points out the common avian relationship between mating system and size/coloration: Monomorphism is associated with monogamy and shared sex roles, whereas dimorphism in size and dichromatism are associated with nonmonogamous systems and divided sex roles. The example is particularly compelling because the differences occur within a single species. Other redwing populations may also have unusual plumages, e.g. the central Mexico population of *A. p. gubernator* is reputed to contain females that are all brown or all gray (Dickerman 1974), but to date there is little information on them. The Cuban redwing is sufficiently different from mainland populations in morphology and behavior that in future classifications it is likely to be considered a separate species.

Morphology and dominance

The potential influence of male and female morphology on social dominance and territoriality will be considered in later chapters. Here it will suffice to say that there is little evidence that size, mass, or coloration of males influence natural dominance relationships, but that the sizes of females may affect their dominance interactions.

GROWTH AND DEVELOPMENT

In the nest

After being incubated for 12–14 days redwing eggs hatch and young emerge. They remain in the nest for 9.5–12 days, or longer if conditions are unfavorable, growing rapidly toward their day of fledging. During the nestlings' first 2–3 days of life they are brooded by the mother for most of each day, but after that time they are increasingly left alone as the mother forages to feed them. Because eggs hatch at least partially asynchronously, there is often large variation in within-clutch nestling size. Size differences may become accentuated because not all nestlings receive the same amount of food (Haigh 1968, Teather 1992).

Mean mass at birth (newly hatched young) ranges from 3.2 g in Washington (Haigh 1968) to nearly 5 g in Ohio (Williams 1940). Sexual dimorphism at birth is weak or nonexistent. At 5 days of age, mean male mass in Washington was 26.0 ± 2.9 g, and for females, 22.5 ± 2.4 g. At 10 days of age, just prior to fledging, mean male mass was 43.7 ± 2.4 g, and female mass, 32.8 ± 1.3 g. Males receive more food than females, at least partly because they can reach higher in the nest to beg (Teather 1992), and so they grow faster than females (Haigh 1968, Holcomb and Twiest 1970). Fiala and Congdon (1983) determined that the average amount of energy assimilated by a male in Michigan during the nestling period was 1014 kJ, and for a female, 797 kJ. At fledging, males weigh, on average, 10–12 g more than females and their tarsi are, on average, 3 mm longer than females'. However, mean wing length of the sexes at fledging, as measured by length of the eighth primary, is approximately the same, because female feather growth starts earlier (Haigh 1968, Holcomb and Twiest 1970).

More female redwings are born than are males. The precise primary sex ratio, the ratio of males to females at conception or birth, is difficult to determine, but estimates can be obtained using broods with 100% survival. Several studies have identified nestling sexes, usually late in the nestling period (e.g. at 8 days of age), by measuring weight or size (Williams 1940, Haigh 1968, Holcomb and Twiest 1970, Laux 1970, Knos and Stickley 1974, Fiala 1981) and consistently found from 51% to 57% females. Fiala (1981) also performed laparotomies (surgical examinations of reproductive anatomy) on 2-day-old nestlings, with the same result. When all studies are combined, the average result is that redwings exhibit a 53:47 female-biased primary sex ratio, apparently departing from a 50:50 ratio owing to events prior to hatching. Data on whether there is sex-biased nestling mortality are ambiguous, e.g. Haigh (1968) found more starvation among male nestlings, but Fiala (1981) found no sex differences in nestling starvation rates. Predation rates are unlikely to affect differently the two sexes because most predators on redwing nests take all eggs or young. A good estimate of the secondary sex ratio for redwings, that at independence (which is at about 2–3 weeks after fledging), is not yet available.

Nestling songbirds are not passive participants in their growth and development. As alluded to above, there is competition for the food brought to the nest by parents. When parents land on or near the nest, all hungry nestlings raise their heads, open their bills, and strain forward. Redwing nestlings that reach higher may get more food than others (Teather 1992). Because males are heavier than females at fledging, it follows that they must either beg more than females or be more successful at getting food from parents. Nestlings also stimulate their parents to feed them by emitting begging calls – high-pitched, continuous, warbling, whining sounds. Nestlings apparently transmit their nutritional needs to parents via these vocalizations. When songbird nestlings are hungry they beg, and when experimentally deprived of food, they beg louder and more frequently (Smith and Montgomerie 1991). When a male redwing is not actively feeding nestlings, begging calls from food-deprived nestlings can stimulate him to do so, provided that he is not tending a fertile female (Whittingham and Robertson 1993).

After fledging, redwings are fed by parents for another 14 to 20 days until, at the age of 25–32 days, they are independent. As juveniles, they gather in small feeding flocks, and begin to replace their juvenile plumage.

Molt

Feathers in good condition are essential for flight and efficient thermoregulation. The importance of feather condition is underscored by the extensive amount of time birds spend preening. Fighting, accidents, and normal wear-and-tear take their tolls, and thus all birds replace their feathers on some regular schedule. If all birds could replace all their feathers twice a year, migrate twice, and still reproduce successfully, they probably would. But molt schedules vary among groups of birds, and sometimes among species within groups, probably as a function of energetic and time constraints (Selander 1972). The natal down redwings and other songbirds are born with is replaced while in the nest with a juvenile plumage. Male redwings then attain their first basic plumage, or subadult plumage, in a molt that begins 45–65 days after fledging and is completed 60–70 days later (Payne 1969). This is the plumage mentioned above that is highly variable in color pattern among males, and often resembles that of adult females. In at least some of the northern subspecies (*A. p. phoeniceus, arctolegus, fortis*), there is also a partial molt just prior to the first breeding season (i.e. a partial 'prenuptial' molt), one that generally does not include replacement of the main contour feathers (long feathers of the wings and tail – rectrices and remiges). During the next, complete, molt, during the summer and early autumn, males finally acquire their adult, jet-black plumage and are thus prepared to attempt territory ownership and breeding the following spring. Once males attain adult plumage they molt only once a year, immediately after breeding, in a postnuptial molt. Females have a molt schedule similar to the males' except that the subadult plumage of females much more closely resembles that of the adults.

The fact that subadult (yearling) males and females undergo a partial molt before their first breeding season, but that adults do not (or do so on a much-reduced basis), supports the idea that there is a trade-off between directing energy to a prenuptial molt or to early, energy-intensive, breeding season activities that are undertaken solely by adults (Greenwood *et al.* 1983). In other words, at this time of year yearlings have extra energy to molt, whereas adults must conserve their energy for reproductive activities. For males, these activities are territory acquisition and defense (subadults do not participate), and for the females, early nesting in unpredictable environments and female–female aggression (yearling females usually nest later, when female–female aggression is less frequent). Molt is energy demanding (food intake during molt is generally raised 30%–40%), takes a long time, and is physically or physiologically incompatible with migration and breeding (except in some tropical species). The time and energy devoted to molt may affect future reproductive opportunities (Lack 1954, Pitelka 1958, King and Farner 1961). Thus, the duration and timing of avian breeding could be at least partly determined by molt requirements. The redwing could be a good example of this. In eastern Washington, nesting terminates in June or, at the latest, early July. Yet insect emergence from breeding marshes, the primary food resource for the young, continues at high rates throughout the summer months (Orians 1980). It is possible that nesting ends earlier than theoretically possible on resource-availability grounds because the birds must shift the bulk of their nutrition and energetics toward molt, the best portion of which must be completed prior to migration.

Subadult males and delayed plumage maturation

One redwing sex-and-age class stands out, presenting in its morphology and behavior a host of ecological and evolutionary conundrums. This is the subadult male, who acquires his plumage during his first late summer/autumn molt, and retains it through his first winter and first full breeding season. Most subadult males do not breed. The subadult plumage, usually intermediate between that of adult males and females, is kept until the second late summer/autumn molt, when it is replaced by a fully adult one. This intermediate male plumage signals a major difference between the sexes, because hatching-year females molt during their first late summer/autumn into a plumage very much like that of adults of their sex, and then breed the following spring. Females begin breeding when they are 1 year old, but males not until they are 2 years or older.

Redwing male subadults are physiologically prepared for breeding. Their testes enlarge in early spring, just as the adults', although they never reach the same size (Wright and Wright 1944, Wiley and Hartnett 1976). They produce sperm. Although subadults constitute the single largest male age-class and are capable of reproduction, very few obtain territories, which is a prerequisite for breeding. For example, Shutler and Weatherhead (1991a) observed in one of their study areas 57 territories owned by adult males and only one by a subadult. At CNWR from 1977 through 1992, we had an annual average of slightly less than one subadult male with a territory on our core study area of 70–85 territories. Subadults will, however, acquire territories, defend them successfully, attract mates, and breed if adult territory owners and any adult replacements are sequentially removed (Laux 1970, Shutler and Weatherhead 1992). The reasons why subadults usually do not breed and why they have a distinctive plumage may be related.

Redwings are members of a large group of passerines in which females assume adult breeding plumage during their first year but males do so only after their first potential breeding season. Delayed maturation in males is most common in polygynous species, but also occurs in many monogamous ones (Rohwer *et al.* 1980). Several hypotheses have been proposed to explain delayed male breeding and plumage maturation in passerines, three of which are most likely to have some applicability in the case of the redwing.

The first is a sexual selection argument especially appropriate for polygynous species. Selander (1965, 1972) argued that, given the intense male–male competition in species in which, for example, males defend resources to attract mates, yearling males, who lack experience, may be at a competitive disadvantage. If so, they should be under selection to maximize survival during their first year, rather than to acquire an adult plumage and pay the costs of male–male aggression associated with it, particularly if they are, in any case, unlikely to reproduce. A subadult plumage, in this scenario, elicits relatively little aggression from adults, thereby enhancing the probability of survival.

Rohwer *et al.* (1980) proposed another hypothesis, that female-like subadult plumages are acquired by males as disguises to deceive adult territory owners about their sex and, by so doing, obtain entry to their territories. Subadult males could obtain several advantages from participating in female mimicry. They could copulate with females on the adult males' territories or, failing that, could at least gain knowledge of the territories for use in later attempts at territory acquisition. The female-like appearance might even permit yearlings to establish their own territories if aggression from adult males were inhibited. The female mimicry hypothesis is an outstanding idea, but is somewhat unlikely to explain redwing plumages, because (1) few subadults obtain territories, (2) subadults could copulate with females when they are off their territories, (3) subadults are recognized by adults and chased from territories, albeit at a greater average latency after their detection than adult intruders, and, most tellingly, (4) female mimicry cannot be an alternative male reproductive tactic because all DNA paternity analyses to date indicate that females do not engage in extra-pair copulations with subadult males.

Later, Rohwer (1986) and Rohwer and Butcher (1988) proposed that many passerine subadult male plumages are not adaptations to first breeding seasons at all, but to first winters. The advantage of a subadult plumage during winter, as it is usually more cryptic than adult plumage, could be in reducing the chances of predation, or in status signalling. The latter function could be particularly important. As Rohwer (1983) points out, social interactions during winter can be frequent and intense because populations that are dispersed during breeding often concentrate in relatively small wintering areas.

Rohwer and Butcher (1988) tested their ideas by evaluating molt patterns of a large number of species. They reasoned that in species with a female-like subadult plumage and an extensive or partial spring molt, if female mimicry during the first breeding season were important, then yearling males in these species in spring molt should remain in appearance female-like or become even more female-like. Of 42 species examined, none did. Supporting the winter-adaptation hypothesis, in species with partial spring molts, including the redwing, the molts tended to make the yearling males resemble more the adult males than females; in fact, Rohwer and Butcher claim that all feathers grown in spring in these species resemble adult male feathers. Furthermore, a number of species have a complete spring molt, and in these cases subadult males invariably change to an adult-like plumage. Thus, for redwings and some other species such as Northern Orioles, Orchard Orioles, the North American tanagers, and American Redstarts, subadult male plumages may be adaptations to their first winter. If so, Rohwer and Butcher suggest that, because an adult-like plumage could be advantageous during the yearling breeding season, the reason that these males do not molt to an adult-like plumage just prior to the breeding season must be one of energetic limitation. Moreover, this line of reasoning suggests that the subadult plumage of yearling male redwings, especially the more female-like ones, are, in fact, disadvantageous and retained only because of the energetic constraints of molt.

Thus, the question of subadult male plumage, like other aspects of redwing morphology, particularly size and color differences between the sexes, does not yield to easy or simple answers. Plumages and molts, it appears, are closely tied to breeding and wintering ecologies and are also under energetic constraints.

Survival and lifespan

Estimates of between-year survival in songbirds can be obtained in several ways, such as recapture rates of live, banded individuals, recovery rates of banded, dead birds, and return rates to breeding sites. Several estimates of annual survival for redwings, based on these methods, are given in Table 3.3. Values range from 42% to 62% annual survival for males and from 40% to 60% for females. Because between-year fidelity to breeding marshes is fairly high in male and female redwings, the best survival estimates emerge from analyses of breeding return rates. At CNWR, where we had excellent information on between-year male territory fidelity and also, when they occurred, on territorial movements, we had about a 55% average annual return rate to territories. This means that, on average, 55% of males with territories in year *x* could be found holding territories, either the same or different ones, in *x* +1. In many studies of birds, males are assumed dead when they fail to reappear seasonally on their territories but, in our extensive trapping during breeding seasons, we captured many male floaters that had previously owned territories; including them, between-year male survival in the study area averaged at least 60%. If we assume that we missed catching some floaters that previously owned territories and that a few males each year dispersed to find territories beyond the range of our censuses, then the actual survival rate was probably slightly higher than 60%. Survival rates of male floaters who never obtain territories are unknown, but may be similar to those of breeders.

Annual return rate of banded females to CNWR averaged 52%, but females

Table 3.3. Annual survival rates of adult male and female Red-winged Blackbirds

Location	Method[1]	Male	Female	Reference
Michigan	A	62%		Laux (1970)
Washington	A	60%	52+%	Beletsky and Orians (1987a), Orians and Beletsky (1989)
USA	B	42%	40%	Searcy and Yasukawa (1981)
	C	54%	54%	
USA	C	52%[2]		Stewart (1978)
Indiana	A	48%		Yasukawa (1987)
USA	B,C	51%	52%	Fankhauser (1971)
Wisconsin	A	55%		Nero (1956a)
British Columbia	A	46%		Picman (1981)

[1]Methods (see text): A = breeding return rate; B = recapture rate; C = recovery rate.
[2]A mixed sample of males and females.

moved more frequently than males. For example, approximately 28% of females changed breeding marshes each year (Orians and Beletsky 1989). Also, although we detected all males during each of their breeding years, about 15% of banded females were not detected breeding during at least one year that they were known to be alive (because their 'missing' years intervened between years for which we had breeding records for them; Beletsky and Orians 1996). They may have moved from the study area during their missing years. If true, this means that a fraction of females may also move from the study area but not return, which would lower survival estimates. Given this potential for underestimating female survival, we believe that annual survival of males and females at CNWR is about the same, 60% (Orians and Beletsky 1989).

Males at CNWR able to establish breeding territories held them for an average of about 2.6 breeding seasons (range 1–11). Males that first obtained their territories at 2, 3, or 4 years of age (which includes 95% or more of territory owners) bred for the same average number of years, indicating that mortality of territory owners is independent of age (Beletsky and Orians 1993). The average territory owner at CNWR, therefore, probably lives about 4 years. Females, who begin breeding when they are 1 year old, bred at CNWR for an average of 2.4 years (range 1–10), and thus, on average, probably live about 3 years. Because we monitored closely hundreds of individuals during breeding, we know that few adult redwings die during the spring breeding season. High mortality probably occurs during autumn and spring migrations, but probably is most concentrated during winter, when combinations of cold stress and the difficulty of foraging in poor weather conditions make survival more difficult. For instance, during two freezing nights in January in Kentucky, with snow on the ground, the estimated mortality of a 45,000 bird roost was approximately 1%. Although redwings comprised only 1% of the roost (consisting mainly of Common Grackles), they comprised 6.3% of birds found dead each morning at the roost (Stewart 1978).

Revisiting a redwing breeding marsh and noticing again males and females interacting to produce young, one still wonders at the variety of selection pressures that have conspired to generate two such different birds. Competition among males for territories has probably acted to drive males toward larger size and such competition, as well as signalling considerations, may have been involved in producing their striking adult plumage. Subadult males have a special plumage apparently to promote overwinter survival. Energetic constraints on breeding and the beneficial effects of appearing inconspicuous during nesting may have contributed to the females' relatively small size and drab coloration. As related in this chapter, a considerable amount of researcher time and energy has been expended to answer the questions posed in the chapter's initial paragraph and to formulate these initial conclusions. All the work was worthwhile because the understanding we now possess of the evolution of redwing morphology – why they look as they do – is among the most complete of any polygamous songbird and, indeed, of any bird.

Now that the reader has attained some knowledge of where redwings are found and what they look like, we can begin, in Chapter 4, to explore behav-

Chapter 4

MARSH LIVING, DIET AND NONBREEDING BEHAVIOR

INTRODUCTION: A YEAR IN THE LIFE OF A REDWING

A good way to begin examining redwing behavior is to review a year in their life. A typical year encompasses both breeding and nonbreeding periods. Precisely when redwings initiate their breeding seasons varies with latitude and climate from February to May; nesting lasts 2–3 months (for examples, see Chapter 6). In the Columbia National Wildlife Refuge (CNWR) population, males begin occupying their territories in February, females arrive and settle in March; nesting commences in early April, and final fledging occurs in late June. After breeding, by late July, redwings assemble into feeding and roosting flocks of from hundreds to hundreds of thousands; they undergo a molt during August through early October. In many north temperate areas, redwings begin southward migrations in late October and November. Most winter in southern regions of the USA in large groups, often mixed with other blackbird species, from December through mid February. By late February they begin moving northwards, arriving back in northern breeding areas in March and April.

For a number of excellent reasons, ecological and behavioral research on redwings is concentrated during spring and early summer, when they reproduce and perform many special behaviors that are not exhibited at other times of the year. To biologists of an evolutionary bent, breeding is of singular import because differential production of young underlies the process of evolution. Also inviting close and frequent study is the fact that redwings during breeding are territorial, and hence, sedentary; they defend small breeding areas in which they nest, roost, feed and, consequently, spend much of each day. At this time, they are easily located, observed and marked with specific color-band combinations so that the same individuals can be observed the next day, the next week, or the next month.

Learning about breeding season activities and the influences of habitat on reproduction are key to understanding much of the ecology and behavior of a species. But the breeding season occupies for redwings at most only 4 months per year. This means that the bulk of the redwings' life is spent pursuing other activities: feeding and roosting in flocks, molt, migration and spending time in wintering areas. We possess much less, and less-detailed information about these activities than we do for breeding both because research is concentrated during the breeding season and because of the almost intractable nature of studying the activities of itinerant animals, especially those in large groups.

Chapter 4 addresses some of the basic 'problems' of daily life for redwings – where to live, what to eat, how to forage, and also how they spend their time outside of the breeding season.

MARSHES AS HABITATS IN WHICH TO LIVE AND BREED

Redwings are primarily marsh-breeders, but also spend much time during non-breeding periods in or around marshes. Much of their behavior is influenced by peculiarities of marshes, from what they eat, to how they forage, where they hide, and even to how they perch – their sharp claws dig into the soft shoots of marsh plants such as cattail and bulrush, allowing perching at all angles (Fig. 4.1). Conversely, redwings are not adapted morphologically to marshes – they do not,

Fig. 4.1. Male redwings clinging to marsh plants. Their sharp claws dig into the soft shoots, allowing perching at any angle.

for example, possess webbed feet (Orians 1985). But marshes have certainly had a profound influence on the evolution of redwing breeding ecology and behavior, and thus, before delving into the species' behavior, one should first have a working knowledge of marshes.

A marsh has a relatively two-dimensional structure (compared with, say, a forest) that provides great visibility, both to outside observers and to its passerine inhabitants. Marshes usually consist of soft, highly repetitive, rapidly-growing and rapidly-renewing vegetation that provides good nest cover amid densely-packed, vertical, stems. Arguably the most critical attribute of marshes for redwings is their role as insect reservoirs. Emerging insects, with aquatic larval but terrestrial adult stages, often constitute the major food for redwings during breeding and are also fed to nestlings and fledglings.

Redwing breeding in many regions is tied to marsh insect emergence. The main characteristics of emergence are that the insect prey is present in great abundance but for only a limited period, which for the most part coincides with the nesting portion of the redwing's breeding season. Emerging insects are also a rapidly renewing food source. The number that emerge in a given area in the morning does not affect the emergence there in the afternoon; Monday's emergence does not influence Tuesday's. As Gordon Orians (1980), the chief purveyor of ideas on the intimate relationships between marsh ecology and blackbird behavior, pointed out, emergence can be so high that one small area, perhaps a grassy patch adjacent to a marsh, can be harvested for insects repeatedly on the same day. It is easy to understand how such patterns of food availability can strongly influence such important aspects of the birds' behavior as the distances and durations of foraging trips. These attributes of insect emergence mean that male and female redwings, although nesting sometimes in high-density colonies that might severely deplete other types of nearby food resources, need never venture far from their nests. They can forage on and near their territories, exploiting the marsh for the rapid replacement of insects that it provides.

An extreme example that illustrates this point is one small marsh at CNWR, McMannamon Pocket, which lies along a stream that drains a large lake. To facilitate water flow, refuge personnel long ago placed a large pipe between the lake and the stream so that all outflow of the lake is concentrated in a small area at the end of the pipe. During my 12 years at CNWR, only three lucky male redwings owned the territory that included the outflow side of the pipe. Their foraging behavior was identical: several times each hour they would fly to the pipe, land either on the mudbank adjoining the start of the stream or on the flotsam that collected in the small pool under the pipe outflow, and forage on the stunned and damaged insects that had washed into the territory during the previous few minutes. In effect, the male that owned that territory, and his harem, had constant access to a food concentrator and collector.

What influences insect emergence? Orians (1980) spent several years during the 1960s finding out. He placed wire-screen emergence traps on lake shores to catch insects emerging at the water's edge, and others he placed over marsh vegetation to catch insects that crawled up stalks to emerge from their larval casings. Traps were visited and emptied daily during the redwing breeding season, and all insects were counted and classified as to taxonomic group. Orians found that various insect types have their peak emergences during different parts of spring. For example, odonate (dragonflies and damselflies) emergence in the CNWR area begins in late April, increases in May, and peaks in June, whereas dipteran (midges, gnats, mosquitoes) emergence peaks in May and decreases in June. Emergence of certain insects is poor in early mornings, but rise to midday peaks; others are earlier 'risers' (e.g. dipterans).

Weather strongly affects emergence -- it is low on cold days. Physical and biological attributes of a body of water determine total seasonal emergence: water chemistry, photosynthetic productivity, depth, permanence, presence or absence of fish, presence of insect predators (other insects, other arthropods, fish, amphibians), and the kinds and density of submerged and emergent vegetation (Orians 1980). Based on variation in these features, even marshes quite close together can have markedly different insect emergences. Moreover, marshes can also vary dramatically in the magnitude of insect emergence from one year to the next. For example, one CNWR marsh, North Juvenile, had the greatest measured damselfly emergence in one year among seven monitored marshes, but the lowest emergence among the same group the next year (Orians and Wittenberger 1991). Such interyear variation affects how redwings make decisions about where to settle and breed (see Chapter 7).

After insects emerge from lakes and marshes, provided that they are not immediately preyed upon, they fly upland, and land nearby to dry and harden their wings and bodies. Termed 'teneral' at this point, they are at a very vulnerable stage – predators such as redwings can simply pick them up. A common scene through a telescope focused on a redwing nest is a female returning to her nestlings, her bill bulging with damselflies, cellophane-like wings glistening in sun (Fig. 4.2). Thus, surrounding uplands, not just the marshes *per se*, are important food areas. Even at CNWR, with its rich, insect-producing marshes, both males and females during nesting regularly leave marshes to forage upland (but rarely do they need to venture farther than about 1500 m from their territories; Beletsky and Orians 1987a).

High insect productivity of marshes affects more than just redwing prey choices and foraging behavior; it also influences polygynous breeding. The insect resource is so abundant and concentrated during the redwing nesting period that one parent can usually capture sufficient food for all dependent young. In the redwing's case, females assume this role, in effect freeing males to attract and court additional mates. Monogamy almost certainly would prevail if redwing young could not survive without feeding assistance from males. Male feeding behavior is

Fig. 4.2. A female redwing delivers a load of insect food to her young.

facultative; males 'know' how to do it, but because young survive solely with female feeding, they often do not participate (see Chapter 8).

Besides food, marshes provide perch sites, roosting cover, escape cover from predators, and, during breeding, nest sites. The protection given redwings by marsh vegetation is clearly demonstrated by their response to a dangerous predator. A typical spatial distribution of redwings on a crowded, noisy marsh in the middle of a breeding season finds many individuals perched high and exposed in the vegetation, in positions to see and be seen. But occasionally, in the blink of an eye, all the perched redwings go silent and dive into the vegetation – an event usually followed by the swooshing sound of a stooping falcon buzzing the marsh.

Redwings sleep down in the marsh vegetation, protected by its denseness and by the underlying water. Once in Michgan I made a point of observing male redwings 'go to sleep' on a marsh in April and then 'wake up' the following morning. As dusk fell male song rates slowed, and then when I could visually just detect their silhouettes but could hear their scratchy movements on the dead cattails, they ceased vocalizing and inched downwards. The following morning, shortly before dawn, I heard the birds first start moving upwards in the cattails, then a first, isolated song or two, then an increasing chorus as the males appeared at the tops of cattails. Meanley (1965) noted that in large marsh roosts redwings sleep usually within half a meter of the water's surface or, in rice fields, within 15–30 cm. Many also sleep directly on the water's surface, on floating mats of vegetation. In shrublands and deciduous thickets redwings sleep anywhere from 0.5–10 m above the ground, often at such high densities that all available perching space is used (Meanley 1965).

DIET, FORAGING, AND FEEDING YOUNG

Diet

Insects and other arthropods are the redwings' main food during the breeding season, and these invertebrates are fed to nestlings and fledglings. However, insects are often scarce or absent during nonbreeding portions of the year, so the redwing diet shifts to seeds. For example, the general diet of a southern Ontario population shifted from mostly insects during early summer to more seeds and corn during late summer, as indicated by the percentage of individuals found with those foods in their stomachs (Fig. 4.3; Hintz and Dyer 1970). Throughout autumn and winter, redwings spend most of each day foraging in fields for seeds from weeds and crops. Even in early spring, when marshes have thawed and male redwings have returned to breeding territories, insect emergence may not yet have started and so males still must subsist on seeds. They spend only early mornings on their marsh

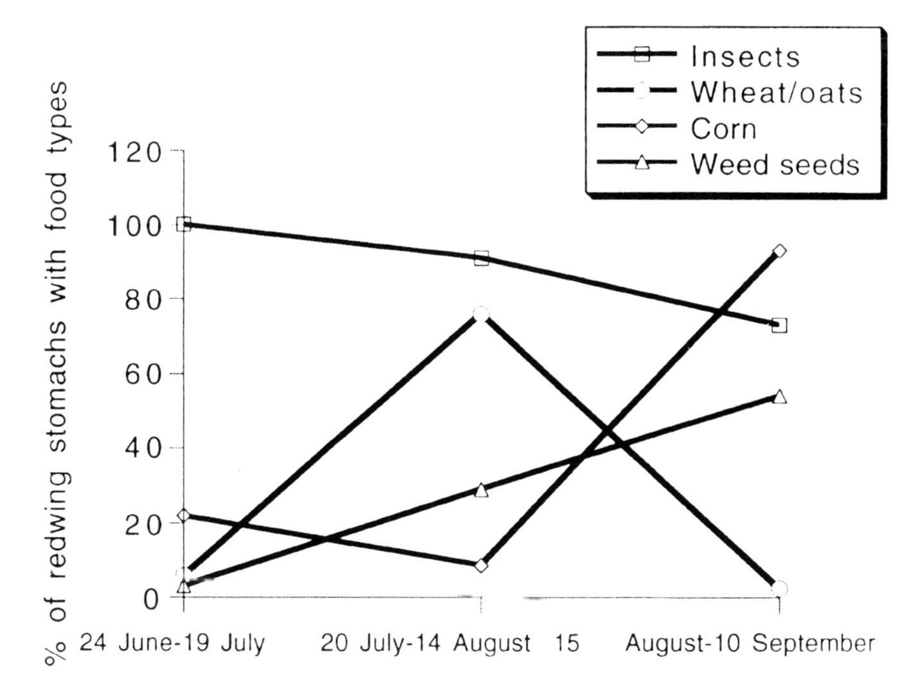

Fig. 4.3. Diet shift of southern Ontario redwings during summer (males and females combined). Values plotted are the percentages of redwing stomachs that contained each food group. First period, n = 218 stomachs; second, n = 160; third, n = 117. (Adapted from Hintz and Dyer 1970.)

territories at this time of year and then leave to forage elsewhere all day, returning only in the late afternoon. The proportion of insects in the diet then increases until, by the time nesting starts, adults are eating them almost exclusively. An exception is in agricultural areas, where waste grain can remain a major part of the diet (Fig. 4.4; McNicol *et al.* 1982).

Many studies have been conducted of redwing feeding and food preferences in summer, autumn, and winter because of the damage the birds inflict when they feed on crops (a subject explored in Chapter 11). Observations of feeding flocks and examinations of gastrointestinal tracts (gullets or stomachs) indicate that, depending on season and region, main foods of redwings outside of breeding are sunflower seeds, corn, rice, wheat, oats, millet, and weed seeds. For example, in Manitoba in autumn redwings feed on cereal grains, weed seeds, and insects (Nero 1984); they feed on corn and weed seeds in Ontario in autumn (Gartshore *et al.* 1982, McNicol *et al.* 1982), on sunflower, corn and foxtail seeds in North Dakota in summer and autumn (Linz *et al.* 1984), on corn and weed seeds in Tennessee in winter (White *et al.* 1985), and on rice (pulling new sprouts and removing ripening grains) in Louisiana in spring and autumn (Brugger and Dolbeer 1990).

Redwings are also somewhat opportunistic, capable of switching diets and even foraging methods when other appropriate prey appear. A good example is that redwings heavily exploit periodic cicadas (Homoptera: Cicadidae: *Magicicada* spp.). These insects live in the ground, emerging often in vast numbers for reproduction in some areas only every thirteenth or seventeeth year. Redwings switch quickly to feed on adult cicadas, rapidly becoming proficient at catching and handling these large insects that they have never before encountered (Steward *et al.* 1988). As a result, nestlings often feast during cicada years, as noted in eastern Washington (where cicadas emerge more frequently; Orians 1985) and in Illinois, leading to increased nestling biomass, decreased nestling starvation, and increased fledging success (Strehl and White 1986).

Although most prey is taken on the ground or while perched, redwings also indulge in a bit of flycatching, especially if the potential reward is great. They are certainly not as adept at it as flycatchers and swallows, but they try. Orians (1961) noted that males in California easily caught emerging dragonflies during their first flights away from the cattails, but gave up quickly when the insects outmaneuvered them. At CNWR, I have often watched adult male redwings chase and sometimes catch flying dragonflies, often making several attempts, but the failure rate is high. The high energetic cost of the rapid acrobatic flying necessary to chase such nimble insects suggests that a rapid 'giving up' time should be advantagous but that the potential reward for success – the energy content of the large insects – supports occasional attempts. On calm, warm afternoons at CNWR, both male and female redwings sally forth to catch on the wing emerging midges and flies. Typically the birds scan upwards and then, presumably spotting a bug, launch themselves straight up, more often than not catching the small prey, and then settle down onto the same perch to repeat the procedure. Redwings also catch butterflies and moths on the wing. There are occasional observations in the literature of redwings taking exotic prey. One male, for example, followed a Spotted Sandpiper on a beach, consuming caddisfly larvae that the sandpiper had excavated but rejected (Shelley 1930). A male redwing in Massachusetts apparently killed and decapitated a Sharp-tailed Sparrow trapped in a mist net (Helms 1962). Among other vertebrates, newts have been reported to be redwing prey (Bendire 1895).

Foraging

Redwings are great gapers (Chapter 2) and use gaping in almost all feeding situations. On the ground they gape to move small rocks, dirt clumps, and even the droppings of grazing mammals, exposing arthropods below that are then consumed (Orians 1961). Gaping is also used while wading in shallow streams to move aside small rocks (up to at least 45 g) 'either by inserting the bill underneath the stone and gaping it up and away from the body with the upper mandible, or by pushing it toward the body with the lower mandible. This exposes (insect) larvae beneath the rocks.' Orians (*loc. cit.*) continues '. . . floating debris among stones and aquatic vegetation is frequently moved by a sideways motion of the head with the bill fully gaped . . . gaping movements are also used in cattails and in grassland, the bill being inserted into the vegetation and then gaped to expose any insects and seeds within . . . gaping is also used when adults are feeding among the foliage of trees. . .' Redwings also use the gaping technique to penetrate corn husks and expose kernels that are then eaten (Bernhardt *et al.* 1987).

Although male and female redwings eat much the same food, they probably have different food and foraging site preferences. In autumn and winter the sexes are often segregated in different flocks, so feeding differences are not surprising; even during breeding, males and females may feed in different locations. At one marsh site in Missouri, males foraged in an adjacent mowed field significantly more often than did nesting females, whose feeding was more restricted to the marsh itself (Wilson 1978). Analysis of stomach contents demonstrated that the redwing sexes in eastern Ontario prefer foods in different proportions before, during, and after breeding (Fig. 4.4; McNicol *et al.* 1982). Feeding differences may be related to the size differential between the sexes, to differing energetic and nutritional requirements, and/or to evolved mechanisms that reduce food competition between the sexes.

Feeding young

Redwings switch from mostly granivorous to mostly insectivorous food during breeding, then back again after breeding. The shifting diet is probably a result of the changing availabilities of various foods and to the special nutritional and energetic demands of rearing young. Also, whereas seeds are excellent food for the maintenance of adults, they are less so for the rapidly growing redwing young, which undergo a 700% increase in mass during their first 9 days of life (Robertson 1973a). The seasonal change to arthropod consumption therefore is also important because it provides a high-protein diet for nestlings and fledglings (Morton 1973).

A simple yet ingenious method was used to determine what redwing parents feed to nestlings. Pieces of wire pipe-cleaner were placed around the necks of nestlings with sufficient slack to allow normal breathing but tight enough to stop passage of most food (Orians 1966). The method is not a perfect way to study

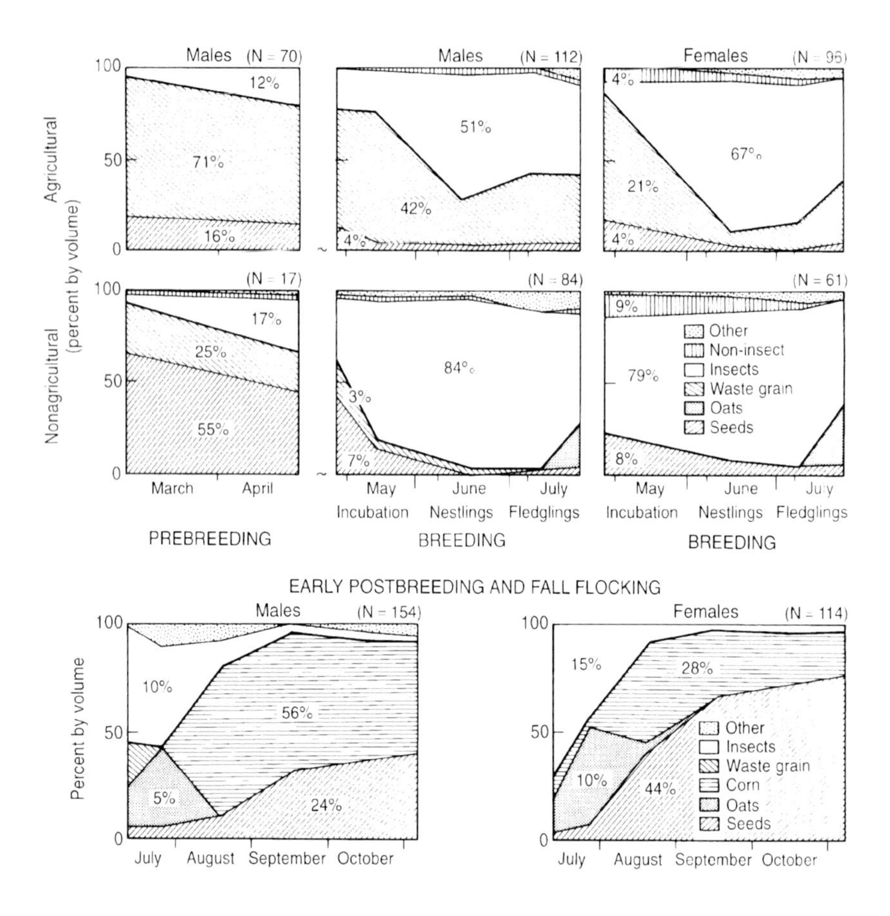

Fig. 4.4. Sex differences in the percentages of types of foods redwings eat in agricultural and nonagricultural areas of eastern Ontario. (From McNicol et al. *1982.)*

parental feeding because not every food item is prevented from passage down the esophagus, and the build-up of food in a nestling's throat probably affects its begging behavior. However, it is an excellent way to determine the types and relative proportions of foods brought to nestlings. Most of the analyses of nestling food samples listed in Table 4.1 relied on this method. As seen in the table, odonates (dragonflies and damselflies) were most commonly fed to nestlings, followed in frequency by lepidopterans (mostly caterpillars) and then orthopterans. Different insect food predominated in different regions. For example, odonates dominated the diets of nestlings in Washington, but orthopterans and lepidopterans predominated in Costa Rica. Note that emerging aquatic insects, although a major part of many nestling diets, were heavily supplemented by more terrestrial foods, including spiders.

Table 4.1. Types and percentages of prey that were delivered to nestling Red-winged Blackbirds[1]

	Washington[2]	Indiana[3]		Iowa[4]	Wisconsin[5]		Connecticut[6]		Costa Rica[7]	Southern Quebec uplands[8]	
		A	B		early	late	marsh	upland		early	late
Dragonflies and damselflies (Odonata)	37	55	25	30.5	13.9	8.7	21	1	3	2	–
Grasshoppers (Orthoptera)	2	–	–	–	–	43.2	–	–	32	4	54
Butterflies and Moths (Lepidoptera)	30	20	12	14.0	73.7	19.0	42	26	35	65	1
Flies (Diptera)	13	23	38	10.4	3.7	6.9	–	–	1	<1	4
Beetles (Coleoptera)	–	2	15	12.6	–	–	–	–	2	12	–
Spiders (Arachnida)	7	–	–	9.8	–	–	–	–	19	2	18
Other insects/items	11	–	10	22.7	8.7	22.2	37	73	8	14	23

[1]Breeding in marsh habitat, unless otherwise specified.
[2]Potholes region of Washington State; Orians and Horn (1969).
[3]Yellowwood Lake; Patterson (1991); A = diet when males helped feed, B = without male help.
[4]Voights (1973).
[5]Snelling (1968); early = 20 May–10 June, late = 11 June–30 July
[6]Robertson (1973a).
[7]Orians (1973).
[8]Bendell and Weatherhead (1982); early = 2–21 June; late = 23 July–2 August.

When feeding nestlings, parents must make repeated round trips from a central area, their nest, to foraging sites; this is generally termed 'central place foraging.' The distances involved per trip, the size and energy content of the items brought to the nestlings, and the number of items ferried per trip, can all vary over a broad range. Because offspring of poor foragers starve or are smaller and weaker when they fledge than those of good foragers, there should be strong selection for adult foraging efficiency. Thus, we might expect redwings, while foraging for their young, to follow a set of decision rules whose effect is to optimize efficiency of energy delivery. If prey items being collected are small, say flies or damselflies, and if it is easy to catch additional items after the first is taken, then we would expect an individual to terminate a foraging trip and return to the nest only after catching multiple prey items. It is common in eastern Washington for female redwings to arrive at their nests with 20 or more insects per load, and up to 36 have been reported (Orians 1985). The average load at one marsh was 12.7 items per trip for females and 22.2 items per trip for males (Orians 1980). Alternatively, when a large food item, such as a big grasshopper or a dragonfly, is caught, a redwing should break off its foraging and return immediately to the nest – a pattern also commonly noted. In Costa Rica, prey, usually grasshoppers, were difficult to catch, so parents usually returned to the nest after catching only one (Orians 1973).

BEHAVIOR DURING NONBREEDING PERIODS: FLOCKING AND ROOSTING

Why live in large groups?

Except during breeding, when adults occupy and defend territories, redwings spend their lives in large flocks, often passing their days gregariously among millions of their fellow blackbirds. On 26 December 1987, one count estimated that there were 53 million redwings together, part of a huge roost of 103 million blackbirds at Miller's Lake, Louisiana, a 2500 hectare coastal wetland (Brugger *et al.* 1992). What drives birds to congregate in such enormous numbers? What are the benefits to an individual redwing or to a small group of either joining a large flock or attracting additional individuals to create such a flock, and what are the advantages of a large roost? Hypotheses advanced to explain the evolution of flocking behavior and communal roosting generally embrace one of two possible benefits: reducing the probability of predation (safety in numbers hypotheses) or increasing the probability of finding food (exploitation of food resources hypotheses). A group cumulatively has, for detection of predators, more sense organs than an individual, and a large group, with thousands of eyes and ears, has a good chance of detecting a predator before it strikes. Groups may also be able to defend against a predator that launches an attack, or the size of the group may itself deter an attack (Lack 1968). Even if a predator is not detected or deterred by the group, the laws of probability argue that the larger the group, the lower is the chance of any one individual being taken by a predator (Bertram 1978). Thus, there is safety in numbers and the degree of safety increases with group size. One can also argue that a large group of birds is more likely to be detected by a predator than is a solitary individual, but the enhanced safety of a group must be decisive.

Enhanced exploitation of food resources has long been suggested to account for the evolution of many kinds of avian social behavior, including winter flocks, communal roosting, and colonial breeding. There is evidence that foraging in flocks improves the ability of individuals to find food (Murton 1971, Caraco 1979). Ward and Zahavi (1973) proposed that a main benefit of roosts and breeding colonies to individuals is that the assemblages function as 'information centers' about food locations. This is not to say that redwings sit around at night and trade tips about the best seed areas. Rather, if food is located in spatially unpredictable patches, then individuals in groups may benefit by taking advantage of others' searching experience. Birds that are new to an area or are unsuccessful at finding food can either follow successful foragers to food locations, or simply note the directions that successful foragers take when they leave a colony. Thus, the time and energy used to find food can be much reduced.

Good support for the information center hypothesis has been found for a few species, such as Cliff Swallows (Brown 1986). Whether redwing roosts function as information centers has not been tested, but Horn (1968) noted that groups of Brewer's Blackbirds at CNWR quickly found concentrations of teneral damselflies and suggested that following each other from the nesting colony was the reason. Gori (1984) tested whether Yellow-headed Blackbird colonies at CNWR serve as information centers by monitoring patterns of colony departures of color-banded individuals and with food placement experiments. His results were consistent with the information center hypothesis. For example, departures from the colony were highly clumped in time and birds that departed within 15 s of each other usually did so in the same direction. Yellowheads arrived at food placed some distance from colonies highly clumped in time, and territorial neighbors of birds who first discovered experimental food sites had a higher likelihood of subsequently finding the food than did non-neighbors. Thus, enhanced food-finding capabilities could be a main advantage of Yellowhead breeding groups, and also of other blackbird aggregations.

Group dynamics

After breeding, redwings assemble into flocks, often in large numbers. Flocks of mixed species are typical, redwings joined with other blackbirds such as Common Grackles and Brown-headed Cowbirds, and with European Starlings. Rusty Blackbirds and American Robins sometimes join the aggregations. The flocks forage in fields during the day and return to roosts – common sleeping areas – to spend the night. Roosts are usually in wetland habitats, such as tidal marshes, freshwater marshes, and rice fields. Otis *et al.* (1986) describe a roost of 750 000 redwings in South Dakota in a 2800 hectare marsh of mixed cattail and phragmites (common reeds). Dense vegetation in such habitats, as well as the water, provides safety from predators. The importance of water to roosting is suggested by the fact that once rice fields, normally flooded to a depth of 10–20 cm, are drained for harvest, blackbirds that have been roosting in them desert (Meanley 1965). Flocks also sometimes roost in drier areas such as stands of coniferous trees, deciduous thickets, and sugarcane fields, all of which provide good protection from the weather. Roosting of enormous numbers of birds damages habitat; for example, the sheer weight of thousands of birds roosting in trees often breaks branches.

The general locations of roosts are probably largely determined by their proximity to food supplies. Meanley (1965) followed the activities of many large roosts in the Mississippi Valley and Atlantic Coastal Plain over a 14-year period. In Arkansas, for example, all 15 roosts of a million or more blackbirds during the 1962–63 winter were located within 25 miles of rice-farming areas. Because there is little food for redwings in and around a roosting marsh in autumn and winter, they must make daily trips to foraging areas. Blackbirds leave roosts at, or right after dawn and return in the evening, usually before sunset. The 'morning exodus' is rapid; an entire roost of 500 000 departed an Arkansas cattail marsh in half an hour (Meanley, *op. cit.*). Often flocks fly considerable distances from the roost before settling to feed, sometimes passing over closer potential feeding sites to do so. Two flocks in Texas flew 46 and 52 miles, respectively, from a coastal roost to inland rice feeding sites. Often, however, feeding occurs within 10 km of a roost (Dolbeer 1982).

Once at a foraging site, blackbird flocks spend the day alternately feeding, flying to water to drink, and resting and preening, often in trees. When feeding, flocks often appear to 'roll' across fields (see illustration, p. 268):

> 'This rolling movement is the result of a particular pattern of movement by the flock members. An individual, finding itself at the back end of the flock, flies over its flock-mates to the front and forages in a small area until it again finds itself at the rear of the group. Each individual thus alternates foraging in a relatively small area with short flights to the front of the flock. The result is a continuous flow of birds across the foraging area. The advantage of the system is that each bird has exclusive use of foraging sites that have not yet been visited by other flock members and yet has to make only short flights between feeding locations.' (Orians 1985).

Starting mid-afternoon, the birds begin return trips, which, unlike the long, direct morning flights, often consist of series of shorter movements. Roosts reform in evenings over longer periods than they empty in mornings, as a consequence of flocks returning from feeding sites located at various distances from the roost. Individuals return to their roosts satiated and are not interested in food. When trying to trap redwings in early autumn at a small marsh roost (about 1000 adult and subadult males) at CNWR, I discovered that the birds ignored my seed-baited traps and even walked over the seeds that I had sprinkled on the ground without feeding on them.

Some roosts are used almost all year round. During the breeding season, roost residents may be predominantly subadult males. Roost numbers increase with the end of the nesting season, with dramatic growth occurring in late summer. Redwings join summer roosts generally within 200 km of their nesting localities (Dolbeer 1982). When molt is completed in late September on the Atlantic seaboard, roosts empty for migration. Some autumn roosts are used for only short periods by migrants, but others that form in autumn are used throughout winter and early spring. Autumn roosts may contain premigratory groups from the local breeding population and incoming, wintering groups. Roosts reach their peak size often in mid- or late-winter, and some continue to be used until mid-April, by which time most redwings have dispersed to breeding sites (Meanley 1965).

The age and sex composition of redwing roosts changes seasonally. In mixed species roosts in Quebec at midsummer, adult male and female redwings, and hatching-year males and females, were about equally represented, but by late summer, the redwing population was nearly 80% male (Weatherhead 1981). Females at that time had presumably begun their migrations. Furthermore, roosts are often spatially segregated by sex or species, resulting from birds spending the day feeding in segregated flocks and returning together, often probably to the same, preferred section of roosts. Meanley (1965) observed several thousand female redwings return each evening for three winters to the 'same low brushy vegetation in the same section of the roost.' In two tree roosts, starlings occupied the highest positions, grackles and redwings were at a mid-level, followed by a stratum of Brown-headed Cowbirds and female redwings, and finally, by Rusty Blackbirds and more female redwings (Meanley, *op. cit.*).

Social behavior

How are roosts socially structured and how do the birds interact? Little is known about the behavior of redwings while in their roosts. Presumably, there is a rich social organization, but how it compares with that found during breeding is not well understood. Pat Weatherhead and his colleagues have the best information. Males in an Ontario roost used many of the same aggressive displays and behaviors that they utilized during the breeding season, such as the exposure of the epaulets, song spread displays, and chasing (see Chapter 5); the use of these in roosts apparently functioned to establish and maintain dominance (Weatherhead and Hoysak 1984). Adult males were dominant to hatching year males, and consequently tended to occupy central roost positions in dense vegetation, locations presumably safer from predators and microclimatically superior. Hatching-year males tended to remain at the roost periphery, in sparser vegetation.

How might the information-center hypothesis and dominance structuring be related in redwing roosts? Weatherhead (1985) thought that a problem with applying the information-center hypothesis to redwing roosts was that it was usually older birds who, with their greater experience, were better foragers. Thus, the hypothesis provided a reason for younger birds to be in large groups, but why should adults, who know where food is, participate? The answer may be that older birds are socially dominant, obtain central roost positions, and so are less likely to suffer depredation (Weatherhead, *op. cit.*). In other words, different age-groups might gather in large roosts for different reasons.

Flock assembly

How do large groups assemble? It is likely that both visual and vocal cues facilitate the formation of flocks? The redwings' conspicuous black plumage probably aids flying individuals to detect others already on the ground (Beecher 1951). Røskaft and Rohwer (1987) demonstrated, after male territory owners were removed and replaced with stuffed mounts, that newly-arriving males preferred to settle near epaulet-blackened mounts rather than near others with red epaulets or empty wood perches. They conclude that the redwing's black plumage may serve as a long-distance attraction signal to facilitate flock formation. Redwings and other blackbirds routinely vocalize while in flocks and roosts, perhaps using their songs and calls to form and maintain flocks. Brenowitz (1981), observing that male redwings sing in both autumn and spring roosts, tested the hypothesis that the songs have a role as an attraction signal. He broadcast redwing songs underneath a commonly used flyway. Males landed within 50 m of the loudspeaker in 11 of 12 presentations, but never when control sparrow songs were played, which supported this hypothesis. The formation of large flocks may be facilitated by vocalizations because, as more individuals aggregate the louder is the cumulative sound produced, thus perhaps attracting even more individuals. The effect is the same as different numbers of people in a football stadium: two people shouting in the stadium is inaudible in the carpark, but 50 000 fans shouting is audible some distance away. Singing and calling in blackbird flocks continues long after all individuals have returned to a roost, sometimes through much of the night (Meanley 1965), so these vocalizations also may have other functions.

Now that we have delved in detail into several aspects of the basic biology and nonbreeding behavior of redwings, we are almost ready to begin an exploration of their breeding. First, though, in Chapter 5, we need to examine visual and vocal signals, which are integral to many aspects of breeding.

Chapter 5

COMMUNICATION

INTRODUCTION

Communication is one of the most intriguing and intensively studied subdisciplines of animal behavior. There are many reasons why so many investigators labor in this area and that great strides have been made in our understanding of animal, especially avian, communication. Foremost is the central role of communication in all social behavior. Whenever two or more conspecifics meet, communication occurs which influences the actions of the individuals, their motivations, and the duration of the association. Even primarily solitary animals at times meet conspecifics and, of course, must get together for reproduction; communication mediates these interactions. Communication also stands at a crossroads of researcher interests; sooner or later, ethologists studying social organization, spacing, dominance, reproduction, sexual selection and other subjects arrive at a point in their work when they must examine signalling. From a practical standpoint, vocal and visual signalling systems especially lend themselves to relatively simple experiments in both field and laboratory. Also, bird signals are often overt – given loudly or obviously – for the sensible reason that their function is to transmit information across some distance; studying overt behavior has obvious benefits. Perhaps the most motivating reason biologists in such numbers study communication is their fascination with the signals themselves, and the very human desire to solve puzzles. Birds clearly convey information to each other by their vocalizations, their often-bright plumage, and their stereotyped movements or displays. But what are they saying? What do the signals mean? Why are they often delivered so frequently? And what are their effects on listeners or observers? The search for answers to these questions at times appears almost to stem from the human compulsion to listen in on other conversations – that it is interspecific eavesdropping appears not to matter in the least.

Communication occurs when one animal, the sender, emits a signal that is received and understood by another animal. Information (messages) encoded in the signal allows the second animal, the receiver, to modify its behavior and make decisions about future actions. Animals give signals because, by sending them, and thus prompting a response in receivers, they benefit; otherwise signals could not evolve. By implication, receivers, by decoding information in signals and acting on it, must also benefit or their responses could not evolve. Some of the complexity of communication systems, and why they are of immense interest to ethologists and evolutionary biologists, is thus readily apparent: communication always involves two or more individuals interacting, with signals mediating the interactions, and athough sender and receiver may have similar or conflicting motivations, they both benefit in some way from the system's operation. Currently there are controversies among researchers and theorists as to whether all communication is 'honest' – whether animals occasionally or almost always try to deceive one another, and indeed, whether all communication can be viewed simply as senders trying to manipulate the actions of receivers for personal gain. However, to research animal communication, the most productive way of viewing signals may be as behaviors like any other, with benefits and costs to the signaller.

Animal signals occur in a broad range of physical forms to take advantage of different transmission media and to stimulate various sense organs. Some animals use as signals chemicals or even electrical discharges. However, birds are primarily visual and vocal creatures and their signals are transmitted via light and sound energy. In fact, this may be one reason for the disproportionate share of researcher interest in avian communication, over other vertebrates. Human communication is also strongly rooted in the visual and acoustic modalities, so perhaps we relate in some ways better to birds than to animals that, for example, inhabit an olfactory world, where chemicals are the signals that mediate most interactions.

Redwings have, to date, contributed enormously to studies of animal communication, enough, in fact, that an entire book could be produced detailing their contributions. In several areas, they have made seminal contributions to our understanding of animal signals. The outcome of four decades of observation and experimental manipulation of redwing communication is no less than what is now arguably the best known and *understood* communication system of any bird. It may well be redwing signals for which we now possess the most complete description: the vocal and visual signals of each sex, how and when they are used, why they are structured as they are, and what they mean.

What sort of signals might redwings be expected to have? At a minimum, the repertoire of males should include signals to mediate all of their main breeding-related activities: (1) acquisition, maintenance, and defense of a territory; (2) attraction and courtship of, and copulation with, mates; and (3) nest protection. Females should have signals to: (1) communicate with mates to establish and maintain a pair-bond, and to coordinate sex and nesting; (2) communicate with other females; and (3) protect nests. Both sexes may have signals to communicate with their young. Because of the redwing's strongly expressed polygyny and colonial breeding situation, they may possess specialized signals that would not be expected in the repertoire of a more typical songbird. In addition, there will be signals for communication during nonbreeding periods and perhaps for communication with other species.

The bulk of this chapter is devoted to describing the diversity and physical properties of male and female redwing signals, and then, in some detail, the observations and experiments that have yielded insights into signal meaning and function. I concentrate on the signals that have been emphasized in research. Male and female signals are considered separately because, although there is some overlap, many of their signals differ both in structure and function. I conclude by considering special aspects of redwing biology that may have shaped their communication system and the potential generality of information on redwing communication for studies of other species.

MALE PLUMAGE AND VISUAL COMMUNICATION

Epaulets

Male redwings are visually striking birds, essentially all black with two splashes of bright coloring, the shoulder epaulets. When we see bright patches on animals, we expect a use in signalling. Certainly, when observing male redwings interact during the breeding season with other males and with females, our attention is often first drawn to their patches of red feathers, as must be the attention of conspecifics. Thus, it is not surprising that the first systematic investigations of redwing communication dealt with male epaulets.

While working at the Columbia National Wildlife Refuge (CNWR) I have been approached more than once by fishermen and asked why the blackbirds inhabiting their favorite fishing spots have so much variation in the size of the red shoulder patches: some have no red, some a little, while others have large red patches. The answer is that all the males they saw had epaulets of about the same size. But the red feathers can be completely sleeked and covered by overlaying black feathers (scapulars) so that no red shows (in this configuration the buff-colored stripe alone is visible), and they can be uncovered with varying degrees of exposure and erection. Fully erect, the epaulets stand out from the wing (Fig. 5.1). Epaulets are made larger and more obvious during many male displays, both those directed at other males and at females; thus, both inter- and intra-sexual functions are likely.

Fig. 5.1. Front and side views of two common postures of male redwings, high-intensity song spread display above, normal resting posture below.

The most important clue to the function of epaulets is that males show their red feathers to each other during territorial and other aggressive contests. Furthermore, when floaters, males without territories, trespass into territories, they do so with epaulets concealed, whereas the territory owners who evict them do so with epaulets exposed. These behaviors suggest that exposed epaulets communicate threat and are thus important for territorial defense. There have been several experimental studies of male redwing coloration focusing on epaulets which, in the best scientific tradition, built on each other and incrementally elucidated epaulet function (Peek 1972, Smith 1972, Morris 1975, Rohwer 1978, Hansen and Rohwer 1986, Røskaft and Rohwer 1987).

At about the same time, quite independently, Douglas Smith (in Washington and Massachusetts) and Frank Peek (in Pennsylvania), decided that a good way to investigate the function of male epaulets was to artificially eliminate their color – 'signal removal' experiments, if you will – and then monitor territorial and reproductive effects. Territorial males were trapped and their epaulets quickly altered: Smith dyed epaulets black; Peek dyed them black, covered them with black 'Magic Marker,' or, in three cases, cut off the red feather areas leaving only white shafts. Results were remarkably similar. During the early parts of the breeding season, when territories were being established and male–male aggressive interactions peak, significant percentages of experimental males lost territories: 62% of 21 males in Smith's study and 50% of 12 males in Peek's. Other altered males had difficulty retaining their territories or maintaining their original size. A few that retained territories were situated on habitat that was marginal for breeding and so were probably exposed to only weak competition from other males. Control males, who were trapped and released after their epaulets were dabbed with alcohol or who were untreated in any way, usually retained territories (92% in Smith's study, 100% in Peek's). A third study of epaulet-blackening with small samples, in Ohio, had a similar result (Morris 1975).

Apparently the main effect of blackened epaulets was that other male redwings, both territorial neighbors and floaters, were 'unthreatened' by all-black redwings and so trespassed freely on their territories, eventually driving them off. Trespass rates were much higher on experimental territories than on the control territories (Peek 1972). Thus, Smith and Peek, replicating each other's work, concluded that epaulets transmit threat to conspecific males and are important for deterring territorial intrusions that could lead to territory loss. Because both Smith and Morris observed that epaulet-blackened males who managed to hold all or parts of their territories were also able to attract or retain females, an intersexual function for epaulets was not indicated. Experimental males in Peek's study were unable to attract mates while their epaulets were black, although one eventually took over a territory on which a female was nesting.

Males take note of epaulets, trespass on the territories of males that lack them, and often are able to drive off these altered individuals and replace them. Is this because males without red epaulets are considered non-threatening or because, lacking red feathers, they simply are not identified as Red-winged Blackbirds? The former hypothesis suggests that neighbors and floaters trespassed because they perceived a weak, non-threatening male conspecific, whom they eventually exhausted and replaced; the latter hypothesis suggests that they trespassed because they perceived a territory without an owner. Hansen and Rohwer (1986) tested these alternatives by presenting to territorial males mounts of redwings with blackened epaulets or normal male Brewer's Blackbird mounts. Brewer's Blackbird males are all black, with areas of greenish or purplish iridescence. Territory owners were much more aggressive toward the stuffed, all-black, redwings, hitting them significantly more often than they hit Brewer's Blackbirds. This result strongly suggests that the live males had no trouble distinguishing between a conspecific intruder (a threat to territory ownership), even one without red epaulets, and a heterospecifc one (no threat). Epaulets, therefore, probably transmit intrasexual threat and, if they do assist in species recognition, they are certainly not necessary for it.

Epaulets are bright, colorful, attention-grabbing, and threatening – attributes that a territory owner can and does put to good use defending his ownership status and territory. But not all males own territories, although we assume that many subadults and all adults would like to. Thus, adult floaters have the same threat signals as owners, the difference being that they rarely display them. Further, there are situations in which even a territory owner does not want to appear conspicuous and threatening. Epaulets are coverable badges – cues of social status that are readily hidden. Hansen and Rohwer (1986) suggest that coverable badges may evolve in territorial systems in which fighting is risky for all participants and floaters frequently trespass on territories. Redwing floaters prospecting for territories or trespassing for other reasons might not (and probably most often do not) want to threaten the owner (see Chapter 10); concealing their threat signals would thus be advantageous. Territory owners also encounter situations in which drawing attention to themselves would be counterproductive, such as when they trespass covertly on other males' territories. They cover their epaulets in these forays. Furthermore, as Hansen and Rohwer point out, because floaters may get territories at any time, males cannot predict before the autumn molt whether in the future they will be owners or floaters. Thus, a coverable badge of ownership status leaves males morphologically equipped to assume either role at any time. It is also possible that coverable epaulets are an antipredation adaptation because males with perpetually exposed red shoulder patches should be detected and preyed upon more frequently than males who can conceal their epaulets for large portions of each day.

Less information is available as to whether males use their epaulets to communicate with females. The fact that males with blackened epaulets in two studies were able to attract females who nested suggests that epaulets are not necessary for courtship or reproduction. Conversely, Peek's experimental males evidently could not attract mates. However, the influence of local floater density, or 'contender' pressure, confounds these studies. As Searcy and Yasukawa (1995) point out, Peek's males had great difficulty maintaining their territories against intruders; thus, it may well have been that females were not repelled by the males' lack of red feathers, but by their very obvious problems with territorial maintenance. If the exposure of epaulets during courtship display is any indication, then epaulets may indeed have some intersexual function: males regularly show them to females (see below). Also, male plumage is known to affect female choice of mates in other species, such as Long-tailed Widowbirds (Andersson 1982) and Barn Swallows (Møller 1988), and thus, it would not be surprising if male redwings signalled females with their epaulets.

Redwings are not all red

If red is threatening to male redwings, how about black, which covers most of their body? Røskaft and Rohwer (1987) conducted an experiment in which territory owners were removed from a long, thin marsh and replaced with four objects arranged in a straight row: a stuffed male redwing with blackened epaulets on a mounting pole; a stuffed male with normal, exposed red epaulets on a pole; a stuffed male with enlarged, exposed red epaulets on a pole; and a bare pole. Trespassers rarely landed near the mounts with red epaulets, but landed often near the blackened mount and the control pole, suggesting that black body coloration is nonthreatening or less threatening than red. Furthermore, in a subsequent comparison on smaller marshes where trespassers had only the choice to land near epaulet-blackened males or the empty pole, they preferred the blackened male. This result suggests that not only is black body color nonthreatening, it may be attractive and is perhaps used as an aggregation or flocking signal.

Apart from red and black, the only other color on the external body of a male redwing is the buff/yellow line that borders the epaulet. It is visible when epaulets are covered but inconspicuous during displays. Because the larger, brighter, red area is the more obvious color signal, the communicative potential of the buff-yellow bar has not been separately studied, but its width is sometimes included in studies of epaulet size versus dominance or territorial status (Searcy 1979d, Shutler and Weatherhead 1991a). The buff patch was presumably darkened in all the epaulet-blackening studies discussed above; only Smith (1972) provided photographs of dyed birds in his published work.

Red bands and Red-winged Blackbirds

Before leaving the subject of male color, I must digress to a seemingly minor issue that relates to the signalling function of red coloration, but one with potentially broad implications for the study of redwing biology. This is the issue of placing red legbands on redwings for studies requiring positive identification of individuals. Until Nancy Burley (1981, 1985, 1986) discovered that band color could influence mate choice, mortality rate, and reproductive success in Zebra Finches, ornithologists had routinely and without reservation placed combinations of colored plastic or aluminum legbands or other color markers on their avian subjects, to distinguish them as individuals. (In fact, more so than any other scientific technique, color-banding, and the direct observation of individuals that it allows, has been a basis of most of the advances we have made in understanding the ecology and behavior of birds during the last 60 years.) Because ornithologists knew or suspected, based on Darwinian principles, that females might choose mates on morphological attributes, they doubtless should have questioned earlier their marking techniques.

The epaulet studies of the 1970s demonstrated clearly that red is an important color signal for redwings. There was thus a possibility that the bands we placed on males at CNWR and that other redwing researchers used influenced mate choice and breeding success. Because our study in particular was designed to measure lifetime reproductive success (LRS) in a natural population, we were especially anxious to determine if, unknowingly, we had been manipulating mate choice and LRS since the study's inception. The danger was that, by marking males with red bands, we had unwittingly rendered them more or less attractive to prospecting females or more or less threatening to other males.

We divided breeding males in our core study area into those with at least one red band (n = 560 territory-years) and those without red bands (n = 231 territory-years) in their color combinations. Males had one to three bands on each leg, with a range of zero to four red bands in their combinations. Other bands were blue, gold or silver. We found that average annual harem sizes and number of young fledged for the two groups of males were statistically equivalent (Beletsky and Orians 1989a). When we looked at males during their first breeding years, when females make their first choice of whether to nest on their territories, the same relationships held. Furthermore, we found no correlation between the number of red bands territorial males wore and their harem sizes, strongly suggesting that females lacked positive or negative preferences for males with red bands. Thus, wearing red bands did not measurably affect annual male reproductive success. Because the average number of years that males held territories was also the same for males with and without red bands, and because territory tenure strongly influences LRS, red bands evidently did not influence LRS in our population. Replicating our analysis in its essentials, Weatherhead *et al.* (1991) found the same lack of evidence for any influence of red bands on male breeding success in their Ontario population.

The matter may have rested there but for a contemporaneous experiment in Ontario by Karen Metz and Patrick Weatherhead. They found a very definite detrimental effect of placing red bands on territorial males. Redwings given all red bands suffered increased aggression from neighbors and lost their territories at a higher rate than control males with all black bands (male redwing legs are black). Metz and Weatherhead (1991) theorized that extra red on their experimental males led to increased trespassing by neighbors because the bands exaggerated aggressive signalling, somehow disrupting the normal balance between a male's actual aggressiveness level or fighting ability, and the levels usually, perhaps honestly, signalled to other males. This makes sense in the light of Scott Freeman's (1987) finding that territorial male redwings may constantly monitor and evaluate their neighbors' aggressive interactions and displays. Further, in a subsequent experiment, Metz and Weatherhead (1993) presented territorial males with a male model with black, blue or red bands. They discovered that the males were equally, quickly aggressive toward the model with blue or black bands, but stayed farther away from and took longer to attack the model with red bands, which provides further evidence that it was the band's red, epaulet-like, color that males found threatening.

Because Weatherhead *et al.* (1991) found the same long-term trends in their demographic data as we had, the work of Metz and Weatherhead was paradoxical: males in an experimental situation with all red bands had difficulty holding territories, yet routine color-banding of males by the same people in the same geographic area produced no discernible band effects. Several factors could contribute to the paradox. The number of red bands may matter; for example, the amount of red may increase to some 'threshold' level that finally stimulates or elicits male–male aggression (Metz and Weatherhead 1991). Experimental males were given five red bands, whereas most normal color combinations contain fewer. Also, the presence of other colors in band combinations may cancel color-based mating preferences (Burley 1985). Furthermore, the shade of red may be critical. Metz and Weatherhead used plastic bands that matched the bright red of natural epaulets. Our aluminum bands are a different, often darker, shade of red, and sometimes a rust color; they usually fade after several years to a pinkish hue. Thus, the experimental and observational studies of red band effects are not strictly analogous. On balance, the preponderance of evidence to date suggests that using red bands in demographic studies of redwings does not measurably affect results. In addition to providing this degree of assurance, the red-band studies, in conjunction with the previous findings with blackened-epaulet males, were useful in illuminating what are apparently finely-balanced epaulet selection pressures – too much red or too little red is socially deleterious.

Male redwing displays

Male redwings give several conspicuous displays that appear to communicate territory ownership and aggressive motivation in territorial contexts. Epaulets are prominently featured in these displays. The three displays described below are given at such high frequency that in any morning hour of breeding season observation, each would be given many times and would be immediately obvious to any viewer.

Song Spread

The most frequent and conspicuous territorial display, by far, is the Song Spread (Fig. 5.2). It is a combined vocal and visual display – the 1–2 s song given concurrently with various degrees of wing positioning, tail fanning, and epaulet erection. Orians and Christman (1968) perhaps best describe the visual aspect of the display:

> 'In its mildest form . . . it consists of a slight lowering of the tail and a slight forward movement of the head when the song is uttered. In more intense versions the tail is lowered farther and spread, the head is thrust farther forward, the epaulets are exposed, and the contour feathers, particularly those of the head and back, are fluffed. In its fullest expression, the tail is strongly lowered and fully spread, the epaulets flared, the wings are fully arched out and down, and

Fig. 5.2. A male redwing on top of a cattail plant, giving a high intensity Song Spread Display (from Smith 1976).

all contour feathers are fluffed and those of the neck, lower back, and flanks fully ruffled. All possible intermediates between the mildest and the strongest forms . . . are commonly used.'

Because of its conspicuousness, high rate of delivery, and presumed role in territory maintenance, the display is often used by researchers in studies of territoriality as a quantifiable measure of aggressive motivation. When so used, song spreads are often divided into low intensity (song delivered with tail and wing feathers sleek), medium intensity (song given with tail feathers somewhat spread and epaulets exposed), and high intensity (song delivered with tail fully spread, epaulets flared, and wings fully arched out and down). The net effect of a medium- or high-intensity song spread is to attract the attention of any observer, both by the stridency of the song and by the contrast a black-and-red bird makes against the (usual) backdrop of brown or green vegetation. The display also, because of wing positioning and feather erection, makes a male appear larger than when in a sleeked or resting posture.

The display is given primarily when the males are on or near their territories, on any substrate – perched on top of vegetation, on or in the marsh, on the ground, or, at CNWR, on the basalt cliffs that overlook many territories. They are given when males are advertising, i.e. when singing to maintain territory and/or attract mates in the absence of immediate listeners, and also when closely interacting with other territory owners, with floaters, and with females. The song spread display rate essentially equals song rate, which, during advertising, averages three to six songs per minute, usually fairly evenly spaced in time. But when a male aggressively engages another male, song spread rate can increase to 10 to 12/min, or even higher for brief periods. For example, males at CNWR in March gave song spreads at an average rate of 38.5 ± 20.5 per 15-min period (n = 161 periods), or about 2.6/min; however, those periods included males observed during all territorial activities, including maintenance activities such as foraging and preening, when display rates are lower.

An investigation of the song spread presented Indiana territory owners with both a redwing mount and recorded redwing song to simulate territorial intrusions; the owners' reactions were monitored (Yasukawa 1978). They responded to the mount with varying degrees of aggressiveness, which were measured as how close they approached the mount and their latency to land near the mount and strike it. Males that gave higher-intensity song spreads when more than 5 m from the mount were more likely to attack it than males who gave lower-intensity song spreads. Thus, the song spread is a graded display: the greater its intensity (degree of wing spread and feather erection), the greater the degree of threat or aggressive motivation communicated to onlookers (Yasukawa, *op. cit.*). A graded aggressive display would be a plausible adaptation to situations in which it is in a male's best interest not to signal an immediate high likelihood of attack, e.g. if signalling strong aggression always leads to fighting and risk of injury.

Bill-up Display and boundary disputes

Another frequent territorial display is the Bill-up (see illustration at chapter head). The wings are held close to the body, epaulets are exposed and flared, and the neck is stretched upwards. The bill is tilted toward the sky, at a variable angle – from approximately 45° from horizontal to almost 90° (Orians and Christman 1968). Males in this posture look as if they are watching something overhead, and would appear comical if they were alone. But bill-ups are given only during boundary disputes and other close, male–male, aggressive interactions.

When two territory owners meet at their common boundary, they often enter into a boundary dispute. In a sustained dispute, the two males, presumably setting, reinforcing, or arguing over a boundary, are often situated only centimeters apart, bills tilted upwards; often they face away from each other so that they are positioned back-to-back. In these positions, when in a marsh, they slowly climb adjacent plant stalks, such as cattails. The effect is one of one-upsmanship – one male trying to climb higher than the other to demonstrate dominance or ownership. When they reach the tops of their respective perches, they continue to give song spreads and repeated bill-ups (lowering and then again raising the bill). The dispute ends when one male flies away. Particularly intense disputes terminate in physical combat. One male launches himself at the other and they fall into the marsh, legs and bills locked together, fighting. After the fight, which can be very brief or continue for several minutes, the loser flies off. If no clear winner emerges, the boundary dispute continues; they crawl up the reeds, give bill-ups and start again. Boundary dispute durations vary tremendously; some, for example, between established neighbors, can be of the order of a few seconds and are terminated when one of the males quickly flies off to attend to more pressing business elsewhere, while others last 15 min or more. For example, during the 1983 breeding season at CNWR, we monitored morning behavior of 40 territorial male redwings during 565 15-min periods. The males averaged 0.9 boundary disputes per 15 min observation period, with a range of 0–7. Disputes occurred in 317, or 56%, of the 565 periods. The average amount of time males engaged in boundary disputes was 1.0 ± 1.3 minutes per 15-min period, range 0 to 9 min. When boundaries are unstable or in flux, males sometimes appear to spend much of the day engaged in disputes with one or another of their neighbors.

Bill-ups and boundary disputes occur throughout the breeding season, first when males establish boundaries with neighbors, then at any time when boundaries change or when a persistent floater enters a territory. The displeasure of a territory owner toward an intruding male, whether owner or floater, is initially signalled with high-intensity song spreads. If the offender does not leave, he is approached and shown bill-up displays with fully erect epaulets. Only then, if the intruder persists, is a physical attack launched, which usually terminates immediately with the intruder flying on his own from the territory (being supplanted on his perch by the owner), or with the intruder being chased out of the territory. Again, the territory owner uses a graded response to a potentially dangerous situation.

The function of the bill-up? When given by males at territory boundaries it is directed at neighbors and appears to mark the position of the boundary and signals an aggressive intention to defend the boundary at that position. When given within the territory to an intruder, the bill-up apparently is an assertion of territory ownership and a signal that a more escalated, physical response is imminent.

Crouch, After-song Display, Precopulatory Display

Males direct song spreads to both males and females. Some of the most intense song spreads are directed at potentially sexually receptive females. Males have sexual displays that resemble a song-spread posture or grade into or arise from it (Orians and Christman 1968). The Crouch, often given prior to copulation, resembles a song-spread posture with the legs flexed, head low, back arched, wings partly out and held down, and epaulets flared. Another display, used in similar circumstances, is the After-song Display. A male in the presence of a potentially receptive female gives a high-intensity song spread and then immediately follows it by fully extending his wings to a flight position while crouched. The wings can be held straight out for many seconds. Males on the ground will sometimes stroll around in front of a female, wings fully extended, epaulets fully displayed, producing a slightly comical effect, as if the wings were stuck in the out position. When a female responds positively, a male's Crouch or After-song Display becomes a Precopulatory Display with the addition of the male's precopulatory calls (see below; Orians and Christman 1968).

FEMALE PLUMAGE AND VISUAL COMMUNICATION

A colleague once asked if I had ever published information on redwings that was wholly wrong or later found to be false. My thoughts jumped immediately to a time before I had handled many female redwings, to an article in which I stated that females probably encoded information about their individual identities in vocalizations because such information could be vital to their social organization and because this information certainly was not visually transmitted: they all looked pretty much the same. The latter statement could not have been more in error. Plumage varies tremendously among individuals. First, as discussed in Chapter 3, females have epaulets that vary in size, color, and brightness; some females have only brown or dull rust-color epaulets, whereas others have bright or dark red epaulets (but usually not the bright scarlet of males). Second, there is often bright color, generally pink or orange, to a varying extent on the chin–throat–upper chest area. And finally, the white eyestripe varies a good deal in width and brightness.

What is the significance of this plumage variation? Does it have signal value? A first consideration is that female plumage generally brightens with age. Yearlings generally have brown or rust-colored epaulets whereas older females have brighter, redder epaulets (Nero 1954, Miskimen 1980). We found this to be the case at CNWR: 86% of females that we had banded as nestlings and that we subsequently trapped as yearlings had brownish epaulets. Epaulets then brighten in some females with advancing age, but the relationship is not absolute; some get duller after becoming brighter.

Much less attention in general has been paid to female than to male coloration in birds and the redwing is no exception. However, several possible explanations of the females' bright and variable plumage have been advanced, some of which deal with communication. The bright plumage of females could be used to communicate individual identity or social status, or it could be a neutral trait, present in females only because males have bright, functional epaulets and the sexes share those genes (Muma and Weatherhead 1989, 1991).

The status-signalling hypothesis is most likely to be correct. Females within harems are often aggressive toward each other and even fight (the causes of such fights are still unclear). Females, like males, have a song spread display, used in similar contexts (Orians and Christman 1968): they fan their tails, put out their wings (although not to the extent of males) and flare their epaulets to show them to other females. Older, redder females usually arrive earlier on the breeding grounds each spring than yearlings, settle first on territories, and become primary (first of the harem) nesters. If some resource is in limited supply, then females might benefit by being able to dominate other females and by the ability to signal that dominance.

Epaulet brightness as a status signal could also be related to physical condition: those females that make large energy expenditures in one breeding season which results in poor between-season condition could molt to a duller plumage, signalling a more subordinate status the next year, one that might elicit fewer agressive encounters. This status-signalling scenario could explain why some females get duller after becoming brighter.

Muma and Weatherhead (1989, 1991) presented males and females in the field with mounts of dull or bright females, and monitored dominance relationships among captive females with natural or manipulated dull and bright plumages. They found no significant relationships between dominance status or how males and females interact, and female brightness. They concluded that bright female color in redwings is probably the result of genetic correlations with male plumage and has no function. However, given the extensive variation in brightness and color, the fact that females differ not just in epaulet characteristics but also in the chin–throat area and eyestripe (where males do not have bright colors), and that female epaulets are made more visible during song spread displays, it is probably premature to reject the idea that color variation in female plumage has a function. Moreover, its use in individual recogniton, although unlikely because plumages can change between years, has not been ruled out.

Another display of females that must be mentioned is the Precopulatory Display. To invite copulation, a female flexes her legs, tips forward, and elevates her tail. She flutters her wings and utters soft, distinctive calls. Males usually quickly approach females in precopulatory display and copulation often follows. When treated with estradiol, females can be 'primed' for copulation, and then this display can be used as a test of the sexually stimulatory effect of male behavior (see below).

MALE VOCALIZATIONS

One's initial impression of a visit to a redwing marsh at early-to-mid breeding season is, frankly, of a madhouse. Breeding colonies are centers of frenetic activity and almost constant, loud sound. Females fly back and forth, some gathering nest material, others on foraging trips for themselves or their young, and all vocalizing frequently. Males race around their territories from boundary to boundary, evicting trespassers, disputing boundary positions with neighbors, occasionally flying off to feed, singing often, and calling, calling, calling.

Even more so than the visual displays, redwing vocalizations have been extensively and vigorously examined: physical structure and transmission properties have been dissected, inter- and intra-sexual communicative functions probed, variation among geographic regions investigated. The Advertising Songs and other vocalizations of males have been studied, as have the songs and calls of females. Consequently, the redwing's vocal system is among the best understood of any bird's. This attention to redwing vocalizations results from the relative ease of conducting field studies on the species and from the great investigative power afforded by electronic sound recording, analysis, manipulation, and broadcast equipment.

Male song

The first male vocalization I deal with has, or is suspected of having, the following uses: signalling species identity to redwing males and females; signalling individual identity; attracting females; falsely giving the impression of higher than actual male density; territory defense; male-male threat; and the attraction of conspecifics to flocks. This is the song given during the Song Spread Displays, a 1–2 s burst of buzzy sound energy known variously as male redwing song, Advertising Song, Territorial Song, or, phonetically, the 'Conc-a-ree' song.

Physical properties and geographic variation

A male redwing song consists of a few brief introductory sounds, or 'notes,' followed by a relatively long trill. Over the years, it has been rendered in English in many forms, but for purposes here, 'Conc-a-ree' will do, as long as it is kept in mind that the 'ree' (terminal trill) is two to four times as long as the 'conc-a'. Each male has two to eight (usually five to seven) different types, or renditions, of his song (Simmers 1975, Smith and Reid 1979, Yasukawa *et al.* 1980, Yasukawa 1981a, Kroodsma and James 1994) that vary in the number of introductory notes (usually one to three) and in the trill modulation rate – from fast trills that sound like a buzz to slow, warbling ones. Sound spectrograms (sonagrams) of some typical songs are shown in Fig. 5.3. For readers not familiar with these graphic representations of sound energy, the horizontal axis tracks passing time, and the vertical axis denotes increasing sound frequency (pitch). The darker the area, the more sound energy is concentrated there, i.e. darker is louder. During advertising, males give their songs at anywhere from three to eight songs per minute (Table 5.1), usually fairly evenly spaced in time. For example, and perhaps most typically, one song of 1.5 s duration is given every 10 s, for an advertising song rate of about 5/min. Males usually sing one rendition many times before switching to another (e.g. in one study, a mean of 14 times before switching, range 1–60 times; Smith and Reid 1979).

Redwing songs are loud, evidently an adaptation to transmit information over long distances. Morton (1975) suggested that the physical properties of bird vocalizations are adapted to the physical environments in the which the vocalizations are typically used, which stimulated a rash of studies of the phenomenon in the 1970s and early 1980s. For example, if transmission over a long distance is functionally important, then a species' songs should be given at appropriate heights and with appropriate sound-energy distribution (amplitudes and frequencies) most resistant to attenuation and degradation in particular habitats. Eliot Brenowitz's doctoral studies determined the properties of redwing songs as they relate to the physical environment in which they are used. He measured the sound-pressure level of redwing songs in New York State fields and pastures (upland breeding territories) to be about 90 decibels 1 m from a singing male – the loudness of a pneumatic drill! Brenowitz (1982a) determined that songs delivered at this sound pressure level would be heard clearly, i.e. with minimal sound degradation, at 190 m – in that habitat, a full two territories away. In the highly productive marshes of CNWR, 190 m could traverse 10 or more terrritories, but the songs would have

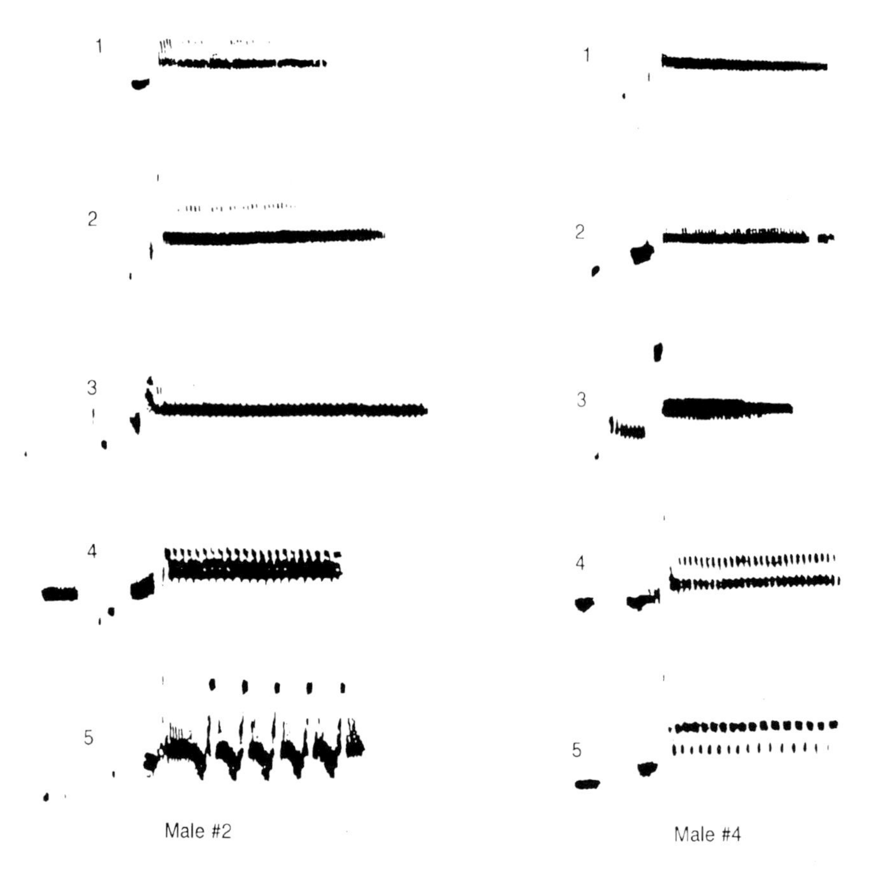

Fig. 5.3. Sonagrams of typical male redwing songs. The complete song-type repertoires from two males are shown, each in order of decreasing trill modulation rate (from Smith and Reid 1979).

to compete with the vocalizations of many near-neighbors. Most of the sound energy of the terminal trill, perhaps the song component best designed for long-distance sound transmission, is concentrated in the 2.5–4.0 kHz frequency range, and in New York upland territories, at 1.5 m above the ground (a height at or below which most redwing males could be expected to deliver their songs) sound traveled best below 4.5 kHz (Brenowitz 1982b). Therefore, the sound energy of redwing trills is appropriately concentrated in the frequencies that will travel the farthest when broadcasting from territories.

Redwing songs are recognizable as such wherever in the USA or Canada that they are heard or, for that matter, at the farthest southern extent of their range, in Costa Rica (Orians 1973); however, there is geographic variation. For example,

Table 5.1. Average song rates of male and female redwings during breeding in various localities

Mean number of songs/min	Location	Reference
Males		
3–6	California	Orians and Christman (1968)
5 (range 3–8)	Massachusetts	Simmers (1975)
4–5	Massachusetts	Smith (1976)
4	New York	Smith (1979)
5.5	Washington	Beletsky (1989a)
3–4	Ontario	Shutler and Weatherhead (1991b)
Females		
3–5	Massachusetts	Simmers (1975)
3	Michigan	Beletsky (1983c)
3	Wisconsin	Yasukawa (1990)

male songs in eastern and western Washington sound different to my ears, the latter almost always having a more nasal quality. As redwings are of different subspecies on either side of the Cascade Mountains, which separate eastern and western Washington, and because vocalization is thought to be a highly plastic behavior, song differences between these populations are not surprising.

Kroodsma and James (1994) detected some geographic differences among songs sampled in different areas in Florida and among songs in California (in relatively sedentary populations), e.g. in the usual number of song syllables used. However, sonagrams published for redwing songs from New York (Smith and Reid 1979, Brenowitz 1983), New England (Simmers 1975), Indiana (Yasukawa *et al.* 1980), Wisconsin (Orians and Christman 1968), Washington, California (Orians and Christman 1968, Kroodsma and James 1994), and Florida (Kroodsma and James 1994) demonstrate that the basic conc-a-ree song is relatively constant over the species' range (Fig. 5.4). Kroodsma and James (1994) suggest that this constancy may be due to the frequent mixing of redwing populations at large winter roosts. The mixing could be genetic, because males born in one area return to other areas to breed, resulting in high levels of gene flow, and/or behavioral (learning), because males sing during autumn and winter. The latter possibility is plausible because it is known that male redwings can learn new songs at least into their second year (Marler *et al.* 1972).

Singing patterns

Male redwings sing throughout the year, although there is a lull in summer after nesting is completed. There is a renewed burst of singing in autumn, perhaps

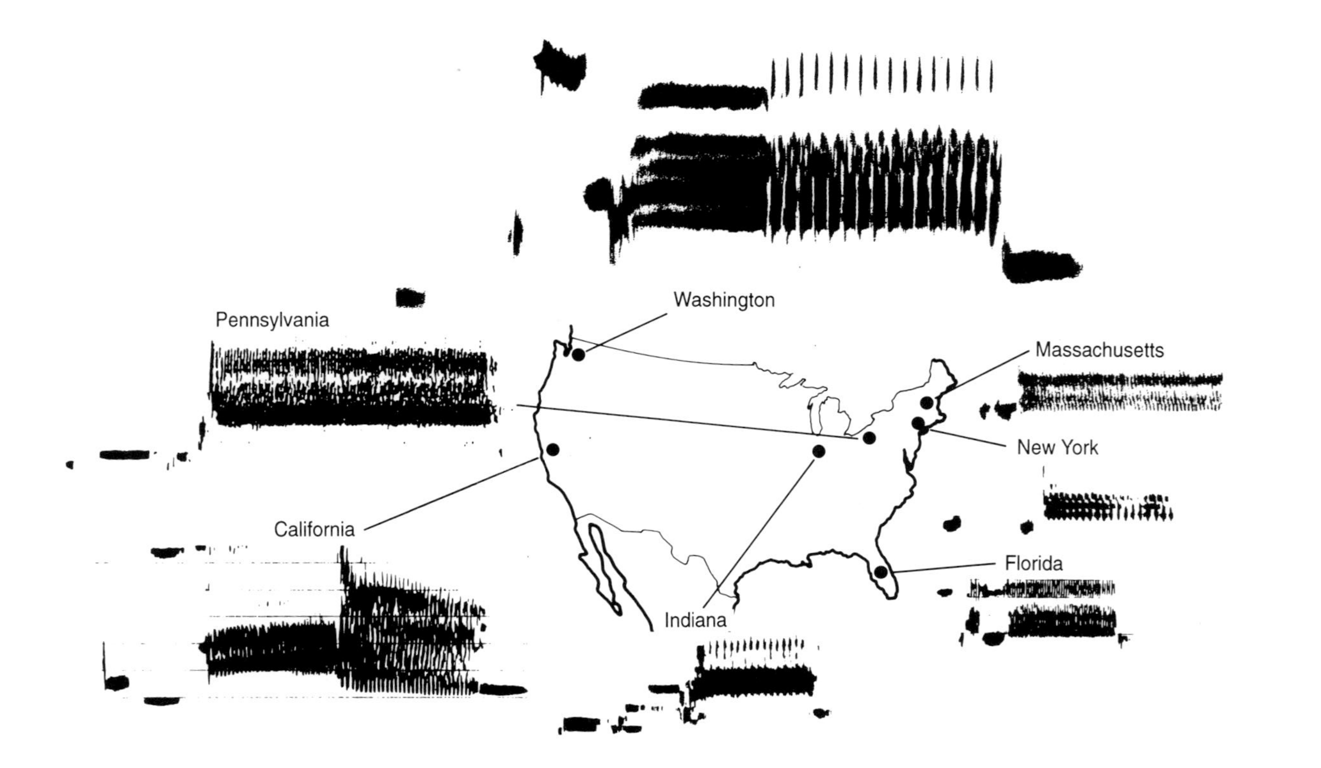

Fig. 5.4. Map of the USA, showing sonagrams of typical male redwing songs that were recorded in different regions. The basic 'conc-a-ree' song is similar among regions.

associated with a temporary increase in testis size and testosterone levels (Payne 1969). Payne observed males in central California singing from December through February. Orians and Christman (1968) noted that males sang thoughout autumn and winter in Central California. Males also sing at high rates during migration stopovers (Nero 1956a, Simmers 1975). One important function of singing outside the breeding season may be to attract other males to establish and maintain flocks (Brenowitz 1981).

During the breeding season, maximum song rates occur during territory establishment and as females arrive on the breeding grounds before nesting starts (Peek 1971, Simmers 1975). Singing continues throughout breeding, usually into July. The daily pattern is for males to start singing 40–45 min before sunrise, continue at peak rates during the 2–3 h after sunrise, decrease their rates during the day, then increase them again in the late afternoon, with a secondary peak at about 15–30 min prior to sunset (Simmers 1975).

Functions

Species and individual recognition

Male song is directed at both males and females and thus apparently has both inter- and intra-sexual functions. Song aids in species (and sex) recognition: parts of the song provide listeners with the information that the singer is a male redwing. When a loudspeaker is placed within a male's territory and the songs of other male redwings are broadcast, the experimental male quickly approaches and behaves aggressively toward the speaker, treating it as an intruding male redwing that must be ejected from the territory (Fig. 5.5). This characteristic response has been put to good use in investigating song function. By literally cutting recording tape in the right places, my colleagues and I dissected a typical male song into an introductory syllable (I), a middle syllable (M; actually a second introductory note of that song), and the terminal trill syllable (T). We played back different combinations of syllables of the dissected song (Fig. 5.6), in various orders, e.g. only I and M syllables or only trills, and discovered that the same highly aggressive response was elicited from territorial males by any of the artificially constructed songs that contained the trill syllable, as long as it was more than 200 ms in duration (the whole trill that was used, which was of average duration, was about 700 ms). The trill could be presented alone or ahead of or behind the I and M syllables and the same male response was provoked. Is and Ms alone elicited little or no response. Thus, it is the trill syllable that encodes species identity (Beletsky *et al.* 1980). (This project constituted my master's degree research; my thesis 'defense' seminar was appropriately titled, on a friend's advice, 'The Red-winged Blackbird: A Trilling Species.') Other studies, using similar techniques, determined that the pulse rate of the trill was also important for species recognition: territory owners respond aggressively

Fig. 5.5. A male redwing responds aggressively to song playback by perching on the loudspeaker in his territory.

only to trills with pulse rates that are typical of their population (e.g. in New York, 40–100 pulses/s; Brenowitz 1983). Introductory notes, demonstrably unecessary for species recognition, could function in other kinds of recognition, such as conveying individual identity, or as a simple means of attracting attention to the important message that follows.

Do male redwings know one another as individuals? And if so, on what attributes is such recognition based? One possibility is that variation among the introductory song notes of individuals allows for individual recognition by voice. Males are able to discriminate between the songs of a neighboring male and the songs of a 'stranger' male played back to them from a loudspeaker, behaving significantly more aggressively toward strangers' songs (Yasukawa *et al.* 1982). It is possible that this discrimination is based on their associating neighbor songs with specific individuals. But it is also possible that the differential response is based only on discrimination between familiar and unfamiliar songs. Furthermore, Yasukawa (1981b) showed that redwing song types within an individual's repertoire are no more similar to each other than song types compared between repertoires, which might be expected if individual identity was encoded in male songs.

Even if we are unsure whether redwings positively recognize each other solely by their vocalizations, I have other evidence that male redwings know each other as individuals. During territorial-male removal experiments (see Chapter 10) we often create situations in which two or more males believe they own the same

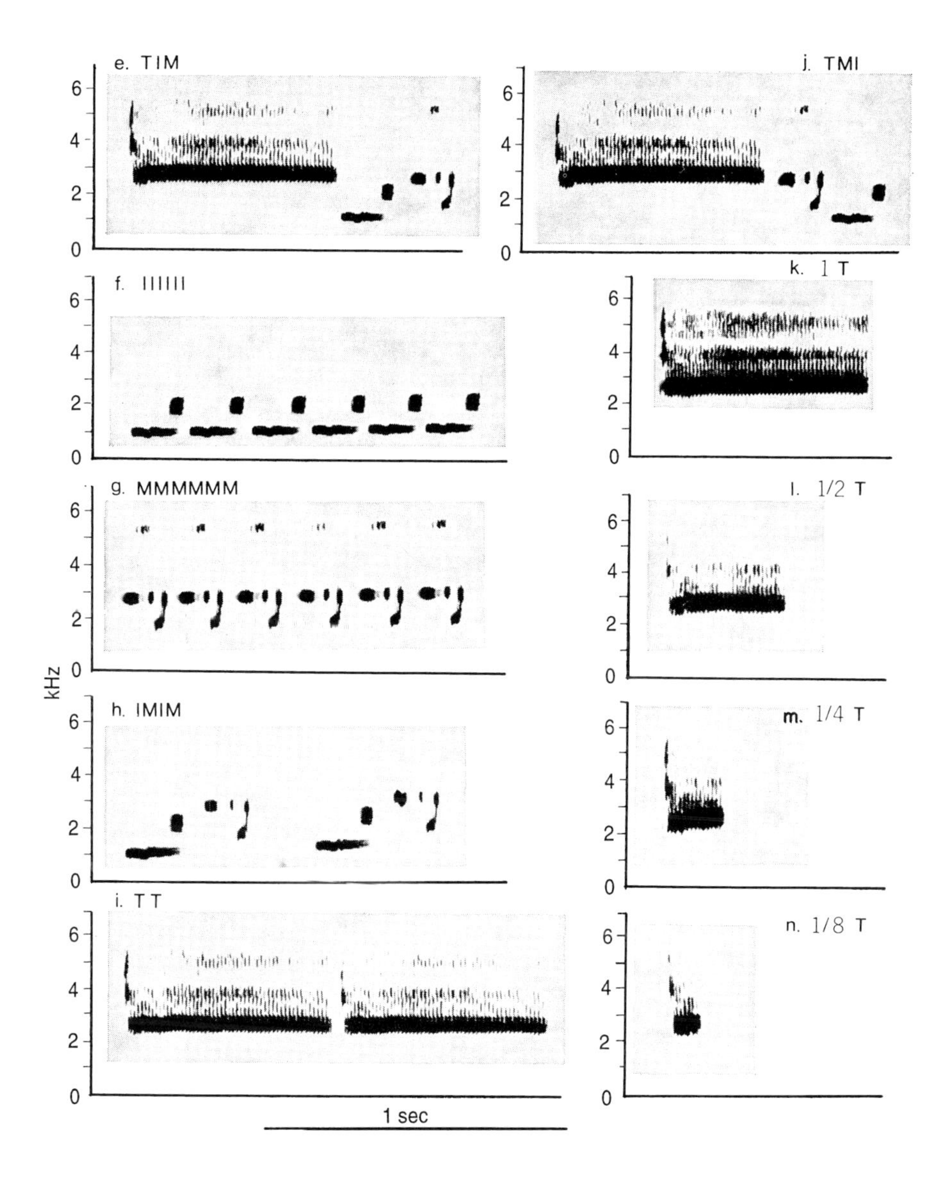

Fig. 5.6. Sonagrams of various songs created for playback experiments to investigate species recognition by voice in male redwings (from Beletsky et al. *1980). A single, typical male song (I, introductory; M, middle; T, and terminal trill syllables) was used to create experimental songs with one or more syllables, in various orders. (e) TIM; (f) only I syllables; (g) only M syllables; (h) Is and Ms; (i) two T syllables; (j) TMI; (k) T only; (l) ½ T; (m) ¼ T; (n) ⅛ T.*

territory. One male wins the territory, but the losers remain in the area, sometimes repeatedly trying to recover their holdings. In these situations, it is obvious that the current owner recognizes the former owners as individuals, because they are treated differently from other floaters. While most floaters that intrude into a territory are simply chased out, the former owners, color-banded for positive identification, are not only chased out, they are not tolerated anywhere near the territory or, for that matter, near the marsh itself. I have watched the current owners chase former owners far from the marsh, until they are out of sight. Further, sometimes after a male song is heard from a clifftop far from the marsh territories, I have observed the current owner dart from the territory and fly a distance to chase the singer from his perch, until he is out of sight. Obviously, the former owners are particular territorial threats and, as such, they are selectively not tolerated within sight of the territory; this selectivity must be based on individual recognition.

Male–male threat

Singing facilitates territorial maintenance and defense for male songbirds, ostensibly by signalling threat. The song as heard by a potential territorial competitor might convey the information that (1) the territory from which the song emanates is occupied by the singer, and (2) if the floater intrudes, he will be evicted by the owner or, if persistent, will have to fight and risk injury. In redwings, several lines of evidence support the territory defense function. First, observation shows that males with territories sing regularly and floaters much less often. When territory owners approach intruders aggressively they direct song-spread displays at them; and song spreads are given during boundary disputes with neighbors. Furthermore, when floaters (or owners, for that matter) trespass, they do not do so in an overtly threatening manner and are generally quiet during the forays. Hansen and Rohwer (1986) created vacant territories by removing the owners and then observed floaters take over the territories. They hardly sang during their first 10 min of occupancy (a period during which a territory owner, if only temporarily absent, might be expected to return), and then song rates rose to normal advertising rates after 20–30 min.

Experimental interference with a territorial male's ability to sing, and the 'occupancy' of a territory solely by a loudspeaker broadcasting songs, provide additional support for a territorial-threat function of redwing songs. Twelve out of 13 territorial males in Pennsylvania that Peek (1972) surgically silenced during the premating or mating period left or lost their territories; eight out of eight sham-operated males maintained their territories. Peek silenced birds by removing 2–3 cm of each of the two hypoglossal nerves in the throat, which provide sensory and motor innervation to the syrinx. Even with their nerves sectioned, males were still able to vocalize to a degree, but their utterances did not resemble male redwing song. The sole male Peek describes made only soft, wheezing sounds after the operation.

Smith (1976) repeated Peek's method on 15 territory owners in Massachusetts (15 others were sham-operated, i.e. nerves were exposed, touched, but not cut). Although Smith's experimental males produced highly abnormal songs (Fig. 5.7), only one of them lost a territory, versus three out of the 15 controls. Given this result, Smith had to conclude that normal songs were not essential for successful territory defense. However, Smith (1976) added two provisos. First, that sectioning hypoglossal nerves under anesthesia often resulted in respiratory distress after the operation, particularly when males were strongly stimulated, as they would be during active territory defense (perhaps explaining Peek's result); and second, that vocalizations may be more important keep out signals in areas where there is greater competition for territories (perhaps Smith's territories were in a breeding habitat that was less desirable than Peek's, resulting in lower competition). So

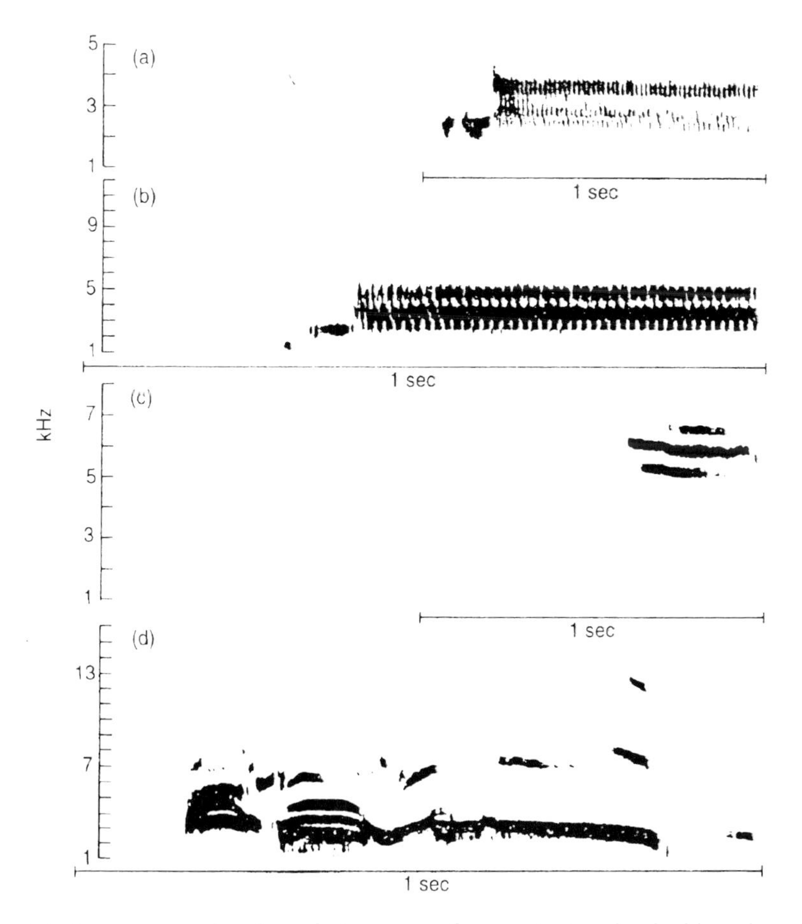

Fig. 5.7. Sonagrams of male redwing songs from intact males (a, b) and post-operative males (c, d) from a devocalization study (from Smith 1976).

Smith (1979) tried again, using another method to mute males. He cut the membrane of the interclavicular air sac, a brief operation requiring no anesthesia. The technique muted males without causing respiratory distress. Moreover, the males regained their ability to sing in about 2 weeks, and thus could be used as their own controls. The experiment was conducted in a New York field earlier in the breeding season than his previous experiment, when there was more active competition for territories. The result was that three out of three muted males left or lost their territories. When they regained their ability to sing, all three recovered portions of their lost territories, even the male who, after the operation, apparently left the area of his own accord. The weight of results of devocalization studies thus supports a territory defense/threat function for male song.

Another method to test song for a territory defense function is to attempt to defend a territory with songs alone. Yasukawa (1981a) removed male territory owners and replaced them with loudspeakers repetitively broadcasting either a single redwing song type or eight different types, at a typical advertising rate (6 songs/min). To simulate an advertising male patrolling his territory, Yasukawa used two loudspeakers, switching the broadcast between speakers every 5 min during each 1-h trial. Significantly fewer males flew through the speaker-occupied territories when the single song type was played than during 1-h control periods (no broadcast). The same result was obtained for playback of eight song types versus control periods. Interestingly, in the latter case, Yasukawa found that although floaters might have been deterred by broadcast songs, territorial neighbors trespassed into the speaker territories at equal rates during broadcast and quiet periods. Neighbors, of course, had more information than floaters – they could see that no real males occupied the experimental territories. Thus, songs alone emanating from an otherwise vacant territory are, at least for a time, able to defend it from floaters.

Intersexual functions of male song

Ethologists continually try to devise tests that will allow them to determine the functions of signals. For vocalizations, the field playback experiment proved to be an extremely effective tool for evaluating the responses of birds to taped vocal signals. The playback technique also has uses in the more controlled conditions of the laboratory. Searcy (1988) demonstrated that male songs have an intersexual function by presenting caged female redwings with 64 repetitions of a male redwing song or the same number of repetitions of a male Song Sparrow song. The response Searcy looked for was the copulation solicitation display, normally given by a female only during the breeding season, on or near her mate's territory, during the periods she is nest-building and producing eggs. Females were brought into breeding condition by implanting them with a subcutaneous, porous tube containing the sex steroid hormone estradiol. They were thus made highly sexually responsive to male displays that might normally stimulate them to copulation. The female copulation solicitation display was then used as a test of male behavior. Female redwings gave solicitation displays at much higher rates during playback of the conspecific song, showing that the song encodes sex and species information that females use, and suggesting an intersexual function for male song in stimulating copulation. That females attend to male songs and that it has a stimulatory effect on them is not surprising because males give their song-spread displays in the presence of females, apparently directing the display at them.

An additional set of playbacks to males and females by Searcy and Brenowitz (1988) demonstrated an interesting sexual difference in response to male songs: females are the more choosy sex when it comes to giving overt, positive responses to playback song. Brenowitz (1982c) had shown that territorial male redwings gave identical aggressive responses to both normal redwing songs and to the song of a Northern Mockingbird imitating a redwing. But Searcy and Brenowitz showed that female redwings, given estradiol implants, gave significantly more copulation solicitation displays in response to a redwing's redwing song than to the mockingbird's redwing song. Furthermore, whereas male redwings responded equally aggressively to a full, normal song or to a lone trill syllable dissected from the same song, females did not: they gave more displays after presentation of the full song than after the trill-only song. Thus, Searcy and Brenowitz found that females discriminate more about the male songs they will respond to than do the males themselves (the study incidentally showed that introductory notes may have roles in intersexual communication). They used their results to suggest that one reason that there are features of male birdsong that can be substantially altered during experiments without changing male aggressive response is that those features are concerned with male–female, instead of male–male, communication.

When investigating functions of songs, it is often useful to consider not only the song *per se*, but also singing behavior. For example, the times of the day or the stage of the breeding season when most singing occurs could provide clues to functions. In some species, male song rates drop precipitously after pairing, and in some, males resume high song rates after this fall if their mates die, desert, or are removed (Krebs *et al.* 1981, Wasserman 1977); both patterns provide support for a mate attraction function for male song. Male redwing song rates are highest during the early stages of the breeding season, when territories are established and females arrive to choose nesting areas, but males continue to sing throughout breeding. Searcy (1988) tested for a mate attraction function of male redwing song by removing the first (primary) females that settled on male territories and then monitoring male singing. Song rates of the experimental males during the 11 days following the female removals were significantly higher than those of control, unmanipulated males, strongly suggesting a female attraction function for male songs.

Another singing pattern that is a common attribute of songbird advertising is that males give their songs with typical cadences, or regular patterns of songs and silences. Songs are usually brief, 1–3 s, whereas the inter-song intervals are longer, typically 7–13 s (Beletsky 1989a). Because we know that songbirds are physiologically capable of sustaining much longer songs – 'continuous' singers, such as Northern Mockingbirds or Eurasian Skylarks, give long streams of song without pause – the common song cadences of redwings and other birds may themselves be functionally advantageous.

A number of years ago I suggested that one function of regular song cadences was that they allow any listener, such as a female prospecting for a mate, but initially out of sight of the vocalizing male, to associate all of a male's songs with a single individual, and then to navigate to the singer. The first function is possible because a signal repeated at relatively invariant intervals is most likely to originate from a single source. Thus, even in a noisy environment filled with the songs of other males, a female might be able to pick out the songs of a single male because his song is, for example, given every 10 s. The second function, locating a singing male by his vocalizations, is possible for the same reason. A female could use the regularly-timed songs as a predictable approach reference, like the revolving beacon at an airport. These functions could be especially important in species where males have repertoires of song types. A female trying to follow songs of her liking to a specific male could not use an individual song type as a beacon because the male would be switching types and perches, but perhaps could use the cadence to predict the onset of the male's next song (Beletsky 1989a).

Repertoires of song types

During the past two decades perhaps the preeminent Grail for biologists who study avian vocalizations has been to explain the evolution and function of male song repertoires. The reason is that various attributes of song repertoires position them at the confluence of several fields of inquiry. First, males of most songbird species possess more than a single song type yet, in most cases, the types appear to be used interchangeably. This offends our sense of biological economy because, as far as anyone can tell, species with one song fare as well as those with multiple types. Deepening the mystery, repertoire sizes vary extensively between species, from 2 to 30 or more (North American wrens, such as Marsh Wrens, Rock Wrens, and Winter Wrens have some of the largest repertoires; Kroodsma 1980). Thus, there is a diversity of signals but a seeming redundancy of function that ethologists seek to explain. Second, because repertoires could represent sexually selected traits, elaborated for male–male competition or female mate choice, they are of interest to evolutionary biologists. Third, because bird songs are learned, the questions of how repertoires are learned, from whom, and why particular song types might be included in a young male's eventual repertoire are of interest to an array of biologists, including those who study language and song learning, and territory acquisition.

The characteristics of redwing repertoires are these: most males possess five to seven song types (Fig. 5.3). Territory owners most often sing one type several times before switching to another (Smith and Reid 1979). Neighboring males or many males in close proximity often 'match' types – they give the same type at the same time. Males often switch types just after they switch perches (Smith and Reid 1979, Yasukawa 1981b). And all song types appear to be used interchangeably; no specific types have been associated with specific behavioral contexts (Smith and Reid 1979; Searcy 1986a).

Although there was previous work on repertoires, the burst of contemporary interest may be attributed to a 1977 paper by John Krebs that caught the imagination of many ethologists. He proposed that repertoires allowed territory owners to deceive conspecific males who were prospecting for new territories, and who were beyond visual but within acoustic range, as to the number of territory owners producing the vocalizations, i.e. they signalled false density information (hence the name Beau Geste Hypothesis; *Beau Geste*, a novel by P. C. Wren, includes a plot development in which people use an analogous ruse). The idea was greeted, in my experience, with heavy skepticism (and jokes: the You Jest Hypothesis), but researchers agreed that the paper made a valuable contribution because the idea could be used to form predictions that could be tested experimentally. The paper thus helped move birdsong research beyond the then current stage of observing and recording birds in the field, preparing sonagrams and inferring function from physical variation and context of use.

There are at least six, nonmutually exclusive, hypotheses of repertoire function that could account for their evolution and use in redwings: (1) different song types could transmit different information, be used in different contexts, or be directed at different classes of individuals; (2) matched countersinging, in which a male would benefit by being able to give the same song type as another male; (3) Beau Geste; (4) status signalling, in which repertoire size would correlate with male age, experience, or dominance position; (5) female stimulation/mate choice, in which females would choose males as mates on the basis of their song repertoires; and (6) signal switching, to be explained below.

Although no definitive study of song-type usage in redwings has been conducted, reports in the literature (Smith and Reid 1979) and personal observations suggest that all types in a male's repertoire are used interchangeably. In one test directed at females, Searcy (1986a) presented individual male song types to captive females that had been treated with estradiol, and they responded identically with sexual displays to each of four types. It is known that both males and females attend to song repertoires. Yasukawa (1981a) found significant differences in the trespassing rate of neighboring males when he played single song types or multiple types through the loudspeakers on his speaker-occupied territories, demonstrating that males listen to and make some use of repertoire information. Similarly, Searcy (1986a) presented estradiol-treated females with repetitions of single song types or repertoires. The females gave more copulation solicitation displays in response to repertoires, showing that they, too, listen to and perhaps extract information from repertoires.

'Countersinging' can have two components. A male can answer the song of another male by singing immediately after the first male's song, associating their songs in time and/or a male can answer with the same song type, provided he has the same type in his repertoire (matched countersinging). The adaptive significance of matched countersinging has never been made clear but, anthropomorphizing, we may assume that the answering male benefits by signalling that he is aware of the first singer and directing his answer specifically at him; perhaps there is an implied threat. There is evidence for countersinging in male redwings. Smith and Norman (1979), calling it 'leader-follower' singing, showed that the timing between songs of two singing males is not random, as might be expected if males sing independently, and that the pattern of intervals between songs suggests that one male's songs follow another's with relatively constant delays. (In Smith and Norman's usage, the leader was usually a territorial intruder, whereas the follower was the territory owner bent on repelling the intruder.) Smith and Norman, however, provide no information on song type-matching. Smith and Reid (1979) reported that male redwings often match song types when singing together, or at least appear to select songs from their repertoires with similar trill modulation rates. Yasukawa (cited in Searcy and Yasukawa 1995) found that when male redwings responded to songs played back from their own repertoires, they did not match the broadcast types more often than would be expected by chance. Thus, male redwings may countersing but there is little good evidence to show that they match song types.

The Beau Geste hypothesis predicts that advertising males should give their song types in patterns that mimic the singing behavior of more than one individual, e.g. they should move as they change song types, and that prospecting males should avoid high-density breeding areas. In fact, many of the predictions of the Beau Geste hypothesis are supported for redwings: repertoires (as opposed to single song types) reduce trespass rates (Yasukawa 1981a) and males often switch perches when they switch song types (up to 80% of the time; Smith and Reid 1979, Yasukawa 1981b). However, the hypothesis fails for redwings because nonterritorial males, presumably searching for territories, are attracted to rather than repelled by densely-occupied breeding areas (Yasukawa and Searcy 1985). Further, one might suppose that, other factors being equal, crowded breeding areas are among the best places, not the worst, to attempt to obtain a breeding territory (Beletsky 1992; see Chapter 10). Moreover, the Beau Geste hypothesis predicts that males should give songs at irregular intervals, the better to mimic the actual sound of many males, each singing with its own cadence; however, an advertising male redwing gives songs at very regular intervals, which provides listeners with exactly the opposite information than that suggested by the Beau Geste hypothesis – that the songs most likely originate from a single source (Beletsky 1989a).

The status-signalling hypothesis suggests that repertoire size is positively correlated with male dominance level, or Resource Holding Potential (RHP). As Searcy and Yasukawa (1995) point out, this is plausible for redwings because older, more experienced territory owners generally have larger repertoires (Yasukawa *et al.* 1980). However, the relationship between age and repertoire size is far from perfect and it is not known if prospecting males avoid older territory owners. Furthermore, signalling relative RHP, at least during breeding, may be unimportant because, by all measures of RHP to date (Chapter 10), including potential vocal ones (Shutler and Weatherhead 1991b), it is not a prime determinant of territory ownership and hence, dominance.

If repertoire size in any way indicates male quality as a mate, whether it signals a male's probability of feeding offspring or his superior genes, etc., then females could benefit by selectively mating with males with larger repertoires. Again, we know that females notice repertoires because they respond differently to playback of single or multiple song-types (Searcy 1986a) Moreover, in field experiments, territory owners switched among their song types at significantly higher rates when they were presented with live or stuffed females than when presented with live or stuffed males or when compared with control periods (Searcy and Yasukawa 1990). This suggests that territory owners tried to exhibit their repertoires to females more so than to male intruders (in fact, the experimental males *decreased* their switching rate among song types when responding to males). Thus, evidence exists that redwing repertoires have an intersexual function. One problem with the possible inter- and intra-sexual functions found to date for redwing repertoires is that, if it is advantageous for males to have larger repertoires, selection should favor even larger ones; disadvantages to having a repertoire of more than eight types have yet to be identified.

Repertoires, of course, may have a variety of functions, including one obvious possibility proposed long ago but never fully accepted or discredited: that multiple song types reduce listener habituation to a male's singing, the 'monotony-threshold' hypothesis (Hartshorne 1956). Finally, repertoires may exist because they allow males to transmit information when they switch among different song types, an explanation of which brings us to the subject of male redwing calls.

Male calls

Alert Calls

Marking birds with color-bands helps ornithologists tremendously in learning about the behavior of individuals. But this technique can lead an observer to focus attention so closely on the banded individual that behavior on any larger scale is missed. Some portion of a bird's behavior simply cannot be interpreted by focusing on an individual because the behavior is part of a group display or interaction. I raise this issue here because any investigation of redwing vocalizations may be doomed to error or irrelevancy unless the social and acoustic environments are taken into account. Male redwings do not vocalize in quiet solitude, but insert their voices into a din, into a cacophonous, usually crowded social milieu. Observing how a group vocalizes, for instance all of the males with territories on a small breeding marsh, may reveal more of the functional significance of vocalizations in many cases than observing single males; this is especially true for male calling.

Again, visiting a redwing breeding marsh during the morning hours, one is struck by the constant sound, and especially by the number and variety of male and female vocalizations. For males, songs are immediately apparent. They are loud, frequent, and generally their longest-duration vocalizations; it is easily understood why most research interest is directed to song. By actually counting, however, one would find that while a male was giving six or eight or even 10 songs in a minute, he might also be uttering, interspersed with the songs, twice that many calls.

There has been a strong emphasis on songs in studies of avian vocalizations. Songs are the relatively complex, patterned, and usually melodic, bird vocalizations long-presumed to function in territorial defense and mate attraction, but birds also possess many different calls, usually distinguished from songs by being shorter and structurally simpler (Thorpe 1961). Research concentration on songs instead of calls has come about for several reasons, perhaps mainly because most of the pioneering work on avian sounds used songs (e.g. Thorpe 1958, 1961; Borror 1959, 1961; Marler and Isaac 1960, Marler and Tamura 1962), thus creating research inertia, and also that biologists are doubtless more esthetically attracted to the usually melodic songs. The emphasis on song and singing behavior, however, has produced the paradoxical situation in which the in some ways richer (vocal repertoires generally include greater varieties of calls than songs) and more frequently used vocalizations are the less well-known and understood. Yet the predominance of calls in avian communication systems suggests that they are important in mediating social interactions of all kinds. This is especially true of redwings. When I began working with their calls, there was an entire literature devoted to their songs and singing behavior, but little information on their calls.

I set out to investigate functions of male calls because I had already worked with male and female songs and I wanted new avenues to explore. One day during the 1984 breeding season I sat on a basalt cliff at CNWR above a teeming redwing breeding colony and considered male calling. As I knew that avian calls had to that point been assigned names and functions by the contexts of their use (e.g. precopulatory call, hawk alarm call, mobbing call, contact call) I expected that, with sufficient hours of observation, I would be able to associate specific call types with specific social situations and thereby assign functions to each type. I should have known that it would not be that easy. Very quickly I discovered that the males vocalizing below me had a variety of very brief calls, that they gave these calls almost all the time – sometimes with almost amazing frequency (one male gave 2200 during a 1-h period) – and that, as far as I could tell from some preliminary observations, a host of different types were given in exactly the same contexts. Orians and Christman (1968) published the most thorough description of male calls, terming these brief ones General Alarm calls. But it was immediately clear to me that when males gave the calls they were not necessarily alarmed which, to be fair, was also noted by Orians and Christman. During my research on male songs, I used the calls in analyzing responses to my playback experiments, terming them 'arousal' calls, which I defined as 'assorted "peeps" and "chucks" and other short sounds emitted by males when stimulated' (Beletsky *et al.* 1980). But again, as I peered down on the calling males, many of them did not appear to be 'stimulated' in any way. What was going on?

Listening to the males call, at first I identified seven different types that they gave in a continuous fashion without overt social stimulation (Peet, Check, Chuck, Chick, Chonk, Chink, Cheer; Fig 5.8); now, after years of working with the calls, I distinguish 10 different types commonly used at CNWR. As soon as I could readily distinguish the seven types by ear, it was clear that males with territories on the small marshes regularly matched call types – when one gave peets, they all gave peets (well, most of them did). This pattern suggested either that each type had a different meaning and that the males all gave the same (appropriate) type simultaneously in response to the same stimulus, or that only one male responded to a

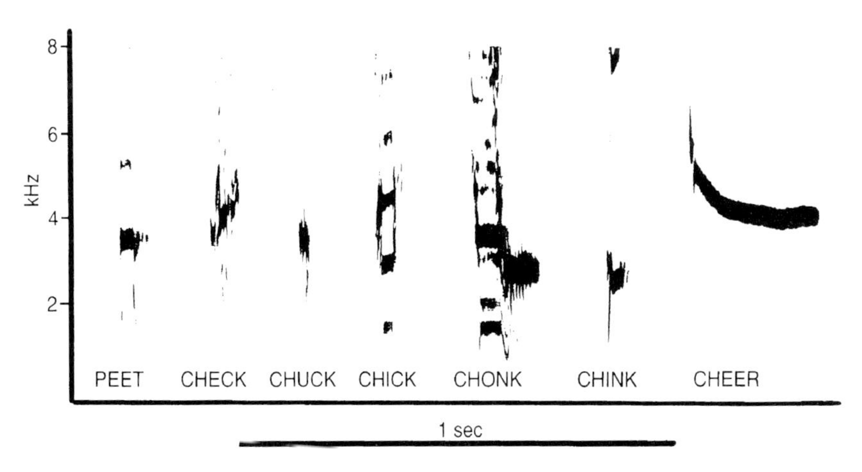

Fig. 5.8. Sonagrams of seven alert calls given by male redwings at Columbia National Wildlife Refuge (CNWR) (from Beletsky et al. *1986).*

stimulus by switching types, and the others, hearing the change, all switched contagiously to his new type. The latter explanation seemed the more likely because, again, the call types seemed to be used interchangeably and in all kinds of contexts. If switches were indeed contagious, then perhaps the calls conveyed little or no information themselves, but were given continuously as a 'background' sound upon which a switch to another type, by its contrast with the prevailing type, might be readily detected. What stimulus would cause a male to change the type of call he was giving? I noticed that when I moved from my seated position above the marsh, the prevailing call type often changed. So males may use their calling as part of a continually operating, general alert system.

The system is based on groups of males on or near their territories each repetitively matching the call type of the others. Against this background new call types are easily detected. An individual shifts call types when he detects potentially dangerous changes in his environment, such as the appearance of a nest predator or the sudden movement of an observer (e.g. me standing up or waving my arms). When other males hear a call type change, their predominant response is to switch to the same call type. The new prevailing type then supersedes the previous one as background, upon which subsequent switches can be discerned and localized, thus resetting the system to a receptive status. The analogy I use is that the males with their continuous calling create a web of sound on and around the marsh, and then listen for changes in the web, just as a spider waits for vibrations, to obtain alerting information. Just as the spider gets information from vibrations that an unfortunate insect is caught in the web and its direction, so too the listening males get the information that something of significance has occurred in or near the marsh and, by the location of the call-changer, its direction.

To convince visitors to the study site that the call system worked as I claimed, I would sit them down near a marsh, wait for male calling to return to a baseline level, and then toss a stone into the water near a calling male. The spectacle of that male changing his call type, and other males, even those across a lake, almost immediately switching to match the new type, usually made believers out of skeptics. To document this communication by signal switching, however, colleagues and I had to demonstrate that, among other characteristics: (1) each of the seven types was given in several general and specific contexts (Table 5.2); (2) adjacent males matched calls; (3) neighbors switched contagiously when one of them switched types; (4) males switched types when detecting the appearance of a predator (a stuffed hawk); and (5) males switched calls contagiously to match artificially broadcast call types (Beletsky *et al.* 1986). Perhaps another explanation of these calls might account for these characteristics, but if so, we have yet to come up with it. (Richard Simmers (1975) documented geographic variation in the songs and calls of redwings in New England and New York. He noted and puzzled over the large number of alert calls he found in each population (up to 15 different alert calls per individual), doubted

Table 5.2. Usage of various call types in different behavioral contexts, expressed as percentage of the total of each call type given (n = 13,206 calls). (Reproduced with permission from Beletsky, Higgins and Orians 1986.)

Context	Peet	Check	Chuck	Chick	Chonk	Chink	Cheer
Aggression	5.5	5.6	6.1	4.0	3.0	9.0	3.0
Sexual	1.6	5.2	5.5	2.5	2.8	8.0	1.6
Flight	13.7	15.0	14.8	10.1	10.1	10.3	7.8
Feeding	2.3	2.2	5.6	1.6	3.4	9.4	6.6
Advertising	24.8	20.3	33.2	25.4	18.5	39.1	4.8
Alarm	52.2	51.6	34.9	56.3	62.3	24.2	76.2
n of column	1903	3892	2822	1294	1073	565	1657
Percentage of total	14.4	29.5	21.4	9.8	8.1	4.3	12.5

that each type had a distinct function, but believed their high number to be an adaptation to reduce listener habituation.) Further, to show that the males gave alert calls even in the absence of potentially predatory biologists, we observed their behavior from a blind (a hide), and also tape recorded a marsh after we had gone.

I performed two other studies to determine what other sorts of information might be transmitted by male alert calling and to identify who might be listening to and benefiting from repetitive male calling. For the latter, I used three methods to test the hypothesis that males direct alert calls to their mates (Beletsky 1989b). When females are on their nests their vision is blocked and it would be plausible that they might listen to male calls for information about activities around the marsh. First, I observed calling males during various parts of a breeding season, predicting that if males direct calls at mates to alert them to potential nest predators, then call rates and call-switching rates should increase as nesting began. I found that call rates and switching rates did increase significantly between the pre-incubation and incubation phases of the female cycle, which suggests that males began to call more just as their first mates began incubating, i.e. when they first began spending long periods on nests situated in dense vegetation, where their ability to detect predators themselves is sharply reduced. Second, I broadcast to nesting females either an experimental playback tape containing four call types (chonks followed by cheers, checks, and chinks) which thus contained three type switches, or a control tape containing only checks. The prediction was that a female placed on alert by call-type switches would behave differently from one hearing only checks, the most common alert call. The females' response to male call playbacks was somewhat ambiguous; they vocalized more often during experimental playback and significantly more females remained on their territories during experimental than during control tape playback. A possible interpretation of the latter result is that, when alerted by call switching, females tend to remain protectively near their nests. Third, I evaluated female vigilance behavior, predicting that if male calling functions as sentinel behavior for females, then females would have to be more vigilant when their ever-calling males were absent from territories. There was a significant positive association between the amount of time females spent on top of the marsh vegetation, where they are in position to detect predators on their own, and the duration their mates spent off-territory, which suggests that when calling males are absent, their mates must assume more responsibility for their own and their nests' safety. Peek (1971) had suggested that male redwing calls serve as an early-warning system for nesting females because he noticed that he could approach their nests much closer without their flushing when their mates had been surgically muted; it appears he was correct. Thus, females may use their calling males as vocal sentinels.

As far as what information, in addition to alert, is contained in repetitive calling, the rate of calling varies enormously so it seemed plausible that call rate might convey additional information to listeners. To test whether call rate encodes information about the distance to predators, I presented to calling males various stuffed predators, which represented differing degrees of potential threat, and then monitored vocal responses (Beletsky 1991).

A stuffed Muskrat, Great-horned Owl, or Black-billed Magpie mounted on a wire cage or, as a control, the wire cage alone, was presented at various distances from the males' territories. Muskrats abound in the study area but they pose no threat whatsoever to redwings of any age. The magpie is the predominant predator on redwing nests in the region but does not prey on adults. The owl is a diurnal and crepuscular predator on adult redwings. The mount to be presented was covered with a gray cloth and placed near a male's territory, after which I hid behind available shrubbery. After a suitable interval, I dropped the cloth by pulling on a string and then recorded the males' responses. The experimental males showed little interest in the muskrat or the plain wire cage, but responded strongly to the magpie and owl, often flying over them, apparently to inspect them more closely. Male call rates were higher in response to the owl when it was placed 2 m from the territory than when it was 15 m away, suggesting that males give higher call rates when predators are detected closer to territories. The best indication of this, however, was when I had a field assistant approach male territories at the slow, measured rate of 10 m/10 s, starting from 50 m away and ending at the territory boundary. Call rates in response were initially low, before the person began approaching, then increased as the person got closer to the territory. Call rate peaked as the assistant covered the final 10 m to each territory, then declined slightly and plateaued as the person remained motionless at the territory boundary (Fig. 5.9). Thus, various tests suggested that males increase their call rates as the distance between a potential predator and their territories decreases, and that lis-

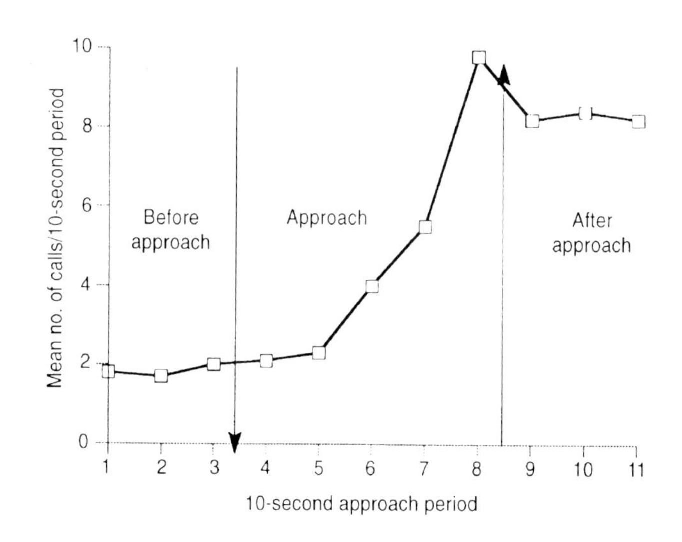

Fig. 5.9. Mean 10-s call rates of 15 male redwings while a person approached their territories (from Beletsky 1991). Periods 1–3, before approach began; periods 4–8, starting from 50 m away, walking toward the territory at 10 m/10 s; periods 9–11, person at the territory boundary, motionless.

teners, therefore, may be able to track approaching predators by listening to male calls.

The sum of my work with repetitive calling to date suggests the functioning of a two- or three-tier alert system. First, males change call type when they detect a predator or other potentially hazardous change in the environment. Second, they progressively increase their call rate as a predator approaches and becomes a direct threat. A third level, indicating alarm, may be used when males hover over predators raiding their nests. In these situations males call at very high rates, usually rapidly alternating between two types (e.g. check-check-cheer, check-check-cheer). Whether other males or mates and young are the primary intended beneficiaries of an individual's calling remains to be seen. What is clear, however, is that listeners, by attending to the waves of calls and call switches that are a constant attribute of redwing marshes, can be alerted to the detection and approach of predators and, by localizing the position of the first male and then the next male to change call types, perhaps also the direction from which the predator approaches and moves.

I have explained the redwing alert call system in detail because I suspect that its understanding may be helpful or necessary for understanding some of the underlying social cohesion of redwing breeding colonies and the costs and benefits of colonial breeding (see Chapters 4 and 12). Also, communication by signal switching has not been documented before, and it is possible that it exists in other animals as well. In situations in which investigators find a plethora of different signals or displays that seem to be used interchangeably, might it be possible that they exist solely so that individuals can switch from one signal to another, transmitting information in the switch, rather than in the different signals? Might songbird song repertoires, in some cases, function in this manner? Individuals could give their songs at the normal rate necessary for mate attraction and territory defense, but simultaneously transmit additional information when they switch among types in their repertoires.

In redwings, there must be an advantage to signalling alert or alarm by vocalizing continually and then transmitting information when making changes in the signals, relative to being silent and then giving alarm signals only when danger is first detected. When this advantage is identified, we might know better where else to look for communication by signal switching. Also, from the very beginnings of the scientific study of animal communication, researchers have been puzzled by the high frequency with which some animals give their visual displays and vocalizations, and with the seeming redundancy of signals (Busnel 1968, Schleidt 1973, Krebs and Dawkins 1984). Redwing alert calling shows that high signal diversity and high repetition rates need not necessarily indicate redundancy; rather they are the necessary attributes of a completely different kind of communication. Finally, it probably has not escaped the reader's attention that if all the males on a breeding marsh participate in call-switching, truly cooperative alerting behavior may be occurring, which is of interest in how we evaluate the benefits and costs of colonial breeding.

Other calls

Alert calls or similar calls are also uttered as males leave their territories, and it is possible to track them in flight by ear, until the calls fade in the distance. These departure calls may allow males to remain in vocal contact with mates or other males, or they may leave an acoustic trail to indicate the direction of travel of the territory owners when they leave the marsh. Females also vocalize when they leave, and I noticed, when studying the female vocalizations, that, although males and females may leave the territory independently, they often return together. Thus, while still on territory they may track the movements of their mates by their songs and calls, making it possible to meet them later, off-territory. Alert calls, intriguingly, are also given by males flying by and over breeding marshes, when they are not obviously interacting with the marsh below. Whether traveling males like these (territory owners, floaters, both?) are using the calls as an alert system, or whether they take on other functions in flight or in other contexts (e.g. contact calls?) is unknown. Further, males visiting a breeding marsh often match the prevailing call types as they land on or near territories. It is possible either that whenever males are close to each other they participate in the call-alert system or that strangers quickly join in continuous alert calling so as not to draw attention to themselves by being silent.

Male redwings appear to announce their arrivals and departures to and from their territories. In addition to giving a string of alert calls as they leave (Orians and Christman (1968) distinguished alert calls given in this context as the 'Flight Call Complex; indeed, these calls may differ structurally from standard alert calls), males almost always give a single conc-a-ree song as they arrive back on their territories, just prior to landing. These flight songs appear to serve as a greeting, perhaps informing listeners that the rightful territory owner has returned. If females need to be more vigilant when their mates are off-territory, then the male flight song could signal that they may now relax; however, the intended receivers of conc-a-ree songs given as males return to territories are uncertain. There is another call males possess and often give in flight (but also when perched) that, although all give it intermittently, I have never been able to associate with specific contexts. Nero (1956a) called it Flight-song and rendered it as 'tseee . . . tch-tch-tch-tch . . . chee-chee-chee-chee-chee'. Allen (1914) described it as 'check,check,check, t'tsheah.' Beer and Tibbitts (1950) thought it might be a victory song, because they observed it given after males chased intruders from territories.

Male redwings have other calls, in addition to Alert Calls. As in many songbirds, there is a Hawk Alarm Call, which is very similar to the Cheer alert call, but pitched higher and covering a narrower frequency range (which makes it physically more difficult to locate its source). It is a clear whistle that declines rapidly in pitch. When this sound is heard about the marsh, one can generally, by quickly scanning the area, detect its stimulus – usually a hawk overhead or hunting nearby. There is little doubt that this call alerts listeners specifically to the presence of a flying predator. Simmers (1975), in his spectrographic analysis of geographic variation in redwing calls and songs, classified from three to seven different 'whistles,' as he called cheers, in each male's repertoire. Some of these cheers may be used interchangeably with other alert calls, while others may be used exclusively as hawk alarms. Again, when males are highly alarmed, as when mobbing a predator raiding a nest, they usually utter more than one call type, often rapidly alternating between types. Cheers are often included: they appear to communicate a higher degree of alarm than many of the other call types.

Some calls are used in highly specific, intersexual contexts. The Growl is a low-pitched, harsh sound that males give when near females. The sound is like a throaty 'haaaah', held for several seconds. Orians and Christman (1968) suggested that it is used during a 'Nest-site Demonstration Display', because a male often utters growls soon after he and a prospecting female disappear into the vegetation of his territory – the idea being that the male, with his call, is perhaps indicating to the female a good place where, should she decide to remain, she could place her nest. Orians and Christman observed that in California males also gave growls in aggressive contexts, both intra- and inter-specific, but I have noted only the sexual usage.

Males also have a precopulatory call, which consists of a rapid series of brief, relatively high-pitched, notes – 'tititititi'. It is given when a male is near a sexually receptive female, especially just prior to copulation. It probably prompts and coordinates precopulatory movements and helps to bring the pair to copulation. Searcy (1989) investigated its function experimentally by playing these calls in the lab to captive females that had been primed for sexual responsiveness by estradiol implants. Precopulatory calls did elicit copulation-solicitation displays from the females, but male songs alone were as good or better. Songs and 'tititi' calls together, however, evoked the greatest number of female copulatory displays, indicating that the two vocalizations in concert have the greatest stimulatory impact; in nature, male precopulatory calls are often interspersed with songs (Orians and Christman 1968, Searcy 1989).

FEMALE VOCALIZATIONS

During the early investigations of avian vocalizations there was a research bias both towards songs over calls, and toward male vocalizations over those of females. That biologists were relatively slow to begin systematic investigations of female vocalizations was at least partly due to the fact that, although females of many species vocalize frequently, few temperate-zone species have females that regularly utter long, melodic songs of a type analogous to male advertising songs. Also, the behavior of female songbirds is, in general, more cryptic than that of males, and so more difficult to observe and study. In addition, there was a common tendency to consider male behavior as proactive and hence of more interest than the supposedly merely reactive behavior of females. While conducting earlier fieldwork with male redwings, I had noticed that females vocalize loudly and often. When choosing a subject to plumb to great depth for doctoral research, I remembered these female vocalizations and, after reviewing the extant literature, realized that very little was known about female singing patterns or the functions of their songs. The field was wide open. I decided to study female redwing vocalizations for this reason and because of a suspicion that their vocal activity would relate to their unusual, strongly polygynous, mating system.

Female redwings have two types of songs. As is the case with male redwing songs, these brief, buzzy or chattering sounds are a far cry from any music-lover's definition of a song. But they are relatively long, patterned vocalizations given in many contexts long associated with passerine song. Therefore, unless one's definition of song includes the word male, the female redwing vocalizations are songs. Orians and Christman (1968) lumped these songs together as 'female chatter', but illustrated their monograph with sonagrams of both types. They thought these 'harsh and rasping' vocalizations might be functionally equivalent to male redwing song, i.e. with roles in territory defense and intrasexual aggression. Orians and Christman noted that the songs were given only when females were on territories, frequently directed at unfamiliar females, and when leaving their nests. Michael Corral (1979) recorded and analyzed female songs in New York, identifying two distinct types: Type 1 songs (also called chit; Hurly and Robertson 1984), which were chattering sounds consisting of up to 20 or more rapidly repeated, brief, notes that averaged about 30 ms each; and Type 2 songs (also called teer; Hurly and Robertson 1984), consisting of repeated growling sounds, with up to 10 or

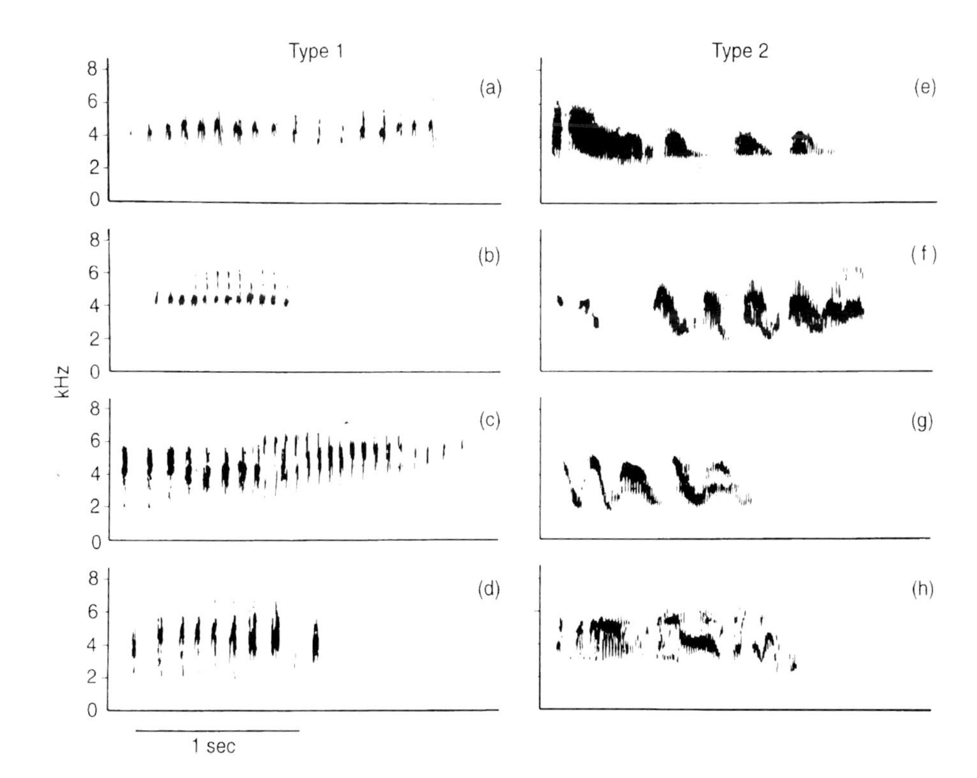

Fig. 5.10. Sonagrams of typical Type 1 songs of female redwings (a–d) and Type 2 songs (e–h) (from Beletsky 1983b).

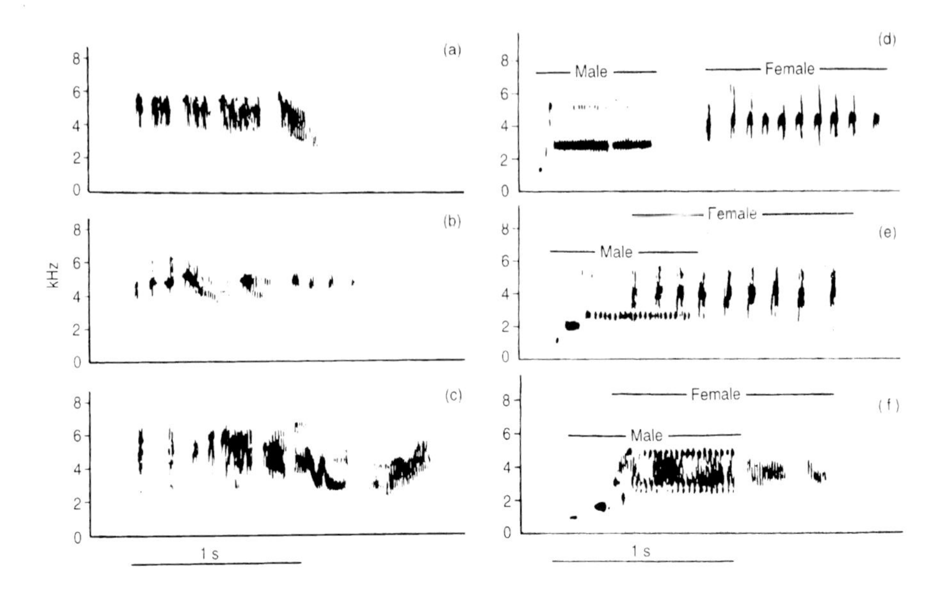

Fig. 5.11. Sonagrams of female redwing combination songs, with both Type 1 and Type 2 elements (a, b, c); sonagrams showing female 'song answers', following and overlapping male songs (d, e, f).

more longer notes averaging about 150 ms each. Both types were, like their male counterparts, typically 1–2 s duration. Females also mixed the types into combination songs, often Type 1 notes followed by Type 2 notes. Corral noted that females frequently gave their songs, especially Type 1, immediately after their mate's songs, their responses so rapid that they often managed to overlap the terminal portion of the male's song with their own (Beletsky and Corral 1983a).

In Michigan, I found female redwings with the same two song types that Corral had described, and combination songs (Figs. 5.10 and 5.11). (The only other common female vocalizations are a brief 'check' call (also termed chet; Hurly and Robertson 1984), given frequently between songs and when aroused or alarmed, and a precopulatory 'titititi,' similar to the male.) To investigate function, I first observed the social contexts in which the songs were used (Beletsky 1983a). I employed perhaps the simplest method – songs were considered to be directed at the closest individual to the vocalizing female. After watching more than 500 song situations for each type, the result was clear: Type 1 was given mostly when the mate was near, whereas Type 2 was used when other females were closest (Fig. 5.12). Thus, one type could be used for communication with males and one for females. This finding would be consistent with work performed on species in which males had two very different song types, one for use in male–male aggression and one for courtship (Morse 1970).

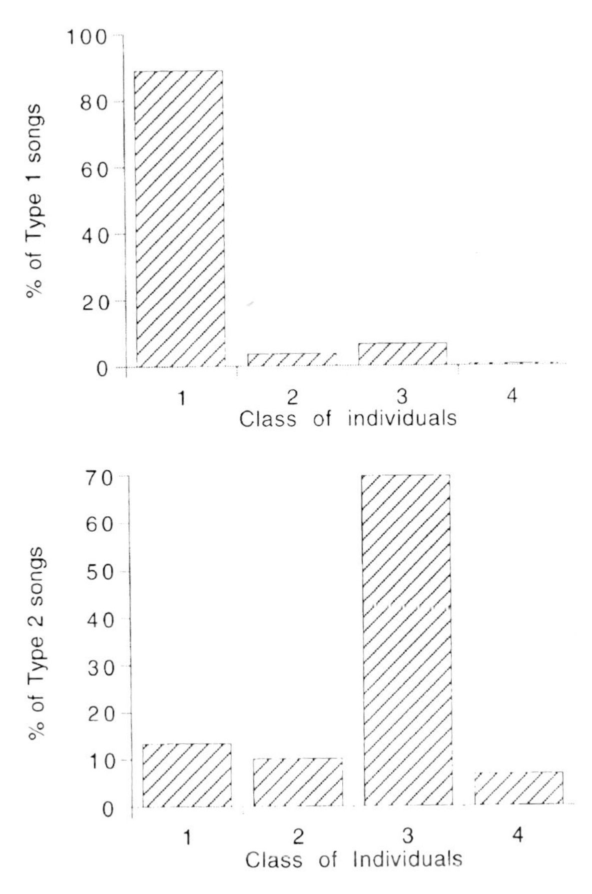

Fig. 5.12. Percentages of Type 1 and Type 2 songs directed at other individuals by female redwings during field observation in Michigan (from Beletsky 1983a). Data are summed from monitoring 10 females. Individual class 1 = mate, 2 = other male, 3 = any female, 4 = interspecific. Type 1, n = 604; type 2, n = 555.

Another fruitful method used to probe the function of a display is to replicate it artificially, present it to live individuals, and observe their responses. Early during one breeding season I recorded local female songs, made playback tapes, placed a loudspeaker in male territories, and broadcast either Type 1 or Type 2 songs to settled females. Their response to both types was immediate. They flew toward the loudspeaker after broadcast of the first or second song, landed nearby, and directed songs at the speaker (Fig. 5.13). Some perched on the speaker. Fully 100% of females approached within 4 m of the speaker playing Type 1 (n = 15 females) or Type 2 (n = 16). Forty-seven per cent of the former and 81% of the latter approached to within 0.5 m of the speaker. Furthermore, during playback the

Fig. 5.13. A female redwing responds aggressively to broadcast female songs by approaching the loudspeaker.

females decreased the rate at which they gave Type 1 songs, when compared with a preplayback control period, but significantly increased their Type 2 rate. Thus, during both experiments, females appeared to interpret broadcasts as another female redwing intruding in the male's territory and they responded aggressively to the potential threat. Moreover, their main vocal response was to give Type 2 songs, suggesting that it is an aggressive signal.

One potential function of signals is that, if they are individually distinctive, they may be used for recognition of individuals. Weeden and Falls (1959) demonstrated that territorial male Ovenbirds gave different behavioral responses when they heard broadcasts of the songs of territorial neighbors versus non-neighbors, which suggested that the birds could at least distinguish familiar from unfamiliar songs, if not actually recognize the songs of particular individuals. Since that pioneering work, similar results have been found for several other songbirds. To test whether male redwings might recognize their mates by their songs, I played to territorial males Type 1 or Type 2 songs of females recorded some distance away or Type 1 or Type 2 songs recorded from one of their own mates. Males gave different behavioral responses to 'stranger female' and mate songs, indicating that they recognized their mates' songs (Beletsky 1983b). In a similar experiment, I sought to determine if females recognize each other by their songs. I played neighbor and stranger female songs to individual females, but their behavioral responses did not differ statistically; thus, they may not recognize each other by voice alone (Beletsky 1983c).

I had identified Type 2 songs as aggressive vocalizations that females directed at other females and at other species (at my Michigan study site, particularly at Common Grackles, occasional predators on bird eggs and nestlings). In a series of observational and experimental studies, Ken Yasukawa confirmed the aggressive meaning of Type 2 (Yasukawa *et al.* 1987a, Yasukawa 1990). Although he found, in more of his 'speaker-occupation' experiments, that simply broadcasting repeated Type 2 on a territory did not prevent female settlement (in fact, it may have attracted new females), he suggested that these songs might function in establishing and maintaining dominance hierarchies within harems.

It was Type 1 songs, however, that, considering their usage, had the greater potential to yield new insight into the redwing social system. Females use Type 1 to answer their mate's songs, and they give Type 1 when they leave and return to their nests. When I began work in Washington, I turned my attention to Type 1 songs. Corral had first documented female song-answering, defining an answer to be a female's song initiated within 1 s of the termination of her mate's song (Fig. 5.11). He found that within 4 days of settling on a male's territory, females consistently answered at least 30% of the male's songs and sometimes up to 65% of them (Beletsky and Corral 1983a). Interestingly, during playback to individual females of mate or nonmate male songs, they answered at equivalent rates. Thus, either females cannot recognize males by their songs, or such recognition is irrelevant to them; they answer the songs of the closest (loudest) male they hear, which is, in most cases, their mate (Beletsky and Corral 1983b). This behavior is consistent with findings that females may select breeding sites by choosing on the basis of habitat or territory attributes rather than on male characteristics – the particular male that owns the territory on which they nest may not be important to them (Chapter 7).

At CNWR, I observed song-answering behavior for about 60 h on 10 male territories (Beletsky 1985). Of 3700 male songs given with at least one female present on the territory, 46% were answered. Individual females answered up to 60% of mate songs during the pre-incubation and incubation phases of breeding. I noted the social contexts of about 2600 male songs, classifying them as: sexual, the male clearly interacting with a female; male–male aggressive, the male clearly interacting with another male; and advertising, the remaining contexts, the male usually perched or on the ground, singing with no other individuals nearby. Females answered 76% of sexual male songs (n = 358), 34% of male-male aggressive songs (n = 420), and 34% of advertising songs (n = 1798). Sixty-two per cent of male flight songs (n = 368), which males give as they return to their territories, were answered by at least one of the vocalizing male's mates. Thus, because it binds male and female songs in time and occurs at its highest rate in sexual contexts, song-answering probably has a role in influencing, perhaps coordinating, interactions between males and females. It could also inform other listeners, such as prospecting females, that answered males are already mated. In a basic way, song-answering is analogous to the duetting behavior of some birds. But duetting is usually confined to monogamous breeders; the simultaneous or alternating mutual singing of male and female correlated with mutually large contributions to the single nest. In the strongly polygynous redwing, however, the contributions of male and female to reproduction and to song-answering are not mutual or equivalent. Rather, the male sings, presumably to maintain territory and attract additional mates; the female answers, associating herself with the male.

Females also give their Type 1 songs when approaching and leaving their nests. This usage provides additional insight into the functions of the vocalization and also into the dynamics of the male–female pair-bond. Because of the acoustic attributes of Type 1 songs, females giving them are easily located in space. Therefore, the locations of their visually hidden nests are almost continually divulged to listeners, to conspecifics, and also to potential predators. Given that a female risks informing predators not only of her nest location but also of her imminent absence, giving these nest-associated songs must be particularly advantageous to females. Gordon Orians and I tested two hypotheses of the function of Type 1 nest-departure songs: (1) that a female directs them at her mate, so that males and females can coordinate their off-territory trips, minimizing the time that nests are left unguarded; and (2) that a female directs the songs at other females, to coordinate or stimulate group departures. Females often seem to leave territories in groups, in a contagious manner, and there might be advantages in leaving in groups (confusing predators?) or foraging in groups.

During 88 h of observation, we found that females vocalized during 89% of nest departures (n = 1130) and during 68% of their returns (n = 772). Most of the songs were Type 1 or Type 1–2 combinations. Individual females gave these songs frequently during the nest contruction, laying, incubation, and nestling phases of breeding, but rarely after their young fledged (Beletsky and Orians 1985, Yasukawa *et al.* 1987a). Our first prediction, that, if females gave nest-departure songs to inform mates of their leaving, they would sing only when males were present, was not supported. Females sang upon leaving regardless of male presence. Our second prediction, that, if females use nest-departing songs to coordinate group departures, then females should be prompted to leave nests by the playback of female songs, was also unsupported. Females left their nests within 5 s about 20% of the time following playback of Type 1 songs or playback of control (American Robin) songs – statistically at the same rate. Furthermore, we found that, although many females appeared to leave territories in groups, the great majority of departures (85%) were actually of single females. Thus, we could accept neither of our hypotheses to explain nest-departure songs.

We then speculated that song-answering and nest-associated songs are components of an intrapair communication system that continually informs males of female activities and reproductive states. After females settle on male territories, and especially following copulation, interactions between mated redwings can be few and irregular. Furthermore, males are ensconced in a female-rich environment, constantly exposed to their own harem, prospecting females, and the harems on adjacent territories – different classes of females toward each of which they must act differently. For example, they must court prospecting females, perhaps chase or attempt extra-pair copulations with intruding females from other territories, and leave their own mates alone. Type 1 songs may thus identify females to their mates as they leave their nests and fly through the territory, to (1) maintain pair-bonds during nesting, when direct interactions between males and females are few; (2) inform males of nest-associated activities and repeatedly demonstrate the location of the active nest; and (3) inhibit sexual advances. This last function could be important because males chase most females that fly through their territories. Chasing their own females to court them when they were already engaged in nesting would serve no purpose and, in fact, would waste time and also delay females in their foraging trips for themselves and their young. If females give nest departure songs to inhibit sexual harassment, then females who vocalize in these situations should be chased by males significantly less often than females who depart their nests silently. We found this to be the case not only for redwings (Birks and Beletsky 1987), but also for Yellow-headed Blackbirds (Beletsky and Orians 1985); yellowheads are also polygynous marsh breeders in which females frequently vocalize when departing nests. Many other female songbirds call from the nest, including several other icterines, and the reduction of harassment function has strong support in several species (McDonald and Greenberg 1991). It is notable, then, that female redwings occasionally depart silently from their nests; might it be possible that they do so to manipulate male behavior when it would be advantageous to encourage male chasing or attention?

In some of his most interesting work, Ken Yasukawa (1989) sought to determine the costs and benefits to females of giving nest-associated Type 1 songs. To test for a cost, Yasukawa placed mock nests containing fake eggs on upland Wisconsin territories and broadcast Type 1 songs from a loudspeaker placed nearby. Ten out of 15 mock nests were lost to predators within 2 days, but none of his control nests (mock nests with fake eggs, near an always silent loudspeaker) were lost during that period. After 10 days, 11 mock nests with Type 1 playback had been destroyed compared with only two control nests. Yasukawa concluded that nest-associated vocalizations have the potential to significantly enhance nest predation rates. To test for a benefit of the songs, Yasukawa compared the vocal behavior of females that successfully fledged a brood and those whose nests were lost to predators. He found that successful females had given significantly more Type 1 songs from their nests. Furthermore, Yasukawa presented a model of an American Crow, a predator on redwing nests, near his mock nests and discovered that the male redwing territory owners mobbed the crow more vigorously at nests located near loudspeakers broadcasting Type 1 songs than at silent control nests. As demonstrated by Yasukawa, the costs to females of giving nest-associated vocalizations include an increased risk of nest predation. The benefits, presumably of greater average magnitude than costs, appear to include the stimulation or maintenance of male nest defense behavior, which must result in an increased chance of successful fledging.

SEXUAL SELECTION AND REDWING COMMUNICATION

Sexual selection is the term evolutionary biologists use to describe the evolution by natural selection of traits concerned with competition for mates. Animal signals that are used during breeding seasons are particularly subject to sexual selection because they often mediate male–male competition for territories and mates and perhaps also mate choice in females. Many studies of redwing signals, particularly those of William Searcy and Ken Yasukawa, have concentrated on testing for the operation of sexual selection in this species, and others, while not performed for that express purpose, nonetheless provide relevant data. Recently, Searcy and Yasukawa (1995) reviewed the contributions of redwings to the study of sexual selection and, after systematically sifting the sometimes contradictory data, concluded that sexual selection is probably responsible for the forms and patterns of use of several redwing displays, notably in the male. For example, they believe that the preponderance of evidence suggests that red epaulets in males are sexually selected attributes with an intrasexual function; in particular they cite experimental evidence that males with blackened epaulets tend to lose territories and thus their chances for mating, and that stuffed male mounts with larger-than-normal epaulets are better at 'defending' a vacant territory from conspecific intruders than mounts with normal-size or no epaulets. For male song, the experimental evidence they cite is even stronger: song has been shown to be used in territory defense (in particular, by devocalization studies), courtship, stimulation of female receptivity, and probably mate attraction. In total, work with redwings has to date provided some of the best evidence for sexual selection operating in birds, particularly because studies have addressed multiple aspects of redwing morphology and behavior, with displays prominent among them.

CONCLUSIONS

Two aspects of redwing biology in particular have apparently shaped vocal and visual display repertoires. These are the strongly polygynous mating system and colonial breeding. Besides other influences, what these factors share is that they both contribute to a crowded breeding situation. A mated male and female in a species that is monogamous and territorial may spend most of the breeding season interacting with one another around their nest. The number of different displays and their frequency of use necessary to successfully coordinate breeding may be less than that required in crowded redwing marshes, where not only does each male have up to a dozen females simultaneously on his territory, many in different stages of breeding, but there can be up to four or five males and harems on immediately adjacent territories and also other nonadjacent but near neighbors. In a different context, Beecher (1989) and others pointed out the need for more highly developed kin-recognition systems in species with colonial versus solitary or low-density breeding habits; for example, abilities of parents to recognize their young. It is not surprising, therefore, that some special signals have evolved to coordinate male and female breeding activities among redwings in their crowded breeding environments, or that others may have developed that take advantage of the crowd.

Songs and visual displays of male redwings are, in many ways, analogous to those of many typically monogamous males, probably because they have much the same main functions: facilitation of territory maintenance and defense, and mate attraction and stimulation. Where major differences occur, they may lie not so much in structure or function, but in patterns of use; for example, in monogamous species, such as White-crowned Sparrows, male song rates decline dramatically after pairing. However, male redwings continue singing at relatively high rates throughout the breeding period, a behavior associated with the long period over which multiple mates are attracted.

Female redwings, however, find themselves in a very different breeding social environment than do monogamous females. Studies of their displays have to date illuminated at least four contexts in which females in the crowded, ploygynous and colonial, breeding situation have need of specialized communication to inform and manipulate their surroundings. First, all females, except those on very poor-quality territories, are members of a harem and, as such, members of an all-female social system, whose organization may be that of a dominance hierarchy, territorial system, or otherwise (see Chapter 7). Females must interact with each other and there is at least one special display, Type 2 song given in a song-spread display, for this purpose. Second, females regularly need to reinforce their pair-bonds because, although they themselves perform most nesting duties, mates are still responsible for nest defense and, in some populations, for assistance with feeding young. Third, because males often have many active nests on their territories, in addition to many failed, inactive ones, females apparently need continually to inform males of the specific locations of their nests, to stimulate, or perhaps simply to coordinate, the males' defense response. And perhaps because males keep trying to attract additional mates throughout much of the breeding season and also because of the large numbers of females on and near territories, females need to prevent unwarranted male chasing and courtship. Type 1 vocalizations have a role in each of the last three interactions. During song-answering, the Type 1 song maintains the pair-bond during periods when mates are usually physically separated. When delivered from the nest it repeatedly reinforces for the male the location of a nest and its active status. When delivered during nest departure or when flying through a territory, Type 1 song reduces sexual harassment. This last function may be particularly important because females seem to have an additional, visual, display to reinforce it. Often when approaching or departing a nest containing nestlings, females move in 'flutter' flight, a slow, wing-fluttering flight, that appears to further identify nestling-feeding females to their mates (Beletsky and Orians 1985).

Crowded breeding colonies also provide raw material for a cooperative alert system. Multiple males stationed on their marsh territories provide a large number of eyes and ears, some of which are constantly scanning for predators. The call-alert system provides for instant alerting of all listeners on the marsh, males, females, and perhaps even young. Thus, although a large, noisy breeding colony obviously attracts more predators than would a quiet, dispersed breeding situation, the redwings' cooperative alerting and mobbing behavior apparently can deal successfully with the threat.

So, redwings produce some highly specialized signals, adaptations to their specific and unusual breeding ecology. It would be a pity, however, if all the effort spent studying redwings did not produce some information of general use. I would suggest that one important lesson for the study of other species that arises from our knowledge of redwing communication is in the areas of the number of displays a species possesses and their frequency of use. Some investigators have attempted to detect relationships between the number of different signals a species has and its social or mating system. For example, Irwin (1990) correlated mating system and song repertoire size in males of the New World blackbirds, but found no significant associations. Orians (1985) examined the number of different signals, by sex, for 18 different species of blackbirds, and found that in polygynous species, males generally had more vocalizations than did females. It may well be that, on average, polygynous species have more signals than monogamous ones, or that in some groups males have more than females. The lesson that redwing alert calls brings to these analyses is that, although one species may have a large number of signals, they do not necessarily have different meanings or functions. This suggests that each species' signals needs to be studied carefully before gross comparisons are made using number of displays. Moreover, the high rate at which some displays are given may need to be considered, in some cases, in the light of what we now know about redwing repetitive alert calling.

Finally, the uses female redwings make of their Type 1 songs for coordinating, in a broad sense, male–female interactions demonstrates one reason why strongly polygamous systems may have more displays than monogamous ones. At first guess, one might predict that because male and female behavior is to a degree disengaged in polygamous systems the sexes should need fewer signals to communicate with each other. But the opposite may be true. Whenever males and females do meet, they will be more or less unaware of each other's immediate situations and, hence, motivations, and thus, must communicate. Further, the unequal contributions to the nest and the resulting different daily activity patterns of the sexes probably make frequent communication between them, in some cases, more important, rather than less.

There have been probably more investigator-years spent in the field studying redwing communication than that of most any other bird and still there are gaps in our knowledge. Most conspicuously, our information is mostly limited to displays given during the breeding season. Although there is some information (e.g. in Orians and Christman 1968), we know relatively little about communication during nonbreeding periods, although they comprise two-thirds of each year! To interpret all redwing signals by their breeding season usage I fear is analogous to beings from Mars arriving on Earth and trying to decipher human language by attending only property transactions, professional boxing matches, seductions, weddings, and births. Just as recent theory suggests that some songbird plumages are adaptations not to breeding but to wintering, so too, perhaps, are some signals. Smith (1991) argues that animal signals are probably often given in many more contexts than we give them credit for, and because we limit most of our study to the breeding season, he is doubtless correct. Another gap in our knowledge concerns the function of male redwing song repertoires which, unfortunately, despite years of study, remains a mystery. Finally, factors that may have been selected for repetitive alert calling are largely unknown; more work in this area, perhaps using a comparative approach, would be invaluable.

Chapter 6

BREEDING CYCLE AND MATING SYSTEM

INTRODUCTION

Male and female birds do not need to associate outside of the breeding season, but to produce offspring they must associate and cooperate. But how, and how much? Except for the shared purpose of producing young, the interests of the sexes do not necessarily overlap. A male redwing seeks to establish and defend a territory from other males during the breeding season, attract to it as many females as he can, and sire as many offspring as possible with his 'own' females and also with those on adjacent territories. A female redwing seeks to select a territory with secure nest sites that is near to rich feeding areas and, in some cases, perhaps a specific position within a territory's breeding hierarchy (first to nest, second to nest, etc.) so that she can garner a large share of the male's help in feeding her young. Thus, with their polygynous mating system, redwings show very little overlap between males and females in breeding interests and roles, and much less than is found in monogamously-breeding birds. The sex differences make the redwing breeding system both very interesting and more laborious to study: because the interests and breeding behavior of males and females differ so broadly, the sexes need to be considered more independently than would be the norm with monogamous birds.

In this chapter, the first of several on breeding and reproductive success, I have two objectives. First, I will in some detail introduce the reader to the redwing's breeding cycle and behavior; second, I will discuss the redwing's mating system, which has drawn so much attention from researchers.

THE REDWING BREEDING CYCLE

Because I know best the Columbia National Wildlife Refuge (CNWR) population, I refer to its breeding information at length in my descriptions. However, I should point out that many aspects of breeding vary regionally across the redwing's range and that such variation will be apparent in this chapter's tables and will also be described where appropriate elsewhere in the book.

Overview

Although it is spring before a male redwing's fancy turns to thoughts of love, or whatever passes as love among redwings, there is much preparation beforehand. Adult male redwings in eastern Washington are resident; they do not leave their breeding areas seasonally to migrate elsewhere. They spend the winter in feeding flocks, foraging in agricultural fields and cattle feed lots, returning occasionally to their marsh territories. Beginning in February, during late winter, males begin roosting each night on their territories, and then each morning advertise and argue territorial boundaries for a brief period before flying off for the day to feed in the fields. This late winter period is apparently the time when most 'new' territory owners, i.e. those that did not have territories during the previous breeding season, establish themselves. In other regions, where males are migratory, they arrive back on the breeding grounds in March to May, depending on latitude, and establish territories (Table 6.1); as is typical for territorial songbirds, there is a high degree of year-to-year fidelity for individuals to specific sites.

At CNWR, as the weather moderates in early March (usually, but by no means always!) and the first groups of females arrive from their migration, males begin spending more time on their territories. On warm February mornings, for example, males rise shortly before dawn, but do not leave for the day until 2 or 3 h later. During these early hours the males patrol their territorial boundaries, flying from song post to song post (generally cattail stalks), defend their territories from other males, and try to attract mates. Females that enter the territories are approached by the males and courted, usually quite energetically.

After a female selects a territory on which to nest, she spends the morning hours there, moving through the marsh, foraging, looking at potential nests sites, and participating with the male in courtship activities, including Sex Chases. She roosts on the territory, but like the males departs in the morning to spend the day foraging elsewhere. Again, at this time of year, there is little food for the redwings on their marshes. At some point, probably in response to a combination of endogenous (hormonal) stimuli and environmental information (weather, food availability, perhaps the activities of other females), a female decides to start nesting. Some females delay up to 30 days or more after settling before beginning their nests. A female constructs her nest alone, copulating during this period with her mate and, often, with other territory owners as well. In early spring there is usually a delay between the end of construction and the beginning of laying – up to 7 or more days; later, laying often begins within a day or two after the nest is finished. Males normally start spending their entire days on or near territories when eggs first appear, presumably because of the need at that time to defend nest contents. Copulations continue during laying. With the appearance of the penultimate egg the female initiates incubation, which is also done without male assistance. As first (primary) females begin to nest, males continue to advertise for and court additional mates. In some areas secondary females do not settle until primary females begin incubation (Langston *et al.* 1990) but in other areas the settling patterns are more complex .

Very young nestlings need little food each day and, being naked, they need to be kept warm in cold weather. Therefore, after eggs hatch, females spend much of their time brooding the young; as nestlings grow, they increase the amount of time they spend foraging for and feeding them. Most males at CNWR do not feed their nestlings, but large percentages of males in other regions do. After young fledge, they continue to be fed by their parents for about 2 weeks. Whether a nest fledges young, is destroyed by a predator, or fails for some other reason, a female generally quickly nests again, provided that it is not too late in the season. Most nesting at CNWR is finished by the end of June or, latest, mid-July, when males, females, and young of the year gather into flocks.

Table 6.1 gives approximate breeding dates for CNWR redwings and, for comparsion, the same for other selected populations.

Table 6.1. Redwing breeding chronologies in various areas[1]

	North-central California in 1958-59[2]	Central Ohio in 1974[3]	South-central New York in 1988–90[4]	South-central Washington in 1980s[5]	South-eastern Alaska in 1982[6]
Males arrive on territories	January	Early	Early April	1–15 February March	14 May
First females arrive	March	–	April	1–15 March	22 May
First nests	Mid-April	Early May	Early May	1 April	Late May
Peak of nesting[7]	Late April/ early May	8 June	18 May	Early May	13 June
End of nesting (final fledging)	Mid-June	17 July	20 July	Early July	10 July

[1]Precise dates are given when provided by authors.
[2]Orians (1961), Orians and Christman (1968).
[3]Dolbeer (1980).
[4]Westneat (1992b).
[5]Orians and Beletsky (1989).
[6]McGuire (1986).
[7]Maximum number of active nests.

Female arrival, courtship, sex chases, and copulations

Females arrive at breeding marshes either singly or in small groups. The attention of all territorial males on a marsh is immediately attracted; epaulets are exposed and 'conc-a-ree' songs given. If the females initially land outside male territories, some territorial males may fly up and perch near the females, while others remain in their territories, perched on top of cattails. If the females fly off, males often chase them for some distance before wheeling in the air and returning to their territories. When a female lands in a territory, the owner flies over and lands near her, epaulets maximally flared. After singing, the male may keep his wings out in the After-song Display. Females still interested generally drop to the ground or to water level. We are not sure what they are doing, but they may inspect the physical structure of the marsh vegetation for potential nest locations or the territory's potential for food production; they have even been observed peering into the water, as if to determine the marsh's future quality as a provider of aquatic insects. They are followed by the males, who may strut around on the ground in the after-song display posture, give precopulatory calls, or the throaty 'haaaa' sound (the Growl). Sometimes the male flies first into the vegetation and is followed by an interested female. The male puts his wings out and up, forming a V above his back, aims his bill down into the vegetation, and 'growls'. Orians and Christman (1968), following Nero (1956a), interpreted this sequence, perhaps a bit anthropomorphically, as a 'Nest Site Demonstration', the male suggesting to the female a good place, should she decide to stay, to locate a nest. However, the researchers never found nests in the particular locations where they saw the display occur.

Many females occupy the same male's territory from one year to the next, at CNWR about 40% of them (Beletsky and Orians 1991; see Chapter 7). Do redwings recognize their mates from previous years? We do not know for sure, but we do see things that provide clues. For example, some females arriving at a male's territory for the first time in spring immediately perch quietly near the male (e.g., Nero 1956a), whereas others move through the marsh on their inspection tours. Are the former individuals mates from last year? We suspect so because they act as if they 'belong.'

Once a female settles on a male territory, the amount of time the pair spend together varies, but is often small. Nest-site demonstrations continue and sometimes the males appear to follow the females on and off the territory (Nero 1956a), but for much of the time the pair operate independently, the male patrolling and defending his territory and advertising, the female foraging, preening and interacting with other females. The most conspicuous interaction between males and females during the courtship period is sexual chasing. Males aggressively chase their females in rapid, acrobatic flights around the marsh, over many territories,

Group sex chase. Several territory-owning males chase a female that has already chosen her nesting territory. The reasons for such chases are unknown.

and often out over the lake or into upland areas. The male pursues closely, shifting his flight in response to each of the female's evasive maneuvers, apparently trying to catch her. Sometimes she is overtaken in flight and nipped on the rump, or even caught outright, the two birds at that point tumbling into the marsh (Nero 1956a, 1984). The chase ends when the male gives up, breaks off the chase, and returns to his territory (usually followed by the female) or when the male and female crash into the marsh (sometimes on the wrong territory) or upland vegetation. The stimulus for a sex chase is not known, but they often appear to start when the male, seeing one of his females flying in his territory or even just perching quietly, suddenly dives at her.

Who initiates a chase and why? One possibility is that such rapid, acrobatic flights allow males to assess the physical condition and health of their mates, or for females to do the same. Much interest of late has been paid to the potential relationship between parasite load and mate choice, and perhaps individuals evaluate their mates for possible parasitic infections or general health by observing sex chase performance. Clearly, heavily-infected animals should not be able to chase fast or long. Such information might be of use to females, for example, when they decide whether to remain with a male and whether and how often to participate in extra-pair copulations (Chapter 7). Thus, it is even possible that it is the females themselves who, perhaps very subtly, initiate the chases. Sex chases do not culminate in copulations, so they apparently are not precopulatory behaviors, meant to stimulate or coordinate copulation (Nero 1956a, Monnett *et al.* 1984, Westneat 1992a).

Sex chases occur throughout the breeding period as long as females are settling on male territories and starting nests (at CNWR, during March, April and May). The most intriguing sex chases are group chases. Quite often nearby territory owners on a marsh are induced to join a sex chase in progress. Six or seven males sometimes end up chasing a single female. Occasionally, group chases occur without participation by the female's mate, a neighboring male starting the chase. Group chases are typically noisy, the males giving rapid conc-a-ree songs as they pursue. A single male, usually the mate, gets closest to the female and typically chases her to ground. At that point the other pursuers either land and perch for a few seconds above the spot in the marsh where the couple disapppeared, before returning to their territories, or break off in flight and return home. Group chases are interesting because for their brief duration they suspend the strong territorial system and males are in a free-for-all contest for a breeding resource (the female). They chase her down and then, as if emerging from a sexual frenzy, look around and realize that they are in another male's territory, and almost sheepishly retire to their own. Furthermore, the group chase momentarily reverses the polygynous sex ratio on the marsh in that it involves many males and a single female. In light of recent genetic evidence that many females are fertilized after extra-pair copulations with neighboring males, it is tempting to link the males that chase a female, perhaps those that chase her best, with those she eventually copulates with.

Copulations themselves are brief affairs, the male mounting from the rear after engaging in appropriate precopulatory displays (Chapter 5). He maintains his position on her back for only 2 or 3 seconds, rapidly fluttering his wings (see illustration at chapter head; Nero 1956a, Orians and Christman 1968). The male then jumps down and usually flies off. Females are quiet and motionless during copulation, but immediately afterwards often flutter their wings and preen. Most copulations occur during early morning and early evening hours.

Nests

Early settling females delay up to several weeks before launching into nest-building. Later settlers begin their nests soon after their arrival on a territory. Females build tightly woven, cup-shaped nests, mostly out of long pieces of cattail leaves (Fig. 6.1). They wet the leaves before weaving them. The inner surface of the nest is lined with fine grasses. The constituents of one dissected nest were '142 cattail leaves, up to 21 inches in length, and lined with 705 pieces of grasses. It also contained 34 strips of . . . bark, up to 34 inches in length' (Bent 1958). In marsh vegetation nests are built anywhere from one to 6 feet above the water, but typically 2–4 feet. One year at CNWR we hired as a field assistant a college student less than 5 feet tall. It turned out that she was not of sufficient height to peer into most redwing nests to check contents, the main purpose of her employment. We quickly devised a mirror-on-a-stick apparatus with which she could inspect nests from below.

Nests are also placed in dense grass clumps, in shrubs, in some regions on the ground, and even in trees at heights of up to 30 feet. In cattail, the nest is woven around four or more closely-spaced stalks. The female weaves while bracing herself on a reed with one foot, while using her bill and other foot to manipulate the nest materials. That a female alone, in the span of a few days, can construct such a neat, lined, tightly-woven nest, especially, in the case of yearlings, without ever having done it before, never ceases to amaze. In 1990 a female at CNWR who built a rather loosely-woven, messy nest had but one leg. I still wonder how she manganged to build a nest at all.

The earliest females to breed at CNWR usually take 4–5 days to complete a nest, devoting to it only so many hours per day. But later on, at mid-season, completed nests ready to receive eggs appear in the marsh in a day or two (Table 6.2). Nests that lose their eggs or young to predators are very rarely reused, even if, to people, they still look perfectly good. Over a 14-year period at CNWR, only 58 of 6787 nests were reused, less than 1% (Harms *et al.* 1991), and in one study in

Fig. 6.1. Typical placement of a marsh-situated redwing nest, woven into vegetation such as cattails, above; top view, showing eggs, below.

Oklahoma, only five out of 306 nests were reused (1.6%; Goddard and Board 1967). Females instead build new ones rapidly, sometimes up to five in one season (Picman 1981, Orians and Beletsky 1989) as they replace destroyed ones and try repeatedly for a successful nest. Sometimes, when prolonged inclement weather occurs right after a nest is built, and laying is delayed, the female eventually builds a new nest in which to lay even though the previous one was never used and, again, to us, looks entirely serviceable.

Egg type and clutch size

Redwing eggs are light blue or a pale bluish green, larger at one end than the other, and moderately glossy (Bent 1958). Most often they are spotted and streaked with a darker color: purple, black or dark blue. Eggs measure about 1 inch

Table 6.2. Durations for redwing breeding phases in various regions

Duration (days)	Region	Reference
A. Nest construction		
5–6 (early season) 3 (later season)	New York	Case and Hewitt (1963)
4–8	California	Payne (1965)
5–6 (or less)	Washington (Turnbull NWR)	Haigh (1968)
3–5	Illinois	Strehl and White (1986)
2–6	Washington (Columbia NWR)	L. D. Beletsky and G. H. Orians (unpublished data)
B. Incubation		
10–12	New York	Case and Hewitt (1963)
11–14	California	Payne (1965)
11–13	Washington	Haigh (1968)
10–12	Illinois	Strehl and White (1986)
C. Nestling period		
10–12	New York	Case and Hewitt (1963)
11–15	California	Payne (1965)
9.5–10	Washington (Turnbull NWR)	Haigh (1968)
9.2 (females) 9.7 (males)	Michigan	Holcomb and Twiest (1970)
11	Michigan	Fiala and Congdon (1983)
10	Illinois	Strehl and White (1986)
11	Ontario	Muldal *et al.* (1986)
11	Washington (Columbia NWR)	Beletsky and Orians (1987a)
D. Fledgling period		
15–19	California	Payne (1965)
About 15	British Columbia	Picman (1981)
7–12	Illinois	Strehl and White (1986)

NWR, National Wildlife Refuge.

Table. 6.3. Redwing clutch size in various regions

Mean	*n*	Range	Years	Location	Reference
		2–5			Beer and Tibbits (1950)
3.5	926	1–5	1960–1961	New York	Case and Hewitt (1963)
3.3	537	2–5	1958–1961	Maryland	Meanley and Webb (1963)
3.4	90	1–5	1963–1964	California	Payne (1965)
3.4	243	1–5	1965	Oklahoma	Goddard and Board (1967)
3.7	499	2–6	1965–1967	Washington (Turnbull NWR)	Haigh (1968)
4.2				Missouri	Crawford (1970)
2.2	36	1–5	1968–1971	Florida (captive)	Knos and Stickley (1974)
3.7	211	1–5	1969–1973	Ohio	Francis (1975)
3.4	170	2–6	1973–1974	Ohio	Dolbeer (1976)
3.7	122	2–5	1972–1974	Iowa	Crawford (1977)
3.2–3.3	120	1–5	1970–1971	New York	Allen (1977)
3.6	381	–	1974–1975	Ontario	Weatherhead and Robertson (1977a)
3.0–3.6	321	–	1973–1974	New Jersey	Caccamise (1978)
4.2	73	–	1963–1965	Washington (Columbia NWR)	Orians (1980)
2.6–4.0	224	–	1977–1978	Washington	Ewald and Rohwer (1982)
2.9–3.8	338	2–5	1975–1977	Illinois	Strehl and White (1986)
3.3	6207	1–7	1977–1992	Washington	Orians and Beletsky (1989), Harms *et al.* (1991)
3.7	83	–	1986–1987	Michigan	Whittingham (1989)

long by 0.7 inches wide (average = 24.9 mm × 17.5 mm; Bent 1958), and weigh, at laying, between 3.0 and 4.5 g (Haigh 1968, Muma and Ankney 1987). Females typically lay three to five eggs, one per day, with four being most common (Table 6.3). At CNWR, we occasionally found six-egg clutches (Table 6.4), and once I found seven eggs in a nest. Because some of them were small and one obviously a runt egg, only one-third normal size, it appeared they were all produced by the same female. Why do redwings lay the number of eggs they do? One influence is

Table 6.4. Redwing clutch sizes at the CNWR study site, 1977 to 1992; percentages of each clutch size each month[1]

	Per cent (*n*) of clutches				
Clutch size	March	April	May	June	Total
1	1.5 (1)	6.9 (135)	12.4 (420)	12.1 (96)	
2	4.4 (3)	7.7 (150)	12.1 (412)	17.6 (140)	
3	25.0 (17)	22.0 (429)	25.7 (872)	37.4 (297)	
4	64.7 (44)	56.6 (1105)	44.2 (1500)	29.6 (235)	
5	4.4 (3)	6.7 (131)	5.2 (177)	3.1 (25)	
6	–	0.2 (3)	0.3 (10)	0.1 (1)	
7	–	–	–	0.1 (1)	
Mean clutch size ± SD	3.7 ± 0.7	3.5 ± 1.0	3.2 ± 1.1	3.0 ± 1.1	3.3 ± 1.1
n	68	1953	3391	795	6207

[1]Based on date of first egg.

age. Yearlings have, on average, smaller clutches than older females. Mature females may be better foragers than yearlings and so would be able to accumulate greater energy reserves for egg production. Crawford (1977) found that yearlings in Iowa had an average clutch size of 3.4 eggs, compared with 4.1 for mature females.

Clutch size at CNWR, from 1977 through 1992, which ranged from one to seven eggs, with four being most frequent, averaged about 3.3. Average clutch sizes were largest during the earliest month of nesting, March, and progressively declined each subsequent month (Table 6.4), a trend reported in other redwing studies (Orians 1980, Strehl and White 1986) and, indeed, for many temperate-zone breeding birds (Klomp 1970). The pattern is often attributed to a reduction in food supplies as breeding seasons progress (Lack 1966) but that is clearly not the case for redwings. Several factors may account for the seasonal decline in redwing egg production. First, the earliest nesters are usually older, experienced females; yearlings, who may produce smaller clutches, usually do not arrive in the study area in large numbers until April, and therefore their clutches will be concentrated in the later nesting months. Second, because the nesting season is a long one in this region, and many females renest several times, they may progressively tire themselves. If this is so, the result would be less energy available for reproduction and, hence, smaller average clutches later in the season (Beletsky and Orians 1996).

Incubation, nestlings and fledglings

Female redwings incubate their eggs for 10–14 days (Table 6.2). During incubation each female remains in contact with her mate, perhaps keeping him attuned to the position of the nest and its progress, by answering his songs while she is on the nest and by vocalizing when departing from and returning to the nest (Beletsky and Orians 1985). Because females begin incubating after laying their penultimate egg, hatching is partly asynchronous, the last egg often hatching 1 or even 2 days after the first. Hatching usually occurs shortly after sunrise. Nestlings that hatch first tend to grow faster, fledge at heavier weights, and have higher likelihoods of surviving to fledging than later hatchers (Strehl 1978). Females feed young by themselves during their first few days in the nest, with males starting to assist their mates, in those regions in which they participate, when young are 4 or 5 days old (Muldal *et al.* 1986, Whittingham 1989, Patterson 1991). In different studies, females averaged between 8 and 12 food deliveries to the nest per hour, regardless of whether their mates assisted with feeding (Strehl and White 1986, Whittingham 1989, Patterson 1991, Teather 1992).

Redwings remain in the nest for 9 to 12 days, depending on region (Table 6.2), or longer if conditions are particularly unfavorable. Young then fledge and continue to be fed by their parents for 2 weeks (Table 6.2) until they are independent. More males assist their mates in feeding fledglings than nestlings, even in regions where few males feed nestlings (Beletsky and Orians 1990). The survival of nestlings to fledging age is strongly dependent on rates of nest predation and, in some areas, on food availability. Little is known about survival rates of fledglings to independence, nor of juveniles to adulthood.

THE REDWING MATING SYSTEM: POLYGYNY

A great deal of labor-intensive research conducted with redwings concentrates on exploring one simple facet of their breeding biology, that typically two or more females mate with, and nest seasonally on the territory of, a single male. This emphasis in redwings is consistent with a general preoccupation with polygamy by ornithologists who study mating systems. As pointed out by others (e.g., Gowaty and Mock 1985), the emphasis on polygamy is paradoxical because 90% of bird species are thought to be primarily monogamous (Lack 1968). Thus, polygyny is relatively rare in birds. In fact, in their survey of the literature on North American passerine breeding, Verner and Willson (1969) considered only 14 of 291 species to be regularly polygynous. The disproportionate interest in polygyny elicited from ornithologists stems partly from the system's rarity, partly from the exaggerated behaviors and striking forms and colors produced by strong sexual selection that acts in these systems, but mostly from a need to explain the puzzle of why polygyny exists at all.

The sexes in monogamous birds contribute fairly evenly to each nest. Provided that polygamy evolves from monogamy, we would like to know how mating systems developed in which males and females contribute unequally to each breeding effort. Each species' present mating system is usually considered to be the outcome of an evolutionary conflict between the sexes. Because male gametes, sperm, are small and thus energetically inexpensive to produce in great numbers, males should gain reproductive advantage by mating with more than one female, perhaps the more the better. However, females, whose eggs are relatively large and energetically much more expensive than sperm, should gain by having a single mate who contributes all of his parental care to her nest and young; indeed, females can incur a cost by sharing a male's parental care duties with other females (Altmann *et al.* 1977).

At first glance, male redwings certainly appear to have decisively won the battle of the sexes. They mate annually with several females, and advertise for additional ones essentially throughout each breeding season. While males advertise and protect nests from predators, females perform most of the reproductive chores: nest building, laying, incubation, brooding young, providing the lion's share of feeding of nestlings and fledglings; there is remarkably little overlap in the sexes' breeding roles (Table 6.5). But there is a 'flip side' to the males' apparent advantage: all females attempt to breed each year, but many males cannot obtain territories and, hence, are excluded from breeding, some, throughout their lives (see Chapter 10). Even so, benefits appear clear to males who participate in such a system – chances to produce large numbers of descendants. For example, one of our CNWR breeding males, RYB-AR (red-yellow-blue on his left leg, aluminum-red on his right), held a territory for 11 years, amassed an average of 7.5 females each year, and had 176 young fledge from the 136 nests located on his territory. For RYB-AR, clearly, polygyny was a rewarding system. But why should females participate? The last quarter-century of research on polygynously-breeding birds can fairly be said to have concentrated squarely on this question.

Early ornithologists logically decided that polygamous mating systems simply reflected uneven adult sex ratios – polygyny was the result when the number of females in a population exceeded the number of males – but this has been proved false. Most biologists credit Gordon Orians and his colleagues during the 1960s with first formally describing a testable hypothesis of how polygyny in birds could yield reproductive advantages to both sexes, and thus could have evolved. Orians' ideas grew out of his many years of investigating redwings and other marsh-breeding blackbirds. The breakthrough occurred with the emergence in the 1960s of the new discipline of behavioral ecology, a union of ecology and animal behavior. The new

Table 6.5. Roles of the redwing sexes during the breeding season

Breeding Activity	Males	Females
Establish a breeding territory	X	
Defend territory against intruders/floaters	X	
Defend individual nest sites from conspecifics		X
Advertise for mates	X	
Evaluate and choose mates		X
Choose individuals for EPCs	?	X
Perform courtship, precopulatory displays	X	X
Guard mates from conspecifics	X	
Construct nests		X
Lay eggs		X
Incubate		X
Brood young		X
Feed young (nestlings)	Some	X
Feed young (fledglings)	Some	X
Defend nests	X	X
Defend young	X	X

field sought to understand the behavior of animals in terms of their ecologies – how they interacted with their physical and social environments. Thus, the thinking went, polygynous breeding might be understood if one considered the habitats in which polygynous animals bred and the resources that the habitats provided.

The Polygyny Threshold

The Polygyny Threshold Hypothesis (PTH) of Jared Verner, Mary Willson, and Gordon Orians suggested that where male territories varied widely in the quality and quantity of breeding resources provided, and where females evaluated territories for such resources when making decisions about where to nest, some females, when confronted with the choice of being a sole female on a lower quality territory or a second or third female on a higher quality territory, should benefit by choosing the latter situation. Given the differences in territory quality, these females, on average, should be able to obtain greater seasonal reproductive success as second females on good territories than they would as first females on poor ones. In other words, any cost to a female associated with her assuming secondary status on a territory, say receiving only a small fraction of the male's parental care for her young and also having to compete with other females for territorial resources, is more than compensated by nesting in a superior territory, perhaps one with better local food resources or superior safety from nest predators (Verner 1964, Verner and Willson 1966, Orians 1969). The polygyny threshold refers to the minimum difference between habitats or territories that would justify a female's decision to mate polygynously; when the threshold is reached or exceeded, then natural selection would be expected to favor the development and maintenance of polygynous breeding.

Once proposed, the PTH dominated research into the causes of polygyny, both because it provided a plausible explanation consistent with contemporary ideas about how selection works on individuals and because it made testable predictions. For example, the PTH predicts that females should not suffer any 'noncompensated' costs for mating polygynously, and therefore female reproductive success should not decrease when breeding on territories with increasing harem sizes. Another prediction of the PTH is that if females rank territories by quality and use those rankings, as well as information on the numbers of previously-settled females on each territory, in their settlement decisions, then there should be a definite order in which females choose to settle on male territories. Imagine an isolated marsh containing 10 male territories of varying quality and harem sizes. Under the PTH, the order in which male territories gain females should be related to eventual harem sizes. In other words, the territory with the largest eventual harem on our imaginary marsh, if the PTH is valid, also should have been selected for settlement first by early-arriving females; the territory with the second-largest eventual harem should have been chosen second, etc. Furthermore, the same pattern should be apparent when secondary and tertiary females begin their settlement. That is, after all 10 territories are settled by one female each, the primaries, the first female to choose to be a secondary should do so on the territory that will eventually have the largest harem, the next secondary should settle on the territory with next-largest eventual harem, etc. (Altmann *et al.* 1977).

Harem sizes: estimating the degree of polygyny

Variation in harem size

Particularly suited to field-test the PTH are species that vary in their degree of expressed polygyny. Redwings are as close to an ideal subject as we are likely to find because they exhibit one of the strongest degrees of polygyny recorded for birds. Average harem sizes vary between two and six, depending on breeding habitat and region. The largest harems are usually those found in the West, where marshes are highly productive. This pattern is consistent with the PTH, which predicts that the largest harems should occur on territories that contain or are near concentrated, high-quality breeding resources. Average harem sizes are smaller in the East and Midwest, where marshes generally produce fewer insects, and in upland areas. Moreover, wherever redwings are studied, researchers find that the great majority of breeding individuals, 80%–99%, mate polgynously. In contrast, in most of the other polygynous species that have been examined, polygyny is often limited to only a small fraction of any population, e.g. only 10% of males and 21% of females in an Indiana Indigo Bunting population mated polygynously, and most polygnous males had harems that averaged less than two females, with a typical maximum of two or three (Carey and Nolan 1979). (An exception is the Yellow-headed Blackbird, another marsh-breeding icterine of central and western North America, which breeds in many of the same habitats as do redwings, and often has comparable-sized harems.)

At Washington's CNWR, territories vary widely in quality, but some have supported the nesting of up to 17 different females in a single breeding season (Table 6.6). The average harem size for 474 different males that bred for a total of 1130 territory-years was 3.9 ± 2.5 females, with harems of 3 being most common (Fig. 6.2). Not all females nest simultaneously, but at times, 10 or more females on the same territory may overlap in their nesting. In Fig. 6.3 are illustrations of the nesting chronologies of two males' harems, showing the nesting of all females on the territories during a breeding season. In Fig. 6.3a, one can see that, in a harem of

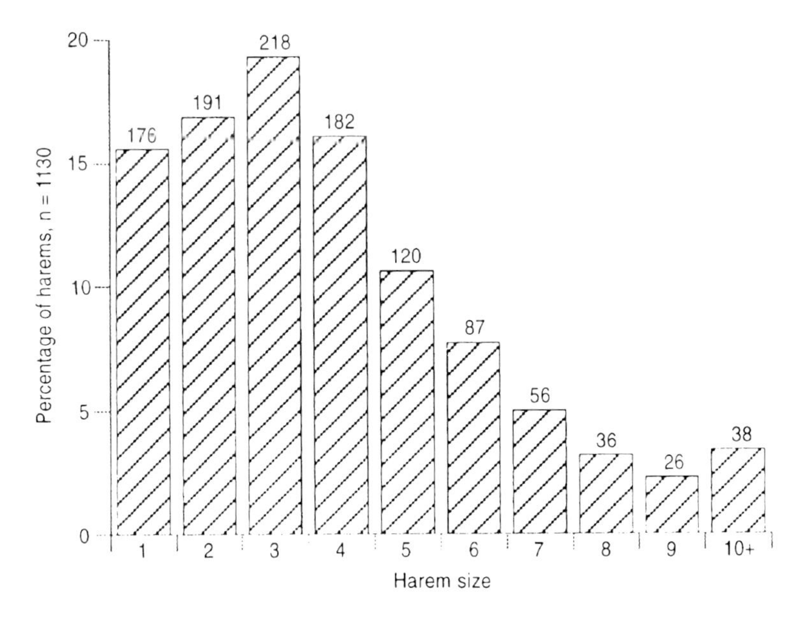

Fig. 6.2. Distribution of harem sizes on male redwing territories at Columbia National Wildlife Refuge (CNWR), 1977 through 1992. Sample sizes are given above bars.

Table 6.6. Redwing harem sizes in various habitats and locations (adapted and updated from Searcy and Yasukawa 1995)

Harem sizes mean	maximum	*n*	Location	Reference
Marsh habitats				
2.2–2.5	–	63	Illinois	Smith (1943)
2.0	3	25	Wisconsin	Nero (1956a)
3.5	6	42	California	Orians (1961)
2.0	3	20	New York	Case and Hewitt (1963)
1.9	–	126	Maryland	Meanley and Webb (1963)
1.6–2.2	–	50	Oklahoma	Goddard and Board (1967)
3.3	8	70	Michigan	Laux (1970)
2.9	6	104	Washington (Turnbull NWR)	Holm (1973)
2.6	9	16	Costa Rica	Orians (1973)
3.0	7	54	New York	Allen (1977)
2.8	9	97	Ontario	Weatherhead and Robertson (1977a)
4.6	8	34	British Columbia	Picman (1981)
4.3	15	107	Washington (Turnbull NWR)	Searcy and Yasukawa (1983)
4.0	–	64	California	Ritschel (1985)
2.1	–	246	Illinois	Strehl and White (1986)
1.7–2.7	6	42	Ontario	Muldal *et al.* (1986)
1.9	–	24	Alaska	McGuire (1986)
2.7	5	22	Pennsylvania	Searcy (1988)
2.8	6	20	Wisconsin	Yasukawa *et al.* (1987b)
3.9	17	1130	Washington (Columbia NWR)	Beletsky and Orians (1996)
Upland habitats				
1.8	2	26	New York	Case and Hewitt (1963)
2.7	4	10	Iowa	Blakley (1976)
4.3	–	31	Ohio	Dolbeer (1976)
1.7	3	27	New York	Allen (1977)
6.2[1]	12	13	California	Ritschel (1985)
2.3	5	18	Wisconsin	Yasukawa *et al.* (1987b)

[1]Harems in upland areas are generally smaller within the same locality than those in marshes, and thus Ritschel's data are unusual; however, she used only four or five territories to compute upland harem sizes.

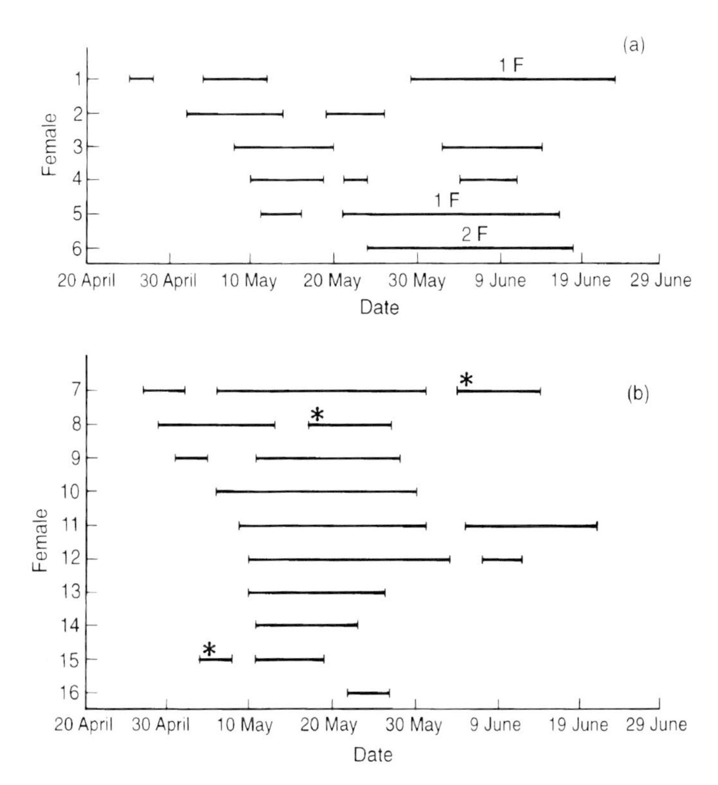

Fig. 6.3. Nest chronologies for two Columbia National Wildlife Refuge (CNWR) redwing harems; (A) A harem of six females with a total of 13 nests, from Juvenile Pocket marsh in 1991. Successful nests are indicated by the number of young fledged (e.g. 2F = 2 fledged); (B) A harem of 10 females with a total of 17 nests (none successful), from the same marsh, same year.
** Indicates that the nest was on another male's territory.*

six females, one built three nests before successfully fledging young, four others tried two or three times each, only some with success, and, at times, there were five nests simultaneously active. Figure 6.3b shows the nest chronologies for a larger harem, one with 10 females, many of which were simultaneously active. In this case, none of the females successfully fledged young. Two of the females moved from the territory after early failures, and one female moved into the territory after first trying her luck elsewhere.

Determining harem size

Table 6.6 presents the numbers of females in harems, but if not all females are present at one time, how can the size of a harem be determined? Determining redwing harem sizes, which is crucial information for studies that test hypotheses of polygynous breeding, is rarely as easy as it sounds. When there are one or two or even three females nesting simultaneously on a territory and each builds only one nest, it is a straightforward matter. But suppose there are five or six or more females, many of them building two or more nests during a long breeding season, some nesting simultaneously on the territory but others sequentially, and more females arriving almost daily from their migrations and settling in the monitored area. Under these circumstances, as Holm (1973) dryly put it, 'it is difficult to determine the number of females (on a territory) from direct observation . . .' Actually, it is almost impossible. Some early studies of redwings estimated the numbers of females in an area by counting nests, resulting in artificially high estimates (e.g., Goddard and Board 1967, Snelling 1968). Picman (1981) described how, had he used nest counts to estimate harems, he would have arrived at average values of 13.2 and 10.4 females per male territory, respectively, during the two years of his study, but by a more accurate method (see below), arrived at average harem size values for those years of 4.7 and 4.5.

Clearly, the only way to determine positively harem sizes is to color-band females so that they can be associated with all their nests. Even then it is not easy. During the CNWR study we banded females but still had trouble counting all the females and assigning all nests to individuals. First of all, some females refused to enter our traps, and so remained unbanded. (Talk about uncooperative study animals! We always believed that birds that *did* acquire bands rather admired their new jewelry.) Second, even with many telescope-wielding field assistants to monitor comings and goings from nests, it often proved difficult to determine which female was associated with which nest, particularly when nests were a meter or less apart in dense vegetation. Third, many nests were destroyed by predators, say at the egg stage, before we could determine their female owners. When that occurred, vital information for determining harem sizes was lost because, for example, it could not be known if a subsequent nest that was built on that territory belonged to the same female who just lost a nest there, or to a new female.

In response to these problems, researchers have resorted to a variety of methods to estimate true harem sizes. One method is to simply use the maximum number of *simultaneously* active nests to indicate harem size, as does Picman (1981). This method assures that females that construct more than one nest are not counted more than once. However, it fails to capture another definition of harem, the total number of females that choose to mate with and nest on a single male's territory during a breeding season. Because no study of redwings is likely to obtain perfect nesting information for all females, a method of estimating total harems is required. A common way is to determine the minimum number of females required to account for all nests on a given territory (Holm 1973, Searcy 1979e). For example, say a territory had two banded females nesting simultaneously and a third simultaneous nest appeared, although its owner was never identified; clearly, the harem size was three. As another example, suppose a territory has two active nests being incubated, one of which is lost to a predator on 20 April, and three new nests are initiated on the territory on 27 April. Because the female whose nest was lost could account for one of the new nests, the probable harem size for the territory was four.

Early in our long-term study at CNWR of lifetime reproductive success we realized that, for the reasons given above, we would not always be able to determine actual harem sizes, nor be able to assign all nests to the females that built them. Given large harems, the large number of male territories we monitored (70–90 annually), and the long-term nature of the study, we decided that our best bet would be a computer program that could take the information that we had on banded females positively associated with nests, and on nest initiation and termination dates, and then estimate harem sizes based on an algorithm of adding to those females definitely known to be on each territory the minimum additional number needed to account for all nests. The program, HAREMSIZE, focused on nest start dates (date of the first egg) and termination dates (actual or estimated). If the known, banded females on a territory, given the known timing of their nests, could have been responsible for all of the 'unknown' nests on the territory, i.e. those for which we did not know the associated female, then no additional females were added to the harem. The first computer-generated female necessary to account for an unknown nest on a male's territory was then considered by the program to be a real female, available to be assigned to other, subsequent unknown nests on that territory, provided that the timing was appropriate. The process was repeated, the program creating a new cyber-space female each time another unknown nest could not be assigned to one of the females already present in that harem. The program also tracked banded females so that if they moved to other territories mid-season within the study area, they could not be credited with additional nests on their initial territories.

Testing the Polygyny Threshold Hypothesis; other hypotheses of polygyny

Orians (1972) used breeding data on redwings from Oklahoma and Washington to support the PTH prediction that the breeding success of females should not be negatively correlated with increasing harem size. The data showed that the average number of young that fledged per nest generally increased as harem size increased. A problem with these early studies of nesting success was that few, if any, females were color-banded and thus reproductive success for each female had to be scaled on the success of only her first nest, or inferred generally from the success of all nests. However, Orians' prediction was that *annual* reproductive success per female should not be negatively associated with increasing harem size. Data from the CNWR study, in which we had good estimates of annual fledging success

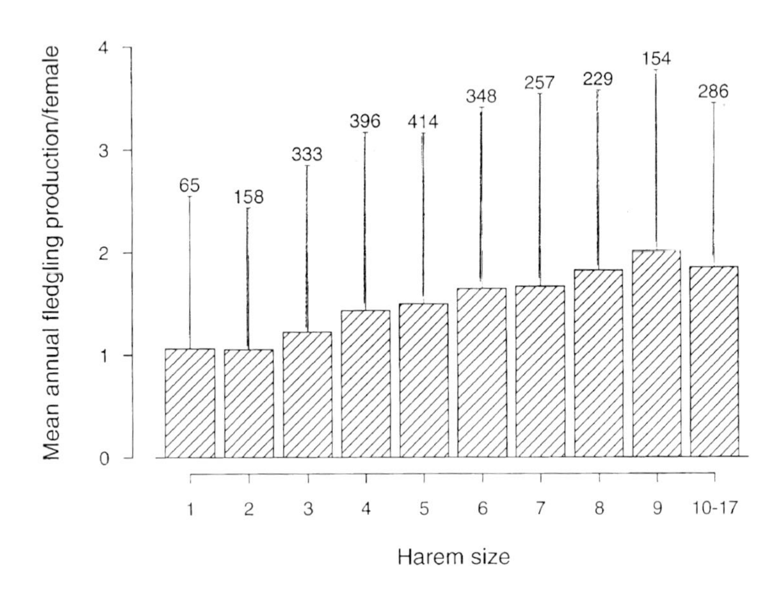

Fig. 6.4. The mean (± SD) annual fledgling production per redwing female at Columbia National Wildlife Refuge (CNWR), 1977 through 1992, as a function of the eventual harem size on the territory each female chose for her first nest each year. Regression of mean annual fledgling production on harem size is significant ($r = 0.07$, $P < 0.0001$).

for each female, supports the prediction: females from larger harems tended to experience comparatively high annual fledging success (Fig. 6.4). A likely partial explanation for the pattern is that male territories with the largest harems tend to be located on 'pocket' marshes. These are large expanses of emergent vegetation where many of the territories do not border the lakeshore, presumably preventing some forms of nest predation, and into which are packed many small territories. Territories on pocket marshes consistently experienced, on average, superior fledging success per male breeder and per female breeder (Beletsky and Orians 1989c, 1996). A significant positive relationship between harem size and annual fledging success per female does not prove the PTH or that females choose sites based on territory quality, but they are consistent with these ideas (Wittenberger 1976, Altmann *et al.* 1977). Another possible explanation for the positive relationship is that the best males attract not only the most nesters, but also the earliest ones, and early nests are usually the most successful ones (Garson *et al.* 1981; see also Chapter 7).

A second PTH prediction, that settlement of females in a given area should follow an order that agrees with the size order of the eventual harem sizes in the territories, was supported for a Washington population of redwings (Orians 1980, Garson *et al.* 1981) but not for Illinois and New Jersey populations (Lenington

1977, 1980). Thus, some empirical evidence exists to support the PTH as an explanation of polygynous mating in redwings, and there is also evidence for it in some other avian species. However, Searcy and Yasukawa (1995) point out that there are problems with the supporting evidence, chiefly that the predictions I mentioned, and others that have been tested, are consistent with more than one explanation for polygyny. For example, the prediction that female reproductive success should not decrease with increasing harem size is compatible with any polygyny hypothesis that assumes no cost to participating females. The second prediction is compatible with any hypothesis that assumes that females choose territories based on male or territory attributes.

The PTH is only one of several hypotheses that might explain polygyny. Some are very similar to PTH, differing only in how females choose breeding situations (Searcy and Yasukawa 1989, Bensch 1993). Some simply elaborate on the PTH, for example, modifying it to include the individual attributes of the territory owner as well as those of the territory he owns as factors potentially affecting female settling decisions (Wittenberger 1976). Others differ more strikingly. The 'Sexy Son Hypothesis' argues that females obtain access to better genes by mating with superior males, which compensates for any cost to them of polygynous mating (Weatherhead and Robertson 1979). Other models have been proposed that suggest different compensations to females mating polgynously, including the cooperative defense model of Picman *et al.* (1988): larger aggregations of females should be better able to defend nests. Two polygyny hypotheses that rely less on females choosing specific breeding situations are the 'Female Deception Hypothesis', which argues that females mate polygynously only because they are unaware of other females already mated to the same male (Alatalo *et al.* 1981), and the 'Neutral Mate Choice' model of polygyny, which suggests that females settle randomly into habitat divided into male territories, but pay no cost if by chance the result of their settling is that they mate polygynously (Lightbody and Weatherhead 1988).

After sifting and comparing all the available evidence, Searcy and Yasukawa (1995) suggest that the most likely explanation for the redwing's mating system is a 'no cost, no benefit' model of female choice. That is, redwings appear to be polygynous because females, when they choose their breeding sites, prefer certain territories and perhaps certain mates. Such female choices will be explored in detail in the next chapter. Last, it is worth noting that no single explanation of polygyny is likely to hold for all species that exhibit the mating system. For example, the Female Deception hypothesis is a likely explanation for the bigamous Pied Flycatcher, in which males often establish two territories, attracting a single female to each, but a poor candidate to explain the strong polygyny on single, small territories that characterizes redwing breeding.

Redwings are strongly polygynous breeders. In fact, they are considered classically polygynous and to typify the mating system known as resource defense polygyny, in which males defend territories containing resources that females require to breed, such as food supplies or nest sites (Emlen and Oring 1977). But are redwings *really* polygynous? That is, do they fit the definition of one male establishing breeding relationships with several females during a breeding season, and each female breeding with only one male at a time? If only it were that simple. Recent DNA genetic testing reveals that many, if not most, breeders, of both sexes, copulate with individuals other than their mates (see Chapter 7). Males copulate with their own females and others, and females copulate with their own males and with others that have territories adjacent to the ones on which their nests are located. Therefore, although *socially* redwings are polygynous, *genetically* they must be classified as polygynandrous – the simultaneous shared access by several males to several females.

Thus, male and female redwings come together in their own characteristic way during the breeding season to produce young. Their polygynous mating has attracted the interest of researchers for three decades and is probably the system most used to formulate and test ideas on how such behaviors develop and operate. Now that we know something of the redwing's mating system and have had a brief introduction to the particulars of the breeding cycle, we can turn, in Chapters 7 and 8, to detailed looks at the respective breeding roles of males and females; specifically, what must each sex do to produce successful offspring?

Chapter 7

FEMALE BREEDING ROLES AND DECISIONS

INTRODUCTION

Animal reproduction can be viewed and studied scientifically from a variety of perspectives. Ecologists and conservation biologists investigate the habitat requirements for reproduction and the demographic consequences of reproductive success. Ethologists study such behaviors as mate location, courtship, communication, and parental duties. Physiologists are interested in the energetics of breeding and the endocrinology of reproduction and breeding behavior. My colleagues and I, behavioral ecologists, believe that one rewarding and productive view is to consider the breeding of birds as a series of seasonal choices, each of which affects an individual's likelihood of achieving successful reproduction that year. We know that large portions of populations try to breed each year but that only a fraction are successful; the successful made better decisions or had better luck. This method of investigating breeding fits well with the job of the behavioral ecologist because it allows consideration of both the behavior of the animal – how and when decisions are made – and of the animal's ecology, or what environmental information, physical or social, is used to base decisions. Decisions can also be considered against the evolutionary backdrop of how they influence individuals' fitnesses; that is, better decision-making, by our definition, results in increased reproductive success and therefore in greater numbers of descendants in future generations. This investigative method assumes that, within certain limits, breeding behavior, although influenced by genes, is modifiable by individuals and that decision-making behavior has been acted on by natural selection so that decisions are geared to increase lifetime reproductive success.

To what sort of decisions am I referring? For polygynous, territorial birds with altricial young (i.e. helpless after hatching), such as the redwing, males must decide where, when and how to establish territories, whether to devote time and energy to attracting new mates or to assisting current mates, and which nests to defend (all of which will be covered in Chapter 8). Females make a host of decisions each year: on which marsh and territory to nest, with which males to copulate, and when to nest. When first nests fail, females must decide when and whether to renest and whether to do so in the same male's territory or to switch territories or marshes. In this chapter I consider in detail four major female decisions: with whom to mate, where to nest, whether to be aggressive to other females, and when to nest. In each case I first consider why the decision may be important for achieving or enhancing reproductive success and then explore the evidence on how female redwings make such decisions. Where appropriate I will support my contentions with information from the Columbia National Wildlife Refuge (CNWR) population.

WITH WHOM TO MATE: MATE CHOICE

Many animals choose their mates carefully and the reason for this selectiveness is clear. Half the number of chromosomes in each offspring of sexually reproducing organisms are donated by the mate. Therefore, the eventual viability, survival, and success of offspring are influenced by choice of mate. Both sexes can exercise choice over mates, such as we believe to be the case in people (although the author's own experience is that female choice is paramount in this species), but usually within a species one sex is generally choosier than the other. Trivers (1972) gave the likely reason: the sex that contributes more 'parental investment' per offspring is a limited resource for the other sex. The sex contributing less parental investment competes for access to the sex contributing more. The sex that contributes more, safe-guarding its larger investment, exercises more stringent mate choice. This explanation of choice correctly predicts that in most avian species females are the choosier sex because only females produce eggs, a considerable energetic investment. Parental investment arguments certainly may explain the redwing system of mate choice, because the females clearly invest more per offspring than do males – females build nests, lay, incubate, and usually provide most food for the young – and also appear to be more selective than males when choosing individuals with whom they mix their genes.

There is no evidence that male redwings exercise choice in which females nest in their territories (although it is at least possible that male choice is involved with selecting partners for extra-pair copulations); indeed, unless more mates seriously deplete local resources for all, theory predicts, and empirical evidence shows, that a male redwing accepts as many females as decide to settle on his territory. Each additional female, after all, provides another opportunity for siring offspring.

It is not just on theoretical grounds that we believe female redwings choose their mates, we actually see them doing it. Male territories are already established when females in small groups arrive on the breeding grounds in late winter or spring. Individual females soon explore territories, often visiting several over a few days before choosing one on which to settle (Nero 1956a, Lenington 1980, L. Gray, personal communication). (For another polygynous marsh-breeder, the Great Reed Warbler, precise information is available: females visit an average of six male territories, range 3 to 11, before selecting one for settlement (Bensch and Hasselquist 1992)). Redwing females spend considerable time exploring but on what basis do they choose their mates? Do they choose individual males based on behavioral or morphological characteristics? Do they choose territories based on physical characteristics? Or is their choice of settling area based on some combination of traits of the male and his real estate?

Early ornithologists often simply assumed that birds chose mates on individual characteristics, as people do, but the actual process of selection was glossed over. For example, Allen (1914) wrote that 'eventually (the male redwing) is ready to select his mate and may be seen following her about. He never allows her to escape from his sight, and as she hunts about near the water's surface, he vaunts himself on the nearest cattail. They now may be considered mated.' In the 1960s, with the introduction of concepts that tied social behavior to ecology, and especially with the enunciation of the Polygyny Threshold hypothesis (PTH, see Chapter 6), the strong possibility arose that, particularly in polygynous species such as the redwing, females might make their mate choices based on territory quality, and research shifted in that direction. By choosing a territory, the thinking went, a female was assured a relatively high-quality mate because males were in competition for territories and some, presumably weak, low-quality males, did not own them. In the 1970s and 1980s, however, mate choice again became fashionable when biologists proposed several mate choice hypotheses that suggested that individual male attributes, either phenotypes or genotypes, play important roles in female choice (described below). And now, in the 1990s, studies of extra-pair copulations and molecular analyses of paternity reveal that whether females choose primarily males or their territories may not be an either/or problem. The behaviors disengage: a female may select a single nesting territory but subsequently choose more than one male with whom to copulate.

Female redwings apparently choose for breeding neither a specific male, nor only a particular territory or marsh, but a 'breeding situation' (Wittenberger 1976). Even though a female selects a particular male's territory, the breeding situation into which she inserts herself has many physical and social facets. The situation determines the general area in which she will place her nest and the availability of hiding places from predators, so physical features of a territory such as vegetation type and density may figure in her decision. It determines where and how much food she will be able to find for herself and her young, so insect productivity of the marsh and surrounding uplands may matter. It determines with which males she will copulate and have fertilize her eggs – the male that owns the territory she has selected and his neighbors. It also determines how many other females may nest or are nesting on the same territory and, as a consequence of that occupancy, the 'rank' of her nest on the territory, whether it is a male's primary nest or secondary, tertiary, etc. All of these factors may affect her settling decision because all may influence her survival and eventual breeding success. But which of these factors are most important to female redwings?

Do females choose individual males?

Do female redwings choose breeding situations primarily based on male characteristics? Females might choose an individual male because of the resources he provides or for genetic benefits. Male redwings do not courtship-feed their mates, so there is no direct providing of resources from male to female. However, males provide resources indirectly: females have access to the resources that males defend, and males, to varying degrees, provide care of young (Emlen and Oring 1977, Searcy 1979e). Females may also select males based on their genes because of potential survival advantages of certain genotypes for resistance to parasitic infections (Hamilton and Zuk 1982) or for maintaining large energy reserves (Zahavi 1975). Finally, some theorists have proposed that females might choose males on arbitrary traits, not related to survival ability, and that the genetic association between the male trait and the female preference is sufficient to maintain the choice system through a special case of natural selection (Lande 1981, Kirkpatrick 1982).

Determining whether female redwings choose territories based on their male owners requires answers to four separate questions: (1) in which male attributes might females take an interest (i.e., which attributes strongly influence female survival or reproductive success), (2) do those attributes vary significantly among males, (3) could females assess the attributes of males by their morphology or behavior, and (4) are the attributes assessable at the time females make settlement decisions?

Paternal care traits

Females might base their decisions on the kind and extent of parental care a male may provide. Male redwings provision young and defend nests, both of which vary among males (Knight and Temple 1988, Whittingham 1989, Weatherhead 1990a, Yasukawa *et al.* 1990) and influence female reproductive success (Muldal *et al.* 1986, Picman *et al.* 1988, Yasukawa *et al.* 1990. Weatherhead 1990a, Patterson 1991). But how could a female determine if a male will be a good father? After all, most females select nesting territories in spring before they can observe a male's parenting skills. They would have to predict future paternal behavior using other traits. Age is one potential predictor because male provisioning of young generally increases with male age and breeding experience (Patterson 1979, Beletsky and Orians 1990). Therefore, females might benefit by selecting territories of older males. Returning females with memories of past breeding seasons could assess male age or experience directly. All females could estimate a male's age indirectly by such cues as song repertoire size, which increases with age (Yasukawa *et al.* 1980), or courtship intensity, which also apparently increases with years of breeding experience (Eckert and Weatherhead 1987a). Likewise, aggressiveness toward potential nest predators varies widely among males, influences nesting success (see Chapter 8), and is positively correlated with male epaulet length. Theoretically, females could choose good fathers for their young based on indicators of age/experience and on epaulet length. But do they?

Weatherhead and Robertson (1977b) found a significant positive correlation between the intensity of male courtship and female settling density, and Yasukawa (1981c) observed a significant positive correlation between male breeding experience and harem size. Both results suggest that females choose males with attributes associated with better parenting. However, these correlations could be explained in a number of other ways that have nothing to do with parenting. For example, Weatherhead (1984) suggested that females prefer older males because they have demonstrated a superior genotype by having survived and held a territory for a number of years. Also, older males might have more females because their territories gradually enlarge with number of years of breeding (Yasukawa 1981c, Searcy and Yasukawa 1995). Moreover, in areas such as at CNWR, in which males provide little or no feeding assistance, females cannot use parenting as a mate-choice criterion. As for females selecting males on the basis of their nest defense abilities, Searcy (1979b) observed no relationship between male epaulet length (associated with nest defense ability) and harem size, but Yasukawa *et al.* (1987b) detected a positive correlation between nest defense and harem size. In summary, although some investigators have successfully identified morphological or behavioral indicators of male parenting ability, we have little or inconsistent evidence that females know about the indicators or use them in making their settlement decisions.

Genetic traits

Female redwings could select mates based on genetic traits, thus acquiring 'good' genes for their offspring, if there is genetic variability and if females can determine males' genotypes or rank males in their genetic quality. No evidence has yet been found that females assess or choose mates in this manner. Searcy (1979b) looked for correlations between harem size and male traits such as epaulet size and color, weight, wing length, and song rate, all of which might be heritable. Most correlations were not significant and some reversed from positive to negative from one year to the next. Yasukawa *et al.* (1980) tested the relationship between song repertoire size of males and their harem sizes, reasoning that females might prefer males with larger repertoires in the way that female peafowl prefer males with larger, gaudier tails, i.e. they might select mates based on exaggerated traits, with their preferences driving the exaggeration. For example, Catchpole (1980) demonstrated that male Sedge Warblers with large song repertoires obtained mates sooner than did males with smaller repertoires. Yasukawa *et al.* found a positive correlation between repertoire size and harem size but, because repertoire size in redwings increases with male age and breeding experience, females could have chosen their mates on a number of other criteria. Finally, Weatherhead (1990b) examined the possibility that female redwings choose males based on their resistance to blood parasites. He observed that although it was likely that parasite load could be assessed by male morphology (epaulet length) or behavior (aggressiveness and courtship intensity toward male and female models, respectively), females did not preferentially choose unparasitized males over parasitized ones as mates, as the harem sizes of the two male groups were the same.

Thus, although research into this question continues, existing evidence provides little indication that redwing females choose breeding situations based on attributes of the males that defend the territories on which they place their nests. Although we do not know if a completely 'naive' female would evaluate males prior to making a settling decision, relatively few females are really naive. Those breeding for the first time, yearlings, generally arrive on the breeding grounds and nest later than older females, so they can use as settling guides the decisions of experienced females. Most older females return each year to the same marsh area or one nearby (Picman 1981, Beletsky and Orians 1991; see below) and so have past assessments of males, territories, marshes, and relative breeding successes to guide them in their breeding situation selections. If the males themselves were important components of those decisions, females should show strong fidelity to the owners of the territories on which they choose to nest, particularly when they have successful nests. However, at CNWR, we found 'divorce' rates to be high. When females' final nests in year *x* were successful, only 49% (147 out of 302) of them nested in year *x*+1 with the same male, even if he survived and owned the same territory. When females' final year *x* nests failed, only 47% (170 out of 365) of them returned to the same male's territory the following year (Fig. 7.1). Of 85 females that returned in a subsequent year to breed on a British Columbia marsh,

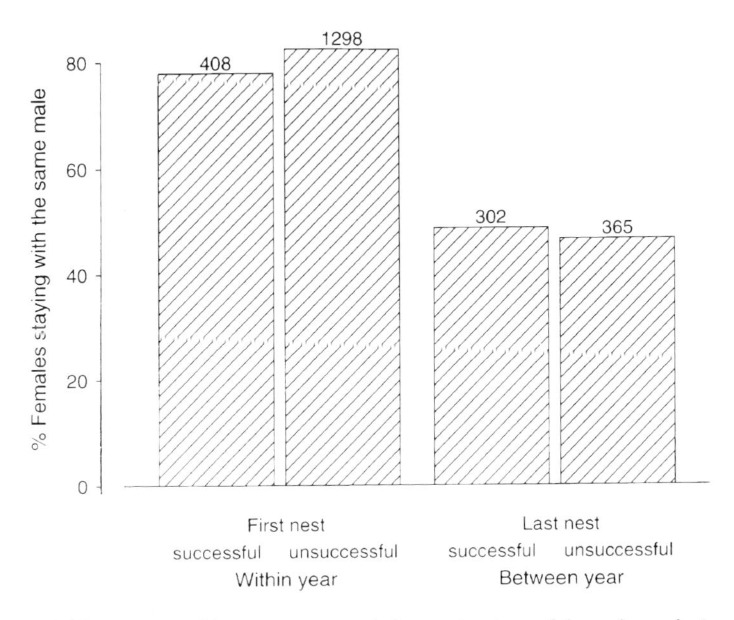

Fig. 7.1. Within-year and between-year 'divorce' rates of female redwings at Columbia National Wildlife Refuge (CNWR), 1977 to 1992. Within-year: percentage of females staying on the same male's territory for their second nests of the year, as a function of the success (fledge >= 1) or failure (fledge = 0) of their first nests. Between-year: percentage of females returning for their first nests in year x + 1 *to the territory on which they placed their final year* x *nests, provided that the same male owned the territory in both years, as a function of the success or failure of the final year* x *nest. Included are the known nesting histories of all color-banded females and nests assigned to them by the* HAREMSIZE *program. Values above bars are total sample sizes.*

62% nested on the same territory as the year before and 31% on an adjacent territory (Picman 1981). Significantly, however, this female 'resettlement' occurred regardless of whether the territories were occupied by the same males as the previous year or different ones.

Also, if females choose breeding sites primarily by the characteristics of males that defend them, then when males change their territory locations to nearby sites, females should move with them. But that is rarely the case. From 1984 through 1991, 21 males at CNWR that switched territories between years to nearby ones (not more than 150 m away) also had one or more color-banded females nest on their territories the year before the move and then also breed the next year. In only 24% of the cases did at least one female follow males from their old to their new territories, even though the old and new ones were located only a short distance away (in four cases, 20 m or less separated the old and new territories, and 130 m in the fifth case). Of the 37 color-banded females that nested both years, only six (16%) followed males to new territories. Clearly, remaining with the same male is not a high priority for female redwings.

Extra-pair copulations

Another perspective on female choice of mate is provided by information on copulation patterns and paternity of clutches. Investigators have long known that female redwings are often unfaithful to their putative mates. Long ago, Allen (1914) and Beer and Tibbitts (1950) observed females copulating with males other than the ones who defended their nesting territories. Bray *et al.* (1975) vasectomized male redwings in Colorado to determine whether interference with reproduction might be a viable method to reduce populations and so reduce the impact of redwings on agricultural crops. To their surprise, up to 69% of clutches (depending on year and marsh area tested) on territories of vasectomized males had fertile eggs. They included only clutches started five or more days after vasectomies, so that fertile clutches were unlikely to result from stored sperm in the females from the treated males. Moreover, nests on territories of treated males that were farthest from territories of control, unvasectomized males, had the lowest percentage of fertile clutches. Roberts and Kennelley (1980) replicated this experiment in a Massachusetts population and obtained the same result: overall, 70% of nests on the territories of vasectomized males had fertile eggs. The strong implication was that female redwings routinely copulate with and are fertilized by nearby territory owners, i.e. males other than their putative mates.

Why should redwings engage in extra-pair copulations (EPCs)? The reason for male participation is clear. Unless there are substantial costs that reduce fitness, males should be able to increase their breeding success by copulating with multiple females. In this view, engaging in EPCs is, for males, simply an extension of polygyny. Possible costs to males are exposure to sexually transmitted diseases and time budget trade-offs, i.e. time devoted to pursuing EPCs cannot be devoted to other reproductive activities such as parental care, guarding mates during their fertile periods from other males, and advertising for additional mates (Westneat *et al.* 1990). In redwings these costs are unlikely to be high (Westneat 1993) and therefore, on balance, EPCs should be beneficial to males for maximizing breeding success.

Potential costs and benefits to females from EPCs are more numerous and more difficult to assess. Females engaging in EPCs may face an increased risk of predation or injury because of greater activity, increased exposure to sexually transmitted parasites and diseases, and possible physical punishment by mates or reduction in their parental care (Westneat *et al.* 1990, Birkhead and Møller 1992). Possible benefits of EPCs to females include obtaining extra male parental care or 'allowances,' such as nest defense by EPC partners or reduced harassment while feeding on their territories, ensuring fertilization of eggs, increasing genetic diversity of offspring, and obtaining matings with males with superior genotypes (Westneat 1992a).

Vasectomy experiments dramatically demonstrate that female redwings engage in EPCs, but they do not prove that females normally do so. The surgery could have altered male behavior in such a way that their reproductive ability was suspect to females, who then could have acted accordingly to ensure fertilization. What was needed were molecular studies of paternity in unmanipulated settings. Such studies with birds began in the 1980s and mutiplied rapidly with the advent of DNA fingerprinting and polymerase chain reaction (PCR) technology. More than 50 species have now been examined and in most of them, even in those long thought to be strictly monogamous, evidence of extra-pair fertilizations (EPFs) has been uncovered. For example, Westneat (1990), one of the pioneers of molecular paternity analysis in birds, used DNA fingerprinting to determine that 35% of Indigo Bunting offspring in a sample from a North Carolina population were the result of EPFs.

Gibbs *et al.* (1990) were the first to report on DNA fingerprinting of redwing broods. They discovered in an Ontario population that about 47% of broods had at least one chick who was fathered other than by the female's putative mate and that overall, 28% (31 out of 111) of young tested were the result of EPCs. The putative mate usually sired at least some of the chicks and sometimes more than one 'extra-pair' father had contributed paternity to a single nest. Gibbs *et al.* identified territorial male neighbors as being responsible for most, if not all, extra-pair paternity. Westneat (1993) reported a similar degree of extra-pair paternity in a New York population: 24% of 235 nestlings were sired through EPCs, and 41% of 68 broods had at least one extra-pair chick. Again, the males responsible for the EPFs were adjacent or nearby territory owners. Westneat's (1992b) females benefited from their EPCs because nests with EPF chicks fledged more young than others without EPF chicks, owing to lower starvation and/or predation rates. Gray (1994) found an even greater degree of extra-pair paternity at CNWR: 34% of 403 nestlings had been sired through EPFs, and fully 54% of tested broods had at least one EPF nestling. Here also, neighboring males were, in almost all cases, the sires of EPF nestlings. Gray identified precise benefits to female redwings from participating in EPCs. She determined with experiments that neighboring males with whom a female had copulated (confirmed by DNA analysis of her nestlings) were more likely to permit the female to stay at feeders on their territories than were other males, and also were more aggressive in helping to defend her nest against a stuffed predator. With these advantages, females who participated in EPCs fledged, on average, about 0.5 more chicks than females who did not (Gray 1994).

The study of EPCs and EPFs in redwings is in its early stages but intriguing variation in female behavior has already been noticed. Most interesting, in the East EPCs are rarely seen by observers. In one of his studies, 94% of 71 female copulations that David Westneat observed were with intra-pair males, yet 23% of those females' nestlings had extra-pair fathers. Eastern females apparently do not actively solicit EPCs, and quite often resist them (Westneat 1992b). But in the West, at CNWR, Gray (1994) frequently observed females soliciting and participating in EPCs. In fact, during her 3 years of monitoring, over 402 hours, she witnessed 375 copulations, of which 66 (18%) were EPCs; no less than 88% of females she watched were observed to participate at least once in an EPC. Thus, in different populations, the costs and benefits to females of participating in EPCs might differ, perhaps subtly, perhaps substantially, and therefore the behavior may be expressed at varying frequencies. Given the regional variation in the energy and time that males, on average, invest in their young (Chapter 8), it is not surprising that patterns of EPCs also differ. Female participation in EPCs could prove to be highly variable within and between populations – pursued only when young will benefit, not solicited and even actively resisted at other times. The main EPF consequence for mate choice is that even if a female selects a breeding situation based on a number of male and/or territory factors, she is subsequently not limited to copulating solely with the holder of the territory on which she nests; she can copulate with his neighbors as well. Little is known of how females choose EPC partners. However, it is possible that neighboring males – potential EPC partners – are part of the equation of female settlement decisions, part of the breeding situation that is chosen.

WHERE TO NEST: TERRITORY AND NEST-SITE SELECTION

Although female redwings appear not to choose their breeding situation based on genetic or behavioral propensities of individual males, there is evidence that the physical features of potential nesting areas figure prominently in their decisions. A female might select a nesting territory because it has 'good' nest sites that will conceal her nests during their egg and nestling stages or, if on-site food will be important for her breeding effort, because of its potential for high seasonal food production. In other words, a female should use in her decision-making available cues that reliably predict future conditions that will influence breeding success. Females do inspect territories before deciding to settle. They usually visit several territories and are often observed moving through the marsh, walking around at the base of the vegetation, and even peering into the water (see illustration at chapter head). What might they be assessing?

Factors to consider when choosing territories and nest sites

Several marsh attributes that females could evaluate in their site selection have been tested, with some positive and some conflicting results (Table 7.1). Most of the tests are correlational ones that relate variation in territory attributes to either harem size or nesting success. If harems are larger on territories in, for example, one vegetation type versus another, female preference for that type is indicated. Comparing territory attributes to nesting success is also an indirect measure of female preference, if we assume that successful females select their nesting territories or sites at least partially by the tested attributes. In regions where redwings breed both in uplands and wetlands, females may also choose gross habitat type, but it is also possible that that is a population characteristic, determined by birth.

Territory size

Sizes of male territories might influence female settling decisions because larger territories could have more breeding resources for females, i.e. nest sites, food, hiding places. However, for some species, high-quality territories are smaller than low-quality ones. This is the case where intense competition for the best areas results in many small territories packed into high-quality sites, called territory compression. The redwing territorial system operates in this way (Orians 1980) and so harems may be larger on smaller territories; however, such a relationship need not

Table 7.1. Territory attributes that have been tested to determine if they influence female redwing settling decisions

Factor tested	Evidence found suggesting use as settling cue?	Types of evidence[1]	References
Territory size	No	A	Holm (1973)
	No	A	Weatherhead and Robertson (1977a)
	Yes[2]	B	Lenington (1980)
Vegetation type	Yes	A, B	Holm (1973)
	Yes	B	Weatherhead and Robertson (1977a)
	Yes	B	Lenington (1980)
Vegetation density	Yes	A, B	Holm (1973)
	Yes	B	Weatherhead and Robertson (1977a)
	Yes	B	Lenington (1980)
	Yes	B	Picman (1980a)
	No	B	Ritschel (1985)
Vegetation edge	Yes[2]	B	Lenington (1980)
Perches	Yes	B, D	Yasukawa *et al.* (1992a)
Water depth	Yes	B	Goddard and Board (1967)
	Yes	B	Weatherhead and Robertson (1977a)
	Yes	B	Lenington (1980)
	Yes	D	Picman *et al.* (1993)
Food production	No	–[3]	Holm (1973)
	No	B	Orians (1980)
	No	B, D	Ritschel (1985)
	Yes	D	Ewald and Rohwer (1982)
	Yes	D	Wimberger (1988)
Previous nests	No	D	Erckmann *et al.* (1990)
Familiar male neighbors	Yes	C	Beletsky and Orians (1991)

[1]Types of evidence: A = correlation with harem size or female density; B = correlation with nesting success; C = correlation with territory fidelity; D = experimental manipulation.
[2]Positive effect found in one of two study marshes.
[3]Little insect food was produced on any of the study marshes.

indicate a female preference for small territories, as each female may have made her settling decision based on other factors. There is evidence that females do not choose territories by their size (Table 7.1).

Vegetation type

Average harem sizes within a locality vary with territory vegetation type, as does nesting success. Cattail is often the preferred nesting vegetation, although nesting success in cattail is not always the greatest when compared with that in other substrates (Weatherhead and Robertson 1977a). Vegetation type affects vegetation density and nest height, both of which may influence predation rates. In one study, nests in cattail were more successful than nests in bulrush because nests could be placed higher in cattails (Holm 1973). But in two other studies, the higher nests were placed in vegetation, the lower was their probability of success (Weatherhead and Robertson 1977a, Ritschel 1985). One cause of poor success of high nests is that they are generally more exposed to bad weather.

Vegetation density

The density of the vegetation supporting nests should strongly influence their success because it affects detectability. What is surprising, however, is that in three out of four studies that identified significant effects of vegetation density on nesting success, nests placed in sparser vegetation fared better. Picman (1980a) speculated that redwings find it easier to defend their nests from predators (Marsh Wrens, in that particular case) when areas around nests are open. However, human intuition suggests that well-hidden nests in dense vegetation should, on average, have better success than those placed in the open.

Vegetation edge

Orians' (1980) work on the relationships between blackbird breeding and aquatic insect emergence revealed that territories that interface with the open water of a lake have very high emergence rates. The reason is that aquatic insects throughout the lake, at the time of their emergence, move to the nearest emergent plant stem to crawl up, and thus stalks at the lake/marsh interface collect the most insects. The amount of vegetation edge a territory possesses should therefore correlate with food production. Also, in large marshes, territories abutting the open water of a lake will be farthest from shore and consequently immune from effects of some land-based nest predators. Vegetation edge, like vegetation type and density, is easily visually assessed by prospecting females.

Perches

Male redwings often sit on prominent perches in their territories to guard their nests and successful nests are often nearer to these perches than unsuccessful ones

(Yasukawa *et al.* 1992a). Settling females could easily assess territories for perches and then attempt to nest near them.

Water depth

Several studies have demonstrated that nests built over deeper water have higher success rates than nests built over shallower water. Presumably deeper water prevents some terrestrial predators from reaching nests. However, the relationship is not universal. Picman *et al.* (1993) observed, in an experiment on nest survival, the expected relationship between water depth and predation rate, but nests above the deepest water experienced heavy predation by Marsh Wrens (which preferred deep water for their own nests). At CNWR, the chief nest predator is the Black-billed Magpie, which flies in and hops through the vegetation; water depth is irrelevant to them.

Food production

When food is gathered only on territories, territorial food production during breeding has a strong effect on nesting success. However, redwings often forage off territory and some of the marshes on which they breed actually have very low insect emergence (Holm 1973). Thus, it is not surprising that some studies have been unable to find a relationship between food production on territories and nesting success. Experiments in which territories are 'enriched' with feeders suggest that females can evaluate food resources when choosing territories because territories with feeders get larger harems than territories without them (Ewald and Rohwer 1982, Wimberger 1988). Orians (1980) speculated that females might be able to assess future insect production of a marsh by simply looking into the shallow areas at lake edges to assess insect numbers.

Previous nests

In Washington marshes, 90% of redwing nests from the previous year's breeding survive reasonably intact to the time females make their settling decisions and humans, at least, can judge whether most old nests failed or successfully fledged young. Therefore, prospecting females could use old nests as indicators of territory quality. However, experimental addition and removal of old nests influenced female settling patterns very little (Erckmann *et al.* 1990).

Familiar male neighbors

At CNWR, between-year fidelity of females to their nesting marshes is positively correlated with the number of returning territorial males (Beletsky and Orians 1991). Annual reproductive success of females is correlated with having familiar male neighbors, perhaps because of benefits of increased cooperative alarm signalling and nest defense. This subject will be explored in Chapter 9.

Fidelity to marshes

Whether females select nesting areas based on physical aspects of territories and marshes can be assessed by their fidelity to those areas. Females generally showed strong marsh fidelity at CNWR, both within and between years (Fig. 7.2), even though their fidelity to particular male territories between years was weak (Fig. 7.1). Females returned to their previous years' marshes in high numbers, even when their last nests of the previous year failed (Fig. 7.2). We expected more females whose nests failed to switch marshes between years in search of better sites. But our long-term nesting success data from CNWR revealed that there were no significant between-consecutive-year correlations in nesting success over a 16-year period, i.e. in the percentages of nests fledged each year on individual

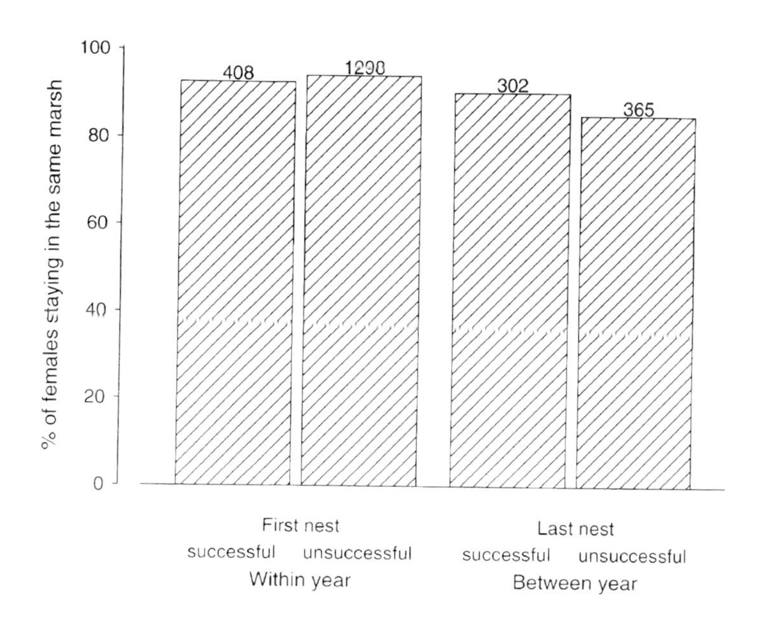

Fig. 7.2. Within-year and between-year marsh fidelity of female redwings at Columbia National Wildlife Refuge (CNWR), 1977–92. Within-year: percentage of females staying on the same marsh for their second nests of the year, as a function of the success (fledge >= 1) or failure (fledge = 0) of their first nests. Between-year: percentage of females returning for their first nests in year x + 1 to the marsh in which they placed their final year x nests, as a function of the success or failure of the final year x nest. Included are the known nesting histories of all color-banded females and nests assigned to them by the HAREMSIZE program. Values above bars are total sample sizes.

marshes (Beletsky and Orians 1996). Therefore, even if a female observed high average nesting success in year *x* on another marsh, that information would have little predictive value for the future; moving to that marsh in year *x*+1 would not, on average, improve her chances for success. In an environment where average year-to-year nesting success varies broadly, strong marsh fidelity makes sense because moving brings no greater likelihood of success than staying, and moving to unfamiliar areas is inherently risky. This variability can also explain why females ignore the evidence provided by old nests.

Our CNWR information on marsh fidelity of females agrees with work done by Searcy (1979e) at Turnbull National Wildlife Refuge (NWR), 150 km away. Searcy determined the numbers of females settling on sections of marsh he termed 'arbitrary territories.' He found highly significant positive correlations over a 3-year period, even though many of the male owners of territories contained in the study area changed, indicating that females were selecting those sites in large numbers and returning to those sites regardless of the identities of the territory owners. Fidelity to nesting sites does not prove that the sites were selected by physical characteristics, but is consistent with that criterion. Fidelity to specific marshes may also relate to benefits of accumulated familiarity with the marsh and surroundings, to a lack of movement options, or to a lack of information about alternative sites.

Once a female redwing selects a breeding area in spring, she is not easily dislodged. Almost all of 27 Indiana females that were trapped and removed from their territories within a few days of settling and held for 3 to 41 days (mean = 17.5 days) returned to the same territories when they were released (Cristol 1995).

WHETHER TO BE AGGRESSIVE: HAREMS, FEMALE RANKS, AND SUBTERRITORIES

Redwing females show aggression toward other females, especially near their nests (Nero and Emlen 1951, Haigh 1968, Jackson 1971, LaPrade and Graves 1982, Roberts and Searcy 1988). Females that settle early appear to use their aggressiveness – approaching female intruders, giving Type 2 songs and exposing their epaulets in song-spread displays, chasing them, and even physically attacking – to discourage other females from settling on the same territories and/or to delay their breeding (Nero 1956a, Lenington 1980, Searcy 1986b). (Indeed, males often attack their already-settled females as they try to evict new, prospecting females, apparently trying to prevent the harassment.) Female–female aggression typically is strongest early in the breeding season, when females are first settling. It then declines until, toward the end of nesting, little female–female aggression is evident (Nero and Emlen 1951, Nero 1956a, b, Case and Hewitt 1963, Orians 1969, Holm

1973). The function of female–female aggression is subject to controversy. Although several reasonable hypotheses could account for it and explain how it has an enhancing effect on breeding success, little uncontested empirical support has been found for them.

Subterritories

Females could use their aggression to defend small subterritories within the male territory, in effect dividing the larger male domain into smaller units in which each nests. Subterritories provide a logical explanation for female–female aggression and have been suggested by many investigators (e.g. Nero and Emlen 1951, Nero 1956b, Orians 1961, Case and Hewitt 1963, Wiens 1965); but the evidence was anecdotal. Advantages accruing to females from defending exclusive spaces could be the usual ones ascribed to territorial behavior: maintaining exclusive access to resources such as food and nest sites and perhaps, by maintaining distance between nests, decreasing the risk of nest predation. Subterritories would be consistent with the observed pattern of seasonally decreasing aggression because as the season progresses the benefits of defending resources diminish. Food becomes more available and, as the marsh vegetation matures, the number of good nest sites increases. If females defend subterritories, then predictions are that activities of individual females on male territories should be more or less restricted to exclusive, nonoverlapping areas; aggression should be space-related; and nests should be dispersed, not clumped, in space.

Hurly and Robertson (1984) and Searcy (1986b) carefully observed the activities of color-banded females, noting their positions during the breeding season on marshes divided into grids. In both studies female activities were fairly well restricted to individually exclusive areas, although the pattern could result from the fact that females spend much of their time near their own nests (Searcy 1986b). The second prediction, that female aggression is space-related, is also partly supported. For example, females attack female mounts placed near their nests (LaPrade and Graves 1982, Yasukawa and Searcy 1982) but not necessarily a female mount placed anywhere within the male's territory. Searcy (1986b) performed an experiment on space-related aggression by placing a female mount or a live female in a cage between two active nests and then measuring the aggressive response of both nest owners; in 85% of the trials (11 out of 13) the female with the nest closest to the intruder was the more aggressive of the pair. Searcy also noted that females advertised in certain, restricted areas on male territories, but these areas overlapped.

Finally, let me add my own evidence for exclusive subterritories. During my doctoral research on the functions of female redwing songs, I conducted more than

100 song playback presentations, broadcasting female songs to settled females. Usually one, two, or three females were settled on the male territories that I used. I positioned loudspeakers in what I believed to be female subterritories based on their use of space (Beletsky 1983b). In no case did a loudspeaker elicit an aggressive response from more than one female. Further, when I moved the loudspeaker on the male territory to another location, another female responded aggressively (but the females were unbanded so I cannot claim that for sure). However, that a single female always attacked the vocal intruder is consistent with the idea of individually defended subterritories.

A third prediction, that nests on male territories should be dispersed in space, has the least support. Nests started within the same 8-day periods in Washington, Indiana, and Wisconsin populations were not spatially separated (Yasukawa and Searcy 1981, Yasukawa *et al.* 1992b). If nests within male territories are not more spatially separated than would be expected by chance alone, then either nest placement was random or clumped, neither of which supports a pattern of nests placed within exclusive subterritories. However, because the best nest sites on a breeding marsh are not necessarily randomly distributed, it is possible that nests could appear in clumped distributions on a territory or a marsh and still be placed within individual subterritories.

Dominance hierarchies

When most females select a male territory, they either start or join a harem, which is an all-female social unit perhaps with its own internal organization. Females within a harem often forage together, nest in close proximity, help defend each other's nests (Picman *et al.* 1988, Westneat 1992b), and may compete for paternal care for their young. In other words, there are both cooperative and competitive aspects of harem life that interact to influence each female's breeding success. In such social units it would not be surprising to find dominance hierarchies – relationships of dominance and subordinance in which the top animals might have priority access to resources, such as the best nest sites and the greater portion of paternal care. Female–female aggression could be the overt sign and regulator of such dominance relationships.

Evidence suggests that if there are dominance relationships among females in a harem, they are not the standard ones determined by size or fighting ability. Searcy (1986b) and Roberts and Searcy (1988) scored aggressive interactions between females on male territories and determined winners and losers. Within a contesting pair, the female who settled earlier usually dominated the later settler, and the female who was closer to her nest usually dominated the female who was farther away. Thus, although they were testing for dominance hierarchies, the

Part of a harem of banded females on one of the CNWR study's breeding marshes. Harems may be cooperative or competitive social organizations, or both. Numbered flags mark the approximate positions of nests, which are a meter or more below the flags.

investigators actually found more support for effects on dominance of spatial position (i.e. for space-related aggression) and settlement order.

Female ranks and temporal spacing of nests

Female–female aggression could be utilized to defend ranks (nesting order) on male territories, and/or to influence the temporal spacing of nests on territories. Females that comprise a harem settle sequentially on a male's territory, often over several weeks. It has long been assumed that the first female to settle (generally an older, returning female) is also the first to nest and that having this primary nesting position carries with it some reproductive advantage. For example, in the closely related Yellow-headed Blackbird, males usually feed young at the primary nest on their territories (Willson 1966, Patterson *et al.* 1980), which is clearly beneficial to primary females. If primary status is likewise advantageous in redwings, then aggression might be used by first-settling females to defend their primary rank from secondaries, and by secondaries to defend their rank from tertiary settlers, etc. In this scenario, once the primary female begins to incubate, her rank is secure, and further aggression on her part would be of no value. For example, if her mate only feeds at the primary (first to hatch) nest, that assistance is guaranteed. The idea of aggression to defend high rank fits well with the pattern of seasonally declining aggression because late settlers have low ranks, and it is doubtful that there are substantial differences in nesting opportunities, paternal care contributions, or nesting success between, for example, fifth- and sixth-ranked females.

Early-settling females could also use their aggression to try to delay the nesting of later settlers. If this were the case, the pattern of aggression shown could be very similar to that generated if females were defending their ranks, but the reason would be different. For example, a primary female might try to delay the settling or nesting of the secondary if food were limited and if the aggression led to partial or total nonoverlap in the periods over which the females gathered food for their young. Females might also try to delay subsequent nesters if higher simultaneous nest densities increased the chances of attracting nest predators.

If early-settling females successfully direct their aggressive behavior to delay the settling or nesting of later-arriving females, then we would predict that the first females to settle on each territory should begin nesting before second females settle, and that nests on individual male territories should be temporally spaced, i.e. they should have a delay of at least several days between their initiations, and that internest intervals should be ordered, not random. As for the first prediction, the results of at least three studies support a direct relationship between settling order and nesting order (Teather *et al.* 1988, Langston *et al.* 1990, Beletsky and Orians 1996) and at least one study (conducted only 50 km from CNWR) found that secondary females did not settle until first-to-settle females began nesting (Langston *et al.* 1990). Langston *et al.* also observed that primary females were more aggressive toward a female mount than secondary females and that their aggression scores declined when they began to nest. However, at CNWR more than one female often settled on a territory prior to the initiation of any nesting (Beletsky and Orians 1996). For example, on Juvenile Pocket in 1990, a marsh with 17 territories, 31 females were settled before the first female nested, an average of 1.8 per territory (E. Gray, personal communication). The settlement order of females on male territories on that marsh before any laying began is illustrated in Fig. 7.3. Therefore, first-to-settle females before they nested did not or could not prevent second females from settling on male territories.

The temporal spacing of nests also provides equivocal support for aggression as a delaying tactic. Yasukawa and Searcy (1981) analyzed the patterns of nest starts within and between territories in Indiana and Washington populations. In Washington (Turnbull NWR) time intervals between nests within territories were

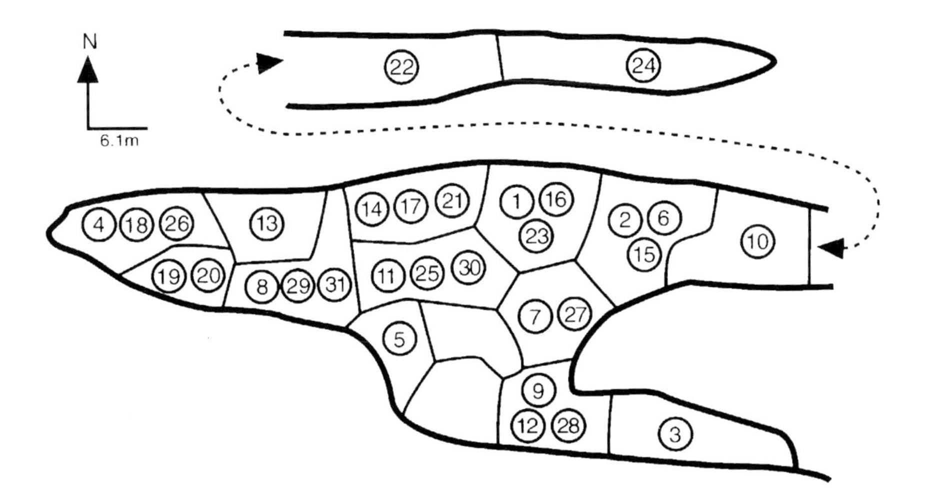

Fig. 7.3. The order in which 31 color-banded females settled on Juvenile Pocket marsh at Columbia National Wildlife Refuge (CNWR) in March and April 1990, before any eggs were produced. Approximate territorial boundaries on 1 April are shown; females are shown on the male territory in which they placed their first nests.

the same as nest intervals between different territories, suggesting a lack of within-territory patterns that would be indicative of females delaying each other's nesting. For the Indiana population, however, there were fewer short intervals than would be expected by chance, indicating that early-settling females may have been delaying the nesting of later settlers. The difference fits with regional variation in paternal care. In Indiana many males help feed nestlings and thus primary females might use aggression to ensure that help; Washington males rarely help feed young. At CNWR, we found that average intervals between primary and secondary nests and between secondary and tertiary nests were both about 6 days (Beletsky and Orians 1996). This supports the idea that intervals are not random, but if primary females used their aggression to delay secondary ones, we would have expected significantly longer intervals between primaries and secondaries than between lower-ranked nests.

If aggression is used to defend nest rank, particularly primary rank, then males should be more likely to feed nestlings at primary nests or defend them. Again, the evidence is equivocal. Most males at Yellowwood Lake, Indiana, provided feeding assistance exclusively to primary nests (in 17 out of 20 cases, 85% of the time; and in two out of the other three cases, primary nests failed before males could assist; Patterson 1991). But in Ontario and Michigan, also in regions where males regularly feed nestlings, investigators observed either no effect of nest rank on male feeding choice (Whittingham 1989) or that males were more likely to feed at secondary than at primary nests (Muldal *et al.* 1986). At CNWR we found that when males fed nestlings they did so at the most advanced nest at the time on their territories, but that rank *per se* was not a determining factor (Beletsky and Orians 1990). There is also some evidence for differential nest defense as a function of rank. Knight and Temple (1988) placed models of American crows simultaneously near primary and secondary nests. Males defended more aggressively at their primary nests, diving at and striking the crows more frequently. One explanation of the males' behavior is that they prefer to defend primary nests, supporting the idea that primary females might benefit by defending their rank. Another explanation is that males defend older offspring more vigorously than they defend younger ones.

A key test to determine if rank is important to females is to compare the annual fledging success of females of various ranks. If females defend their high rank, it must be because such rank is advantageous in some way that leads to improved breeding success. Thus, a critical prediction would be that primary females should have higher success than lower-ranked females. In some redwing studies this prediction was borne out: Crawford (1977; in Iowa), Yasukawa (1989, in Wisconsin), and Langston *et al.* (1990, in Washington) all determined that primary nests, on average, experienced significantly better fledging success than lower-ranked nests. In southern Ontario, Muldal *et al.* (1986) observed than secondary females actually had higher average fledging success than did primaries because males tended to provision secondary rather than primary nestlings. A problem with all of these studies is that few if any females were color-banded, so researchers could reliably report only the success of females' first nests. However, renesting in redwings is very common, following both successful and failed nests and thus, annual reproductive success was not measured. With a large population of color-banded females and good, long-term information on nesting success, we used our CNWR data to test for a positive relationship between female rank and annual fledging success. Instead, we found for high ranks a nonsignificant trend in the direction opposite: primary females had relatively low seasonal success, secondaries did better, and tertiaries had the best average success (Fig. 7.4; Beletsky and Orians 1996). Similar trends in fledging success of females of varying rank have also been identified in some other birds (e.g., in Corn Buntings; Hartley and Shepard 1994).

All the information gathered to date on female redwing aggression still provides no clear explanation for it. There is evidence that females exhibit space-related aggression during breeding and that sometimes female activities are

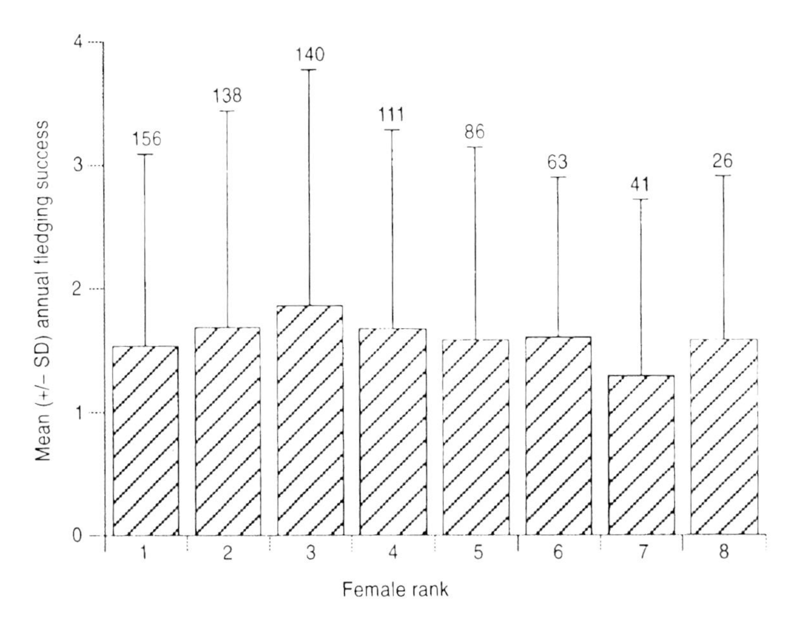

Fig. 7.4. Annual fledging success for Columbia National Wildlife Refuge (CNWR) females, 1977 to 1992, as a function of their nesting rank. Rank was assigned to females by the order of their first nests each year (e.g. a female whose first nest of the year was a secondary nest on a male's territory was considered a secondary female during that entire year). Included here are only those females who nested on pocket marshes, generally a high-quality breeding habitat with large harems. Sample sizes are given above error bars. Individuals could have been included more than once, in different years.

restricted to portions of male territories. However, because defended areas are not exclusive, Searcy and Yasukawa (1995) hesitate to affix the term 'territorial' to female redwings. Instead, they suggest that females defend or are aggressive in 'dominions', defined by Brown (1975) as 'areas of dominance from which submissive individuals are *not* excluded'. Linear dominance hierarchies probably do not exist within harems. As for females using aggression to defend a high nesting rank, on one hand there is some evidence that males preferentially feed and protect primary nests, but on the other hand, primary nesters in our long-term study, the best information currently available, were not the most successful. It is possible that it is all much simpler than that – more aggressive females may simply have enhanced breeding success. Langston *et al.* (1990) observed such a relationship (again, using only first nests), but Searcy (1988) did not. Searcy (1988) and Searcy and Yasukawa (1995) conclude that when all the evidence is considered, no consistent function for the females' aggression is evident. They suggest, as a default hypothesis, that female–female aggression in this species occurs as a carryover from selection for aggressiveness in males. This explanation is, for a behaviorist, far from satisfactory, but until a positive effect of female–female aggression on reproductive success is demonstrated, it may have to suffice.

WHEN TO NEST: STARTING AND STOPPING

When to start nesting is an important decision because starting too early or too late can reduce nesting success. Females beginning their nests early in spring assume the risks that foul weather, even snowstorms, will terminate their efforts or that there will be inadequate food available for their young. Because nest predation rates generally increase seasonally, if a female starts later, her nest is less likely to be successful and, if her first nest fails, she has less time to try again. In Ohio, for example, the earlier a nest is started, the better average fledging success it has (n = 186 nests started from May to July; Dolbeer 1976). Likewise, Langston *et al.* (1990) showed a strong relationship between nest start date and the chance of failure owing to nest predation (based on 1321 nests on 11 different marshes, started in April through June). However, because these investigators did not mark females, they could only report general nesting success and not the annual success of individuals. Our long-term breeding success information for color-banded birds demonstrates for CNWR females that, as found by others, nests begun earliest during the season have the highest probability of successfully fledging young (Beletsky and Orians 1996) and also that a female's total seasonal breeding success is significantly correlated with the week she began nesting – the earlier, the better (Fig. 7.5). Although this is likely to be a widely generalizable result, it may not apply to all populations in all years. For example, Westneat (1992b) reported for his New York population a positive correlation between a female's first egg date and annual breeding success in only one out of 3 years.

Waiting periods

We suspect that females actively decide when to nest because most of them wait for up to several weeks after they have settled before they build nests and lay eggs. Waiting periods of 3–4 weeks are common. Nero (1956a) reported a mean waiting period of 20.7 days (n = 4) in Wisconsin between 'pair formation' and first egg dates. Teather *et al.* (1988) observed waiting periods in Ontario of 14.1 ± 9.3 days (n = 17), and Langston *et al.* (1990) found waiting periods in Washington of 37 days for primary females (n = 14) and 17 days for secondaries (n = 30). There was extensive variation in the duration of CNWR waiting periods. Because we did not

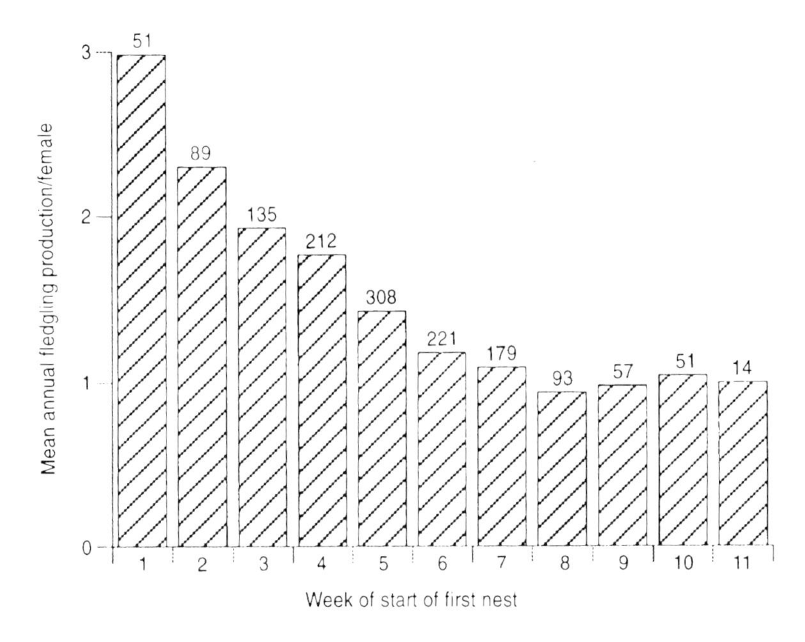

Fig. 7.5. Mean annual fledgling production per female at Columbia National Wildlife Refuge (CNWR), 1977 to 1992, as a function of the week of start of each female's first nest. Week 1 is the last week of March; week 11 is the first week of June. Sample sizes are given above bars.

know settlement dates for most females, we generated conservative estimates of waiting periods by counting the number of days between each female's first capture on the study area and the date each laid her first egg. Many actual waiting periods were longer because few females were captured the day they arrived. Conversely, some of the very long waiting periods arrived at by this method are likely to be due either to females moving into the monitored area after their first nests elsewhere failed or to observers missing females' first nests. The resulting distribution (Fig. 7.6) shows a large number of waiting periods in the 10- to 40-day range. For the 31 females settling early in 1990 on Juvenile Pocket (Fig. 7.3) we knew exact dates of settlement and nest initiation. Those females waited an average of 27.4 ± 13.5 days between settling and laying first eggs.

We are confronted with a paradox. Our data show that starting to nest early is advantageous, yet females wait several weeks after arriving to do so. Why are they waiting? One obvious candidate to explain waiting is the seasonal availability of food resources. Females may need to accumulate sufficient energy reserves for egg production and the rigors of nesting. This hypothesis is supported by several

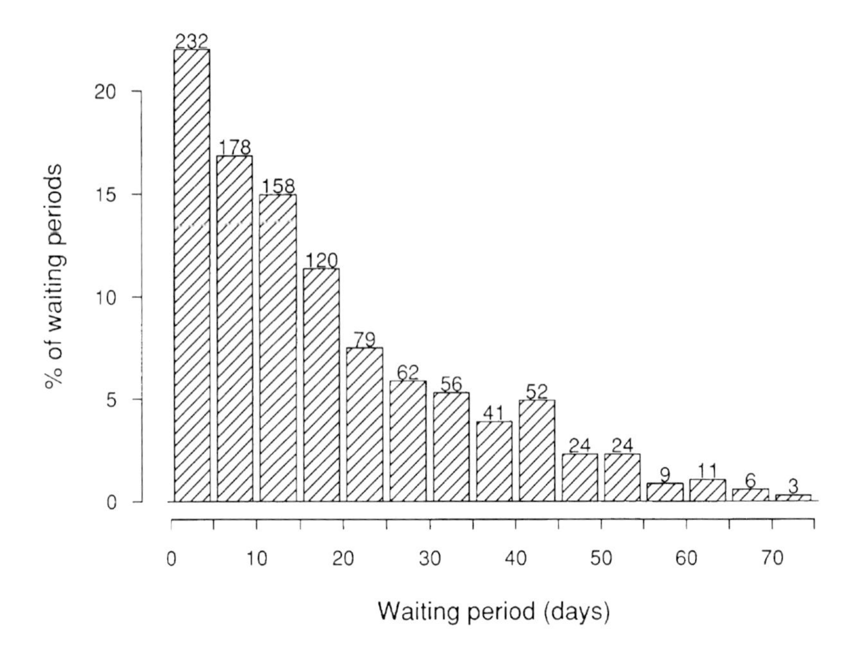

Fig. 7.6. Waiting periods between settling and nesting for females at Columbia National Wildlife Refuge (CNWR), 1977 to 1992, estimated conservatively as the interval between the date each was first trapped and their first egg dates. Sample sizes are given above bars.

studies that demonstrated that seasonal nest initiations in redwings can be advanced, sometimes by several weeks, by providing supplemental food in or around territories (Ewald and Rohwer 1982, Wimberger 1988, Beletsky and Orians 1996). Other factors, such as weather, which influences food production, could also contribute to female nest initiation decisions.

For our CNWR population we considered the factors that might influence nest initiation and tested several of them (Beletsky and Orians 1996). Delay may be influenced by environmental factors such as the weather, the presence of good nest sites in the previous year's growth of vegetation (the current year's growth of marsh vegetation has usually not yet begun or advanced much when females make their nest start decisions) and the availability and quantities of early-season food. Moreover, because the females are migratory, environmental conditions elsewhere, in wintering areas or along migration routes, could also influence nesting start times. Females arriving back on their breeding marshes in poorer condition might have to wait longer to recover and to begin nesting.

Weather apparently does not influence when females start to nest at CNWR.

We found no statistical relationships between nesting success and such facets of weather as mean temperature in February, March, or April, number of days each month that did not rise above freezing, total monthly precipitation, or the number of days of precipitation each month. The only significant association we detected was in the direction opposite to the one we expected: more precipitation in February and March was significantly correlated with earlier nesting. Rain, therefore, did not delay the females.

Social interactions could also influence female nesting times. Males do not delay female settling or nesting, nor is there any reason to suspect such behavior. As indicated previously, females themselves might have interests in delaying the nesting of others in their harems but there is evidence against such a social effect; indeed, some investigators have observed the opposite – a positive effect on reproductive success of synchronous nesting among redwing females (Picman *et al.* 1988, Westneat 1992b).

We succeeded in identifying individual differences among females that contributed to their nest initiation dates but, because they were intrinsic traits, these probably had little to do with decision-making. Age played a role. Mature females began nesting, on average, a week earlier than did yearlings. Also, if a female started her first nest earlier or later than the average in one year, she usually did so consistently in other years. Thus, there may be genetic propensities for nest initiation times, perhaps mediated through differences in hormonal states, different foraging abilities, or different wintering grounds.

In summary, a variety of factors, some environmental and others perhaps intrinsic, probably contribute to female decisions about when to start nesting. It may well be that for each female in each particular breeding situation, there is one best time to initiate seasonal breeding, the better to maximize annual fledging success.

When to stop nesting

The decision to terminate breeding for the year may also have multiple inputs. The decision to stop nesting is especially interesting because the females do it in summer while nesting success is still high and conditions are still conducive to breeding. Most females at CNWR are finished nesting in June, yet insect emergence from their marshes remains high through July and August (Willson and Orians 1963, Orians 1980). Several factors may account for the paradox. First, investing in late nests may be very unprofitable because predation rates tend to rise later in the breeding season (Orians 1973, Caccamise 1978, Langston *et al.* 1990). However, we did not find low fledging success rates for late nests (Beletsky and Orians 1996). Second, female redwings, avid renesters, may tire themselves with their repeated nesting attempts, and exhaustion could hamper migration abilities and also perhaps future breeding efforts. Finally, females may have to stop nesting in order to divert their metabolisms to their energetically expensive molt, which must be accomplished prior to the autumn migration.

NEST DEFENSE

Another activity of breeding females is nest defense. During laying and incubation, females are the main guardians of the nest, leaving it, on average, for only 6–7 min at a time (Ritschel 1985). After the eggs hatch, the nest is not so well attended because females make frequent, extended foraging trips, and males do not or cannot always guard nests during female absences. Males often have several nests simultaneously active on their territories, not all of which can be guarded at the same time. Females do signal their departures by singing when they leave their nests but males do not adjust their behavior in response to those songs. A male is just as likely to leave his territory to forage immediately after a female leaves her nest as at any other time (Beletsky and Orians 1985, Ritschel 1985).

When females detect nest predators, they give alarm vocalizations to attract other redwings and then, alone or in conjunction with males and other females, defend their nests by mobbing potential predators. They use 'Check' alarm calls and 'Screams', long, harsh sounds given by both sexes when attacking or after being caught by a predator (Orians and Christman 1968), to summon help. When Knight and Temple (1988) played these vocalizations from a tape recorder on male territories, redwings approached the recorder and hovered over it, especially in response to screams.

Mobbing is 'a joint assault on a predator too formidable to be handled by a single individual in an attempt to . . . drive it from the vicinity . . .' (Wilson 1975). Both males and females mob, but males, perhaps owing to their larger size and presumed greater effectiveness, generally take the more active role. Nonetheless, females vocalize, hover over nest predators, dive at and even strike them. From long personal experience I can attest that both sexes hover over and dive at people moving through marshes to check nests, but that it is almost always the males that actually strike. However, that is not always the case. One CNWR female was as aggressive and fearless as any male. For several years, each time a field assistant would enter the marsh near one of her nests, she would immediately land on the assistant's head and commence pecking, and even try to rip out hair.

Females increase their mobbing intensity as their nests progress from egg to nestling stage, in accordance with predictions from parental investment theory (Trivers 1972) that parents should increase nest defense as investment in and value of the nest increases. D'Arms (1978) placed a live Black-billed Magpie, the major nest predator in the area, near nests at CNWR and observed the females' aggressive responses (hovers, dives, strikes, screams, attack initiations). She found little or no mobbing before nests received eggs, a 'medium' amount of mobbing at nests with eggs, and the strongest mobbing response during the nestling stage. The efficacy of redwings mobbing nest predators probably depends on the number of mobbers participating, but mainly on the type of predator involved. Large mammals, such as mink, weasel, and raccoon, are unlikely to be deterred. Smaller avian predators, such as jays and grackles, are more likely to be driven away than larger birds, such as crows and magpies (see Chapter 9).

CONCLUSIONS

A great deal of ecological research on redwings has concentrated on the behavior of females. Their breeding season decisions and associations are more complex than those of males. Females choose mates and places and times to nest, and they must maintain pair-bonds with males and relationships with other females in their harems. Also, early work on the behavioral ecology of birds suggested that territory quality could be the decisive factor in female mate-choice in polygynous species and the idea stimulated research on redwing females and their breeding decisions. The question of whether female redwings choose nesting areas by the qualities of the males that defend territories or on qualities of the territory itself has been fairly well resolved in favor of territory. There is evidence that females do not judge individual males, suggesting more that they prefer certain physical traits of territories. Settlement choice is better viewed, however, as females selecting breeding situations – multifactor situations that are probably assessed for both physical and social circumstances, including the numbers and identities of other females already settled on the territory or the marsh, and the number and identities of male neighbors. Recent work on EPCs and paternity analyses have provided another piece of the puzzle of female breeding choices, showing that they can select territories first and then much later decide on the paternity of their clutches; they are not limited to mating exclusively with the male on whose territory they place their nests.

Most of the work on nesting ranks and female–female aggression was stimulated by interest in the workings and social dynamics of harems, but also by a need to test predictions of the Polygyny Threshold Hypothesis. The purpose of female–female aggression is still obscure. There is some evidence that males treat primary nests differently than others, but our data, by far the most complete available, reveal no advantage of primary rank on a female's annual reproductive success. In fact, at CNWR females should prefer to be tertiary nesters, as these have the highest average nesting success (Fig. 7.4). Research to date on nesting choices perhaps allows us to characterize how a CNWR female would choose and use an ideal breeding situation. A female should settle on the same marsh she had nesting success in the previous year, in about the same place, regardless of the identity of the male holding that territory. Surrounding her nesting territory should be many of the same males from the previous year, with whom she will copulate and from whom she will expect nest defense. She should nest in cattail in a spot of moderate stem density, starting her first nest early because early nests have a higher probability of escaping predation than later nests and because, should her nest fail, she will then have time to try again. Because we found no benefits of high rank, she should not be aggressive toward her harem-mates, but should accept primary, secondary or tertiary nesting status.

Chapter 8

MALE BREEDING ROLES AND DECISIONS

INTRODUCTION

In the redwing rendition of polygynous breeding, male roles and decisions are almost completely divorced from females'. There is an almost complete division of labor by gender; breeding roles coincide only over defending young and, sometimes, provisioning them. Males devote their greatest effort each year to establishing and defending territories. To these all-important parcels of land, females are attracted to build their nests. Put simply, males of this species that fail to obtain or adequately defend territories cannot reproduce. Therefore, considerable research attention – mine and that of other investigators – has focused on territory acquisition and retention as the single most important component of male redwing breeding biology. Because of its importance, I devote most of this chapter to descriptions of territories and of territorial behavior, and Chapter 10 to territory acquisition and territorial dominance.

Continuing with the tactic of viewing animal reproduction as a series of decisions made by breeding partners, males, after deciding when, where, and how to obtain territories, court females that arrive to inspect their territories, copulate with the ones that stay, and then decide how to channel their energies among several different activities, each of which influences their seasonal reproductive success. Among these activities are, assisting their present mates by feeding offspring (enhancing the viability and survival of current offspring), advertising for additional mates (providing opportunity for more nests), guarding their already-settled mates from other males (ensuring paternity), and pursuing extra-pair copulations (EPCs) with neighboring females (providing opportunity for extra-pair young).

TERRITORIES

Coming upon a cattail marsh early in the morning of a spring day almost anywhere within the great range of the redwing, one is likely to encounter a group of adult males singing, calling, showing their red shoulder epaulets, chasing, and flying from perch to perch (see illustration at chapter head). With some observation, it becomes clear that individual males are not randomly distributed over the marsh or traversing it entirely with their flights, but that each is restricted to a particular section in its perchings and movements. Males patrol their respective sections, chase intruding males from them, and frequently come into close contact with adjacent neighbors to argue boundaries with bill-up displays (Chapter 5). Male redwings exhibit a classic territoriality: during the breeding season each maintains a defended area from which all other male redwings are aggressively and rigorously excluded. Much of the male redwing's overt behavior that people commonly notice – many of their vocalizations, visual displays, interactions with other males, frenetic activity levels on breeding marshes – are expressions of territoriality.

Average size and usage of the defended spaces vary regionally and among habitats, but in all areas males center their activities during the breeding months within and around their territories. To this jealously-guarded space females are attracted to settle, nest, and raise young, and also here males and females roost seasonally and forage for a significant fraction of their diet. The size and quality of the habitat enclosed by a territory's borders are usually inversely related, and both features may influence the density of female settlement and hence, male reproductive success. Males defend their territories not only from other male redwings, but also in many cases from other species. Although the boundaries can change weekly or monthly and often differ in their approximate locations each year, a male's territory in most cases remains his until death. However, a small fraction of males each year change their territory locations, or desert or are evicted from territories to rejoin the floating population (Beletsky and Orians 1987a).

Functions, use, and size

Birds have various kinds of territories. In some species, individuals defend territories only for the purpose of feeding on them; other species have territories only for mating. Redwings have all-purpose territories. Males roost, advertise to repel other males and attract females, court and copulate with their mates, forage, rest, perform maintenance activities (preen, bathe), and hide from predators there.

When males have active nests, they spend long periods each day, including times during which they are otherwise inactive, on high cattails or in trees on or adjacent to their territories, in good position to inspect their territories and detect approaching danger, be it a nest predator or a potential territory usurper. On high perches the males apparently act as sentinels, positioned to protect nests and give alarm signals to alert their mates and other redwings on the marsh to potential dangers (Linford 1935, Beletsky 1989b, Yasukawa *et al.* 1992a). At the Columbia National Wildlife Refuge (CNWR), many of our marshes were positioned directly below high basalt cliffs that the males often perched atop, surveying their domains (Fig. 8.1). Males often have favorite perches, such as particular cattails or rock ledges, to which they return again and again.

Fig. 8.1. A Columbia National Wildlife Refuge cattail marsh that contains redwing territories and that is directly below a cliff.

During the breeding season, the activities of some songbirds, such as Song Sparrows and Prairie Warblers, are restricted to their all-purpose territories (except for brief absences for water or trespassing on nearby territories). Although the activities of male and female redwings center around the male's holding, both sexes spend a part of each day off-territory. How long they leave and how far they go depend on food availability. Before nesting begins, there is often little food for redwings on breeding marshes, so both sexes leave the territory for large portions of each day to forage on seeds in upland areas, sometimes some distance from the marshes. Later, when insect emergence makes marshes more profitable for foraging, females often seek additional food away from their mate's territory, sometimes on other males' territories but also in upland areas and in undefended wetlands. For example, Haigh (1968), who monitored redwing breeding at Washington's Turnbull National Wildlife Refuge, noted that females on some marshes foraged mostly away from their territories, in upland areas and on adjacent lakes with higher insect emergence rates. Many of these females bred on lakes that dried in summer and hence had poor insect emergence rates. Females, once settled on male territories and once food is abundant there, spend most of their time on or near them. But this behavior varies geographically – Westneat (1994) observed that in his New York population females rarely left territories, whereas at CNWR and in some other areas they do so regularly (Nero 1956a, Wiens 1965, Snelling 1968, Weatherhead and Robertson 1977a, Picman 1981). Females may also seek EPC opportunities while off territory (Gray 1994) or evaluate nesting success on other marshes, but to date we have little direct evidence for these endeavors.

Males must guard their territories every day against whole or partial loss to floaters and from encroaching neighbors trying to expand their own holdings, to protect their nests from predators, and to guard their females from EPC attempts by other males; hence, they leave territories primarily for only three important reasons: foraging, trespassing, and spying. When food is scarce on and adjacent to a male's territory, he must forage elsewhere. We have some information on the distances males travel to find food because we caught redwings at CNWR with seed-baited traps. Thus, we know the maximum distances color-banded males traveled from their territories in search of food, at least as defined by how far away we trapped them. Males on our study marshes in 1983 and 1984 (n = 125 males each year) were rarely trapped more than 1600 m from their territories (Beletsky and Orians 1987a). Significantly, many males who were not captured during those years in large traps, whose positions were fixed throughout the study, had been captured in them during the years before they owned territories, giving the impression that they knew about the traps as food sources but that they were too far from their territories to visit. (Males with territories close to the traps were usually caught annually.) Distances males in different areas travel from their territories must vary depending on how far they have to go to find sufficient food, but long flights to distant areas leave mates and territories unguarded.

Males also leave their own territories to trespass on those of neighbors, and probably also to scout other marshes for potential future territory locations. Males intrude in their neighbors' territories for several reasons. They often enter to assist the resident male repel predators during mobbing, they sometimes end up in the wrong territory following group sex chases (see Chapter 6) and several investigators, myself included, have observed instances during which a male will enter a neighbor's territory to chase and evict a male redwing intruder when the neighbor is temporarily absent (Nero 1956b). In one case, I watched a male fly across a lake to the territory opposite his own to perform this service. Possible reasons for such seemingly altruistic behavior are discussed in Chapter 10. Males have also been observed intruding into neighbors' territories to watch their neighbors interact or fight with trespassing floaters.

Males also leave their territories apparently to spy on other territories, gathering information that may be used for future takeover attempts at those sites when the current resident dies or leaves. At times at CNWR I would spot a male on a territory, raise my telescope to read his color-band combination, then stare in wonder as I realized that he was not the resident that I expected to be there, but an individual that I knew was presently holding a territory on another lake. What males are doing in these trespassing situations is a mystery, but we suspect that they are evaluating the quality of other territories and marshes for future movements to them. Such monitoring of another male's territory has been discovered in other birds that have been observed very closely. For example, Prairie Warbler males, who spend almost all their time on their territories, were observed by Nolan (1978) to depart sometimes for 30 to 60 min to perch in trees quietly to watch activities on other territories. We might expect that during such distant trespassing redwing males would seek EPCs, but molecular evidence demonstrates that females are normally fertilized only by their putative mates and males with neighboring territories, allowing us to eliminate these forays as a reproductive tactic (but see Weatherhead and Boag, 1995).

The amount of time males spend daily on their territories during the breeding season varies with time of year, breeding phase, and geographic region, but again, the proximate determining factor must be availability of food. Because breeding areas often are surrounded by undefended habitat, male redwings, unlike many other territorial birds, have the option of spending time off, but near, their territories – positions from which they can, if need be, speedily return. Several studies have gathered 'time budget' data on male redwings, that is, researchers have monitored their activities, recording their behavior by category. Gordon Orians monitored several males at Jewel Lake, California in 1958. He found a temporal pattern of territory occupancy much like the one practiced by CNWR males. In

February, they spent only 15 minutes or so after sunrise defending their territories before they left for the entire day, returning only as evening fell. During March, the number of hours males occupied their territories each morning rose progressively to 3.5 (again, quite similar to the CNWR schedule). By the first week of April, when nesting began, males spent all day in the immediate area, but physically on their territories only 75% of the time (Orians 1961). This proportion is consistent with redwings studied in Wisconsin, who spent between 73% and 88% of the morning hours on territory during the incubation periods of primary and secondary nests (Yasukawa *et al.* 1992a).

We studied male time budgets at CNWR for various reasons. From 1983 to 1985, we monitored males with varying numbers of years of territory ownership to look for differences in their behavior. Overall, from 06.00 to 10.00 hours, males spent about 73% of the time on territories (n = 62 males observed for a total of 565 15-min periods during March, April, and May). We found a slightly higher percentage in 1987 and 1989; during March, from 06.00 to 09.00 hours, males spent 84% of their time on territories (Beletsky *et al.* 1990). The 1983 to 1985 data can be broken down by breeding phases. The males monitored actually spent, on average, declining amounts of time on their territories as females settled and first nests

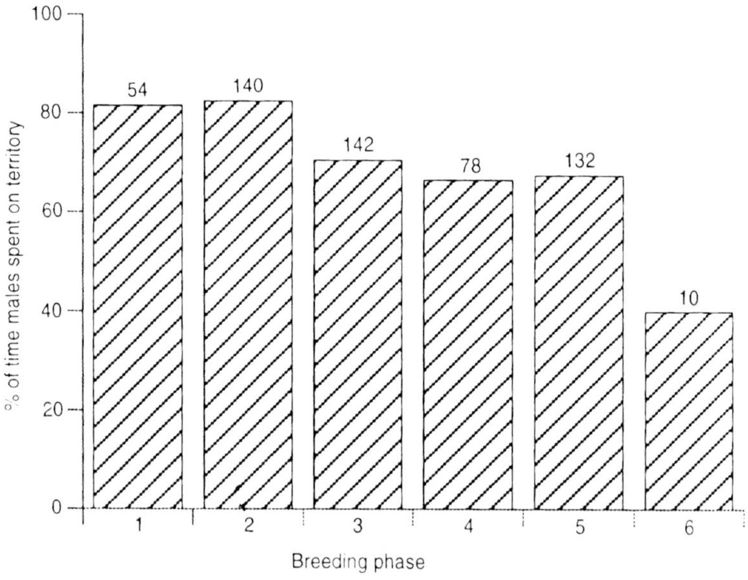

Fig. 8.2. Proportion of morning time that male redwings spent on their territories at Columbia National Wildlife Refuge, by breeding stage. 1 = prior to female arrival; 2 = after female arrival but before the first nest begins incubation; 3 = during first nest incubation; 4 = during first nestlings; 5 = during first fledglings and beyond; 6 = after predation on first nests but before second nests begin.

progressed (Fig. 8.2). A likely reason for this trend is that males had to remain on territories when advertising for their first few mates, but then could afford to leave more often, particularly after the number of prospecting females declined. Also, it should be noted that although these males frequently left their territories, they usually remained nearby, often in locations from which they could monitor their territories. One last factor influencing the time males spend on territory is demonstrated by an instance where, after Black-billed Magpies destroyed all or most nests on a marsh, resident males would be absent during large portions of each day for a few days, until new nests were started; clearly, it was active nests at that period of the season that held them close to their territories.

Precisely what do male redwings do while on their territories? We can use our CNWR time budget information to make a fairly good chart of an average male's morning activities (Table. 8.1). They spend, on average, 10–13 min per 15-min

Table 8.1. Rates of various behaviors[1] (average number per 15-min period) of male redwings while on their territories during spring breeding seasons at CNWR

Year	Number of males observed	Total number of 15-min periods	Time on (min)	Time up (min)	Vocal time (min)	Number of calls	Number of songs
1983	44	345	9.9	9.1	9.0	83.7	13.7
1984	21	158	12.5	12.3	12.0	80.8	35.8
1985	16	62	12.5	11.3	12.3	90.8	44.4
mean ± SD	1983–85		10.0 ± 4.6	10.2 ± 4.1	10.2 ± 4.3	83.7 ± 74.1	23.2 ± 19.5
1987	11	50	11.2	9.5	9.9	95.8	25.9
1989	15	72	13.5	12.0	12.2	107.6	42.7
mean ± SD	1987–89		12.6 ± 3.8	11.0 ± 3.7	11.3 ± 3.8	102.8 ± 69.3	35.8 ± 17.0

		Boundary disputes					
	Number of Bill-ups	Number	Duration (min)	Number of fights	Number of flights	Number of male chases	Number of female chases
1983	2.4	0.9	0.9	0.1	7.8	0.2	0.5
1984	4.7	1.1	1.1	0.3	10.8	0.7	0.4
1985	2.1	0.6	1.0	0.0	6.7	0.3	0.4
mean ± SD	3.0 ± 6.8	0.9 ± 1.3	1.0 ± 1.8	0.2 ± 0.8	8.5 ± 7.0	0.4 ± 1.4	0.5 ± 1.2
1987	3.1	0.7	0.8	0.0	6.9	0.1	1.1
1989	1.5	0.4	0.3	0.1	8.4	1.0	0.3
mean ± SD	2.1 ± 4.8	0.5 ± 0.7	0.5 ± 1.3	0.2 ± 0.18	7.8 ± 4.7	0.6 ± 1.9	0.6 ± 1.0

[1]Time on = time on territory per 15-min. Time up = time male is on territory and visible to the observer per 15 min. Vocal time = the sum of all 10-s intervals/15-min that a male gave at least one call or song. Bill-ups = aggressive displays given when males are engaged in boundary disputes with neighbors and when trying to evict intruding males. Boundary disputes = ritualized arguments with neighbors over the positions of territorial boundaries.

period on territory. Most of that time, 9–12 minutes, is spent in exposed positions ('time up'), from which they can advertise and also monitor their mates, territories and surroundings. During their remaining time on territory they are out of sight of observers, usually in the dense vegetation of the marsh, either foraging or interacting aggressively with intruding males or neighbors, or sexually with females. They utter a vocalization in every 10-s period (vocal time) during almost all their time on territory – they are, in fact, rarely quiet during mornings for more than 20 or 30 s. Alert calls, usually given while on territory, account for much of their vocal activity (Chapter 5). They also engage in border disputes with neighbors, give bill-up displays, occasionally fight, and fly from point-to-point within their territories while patrolling boundaries or chasing male intruders or females (Table 8.1).

As we have seen, during the breeding season the behavior of male redwings is largely geared to territoriality – they need territories to reproduce, and once acquired, they spend most of their time on or near them. But defending space from other males is a costly business, because time and energy devoted to defense cannot be used to pursue other activities, and because activity levels associated with defense must surely render males more noticeable to predators. Might males try to minimize these costs by defending only small spaces? Several factors will determine how large a space males decide to defend.

The main resources redwing territories provide to breeders are food supplies and safe nest sites; ecological factors that affect those resources should influence territory size. First, for nests, territories must provide concealing cover and anchor sites. Because plant attributes vary, e.g. in the concealment they provide, vegetation type may influence territory size. Thus, redwings in eastern Washington prefer territories in cattail over bulrush, probably because cattail provides more safe nest sites per unit area. Territories in one locality in bulrush were, on average, larger than those in cattail (Haigh 1968; Table 8.2). One possible explanation for the pattern is that bulrush territories, to offer the same density of quality nest sites to females, had to be larger than their cattail counterparts.

Second, a territory must be large enough to provide food for the owner, his mates and young. Depending on the availability of undefended feeding areas near territories, the concentration of on-site food supplies could be a critically important determinant of the size of the area males try to defend. Generally, there should be a correlation between territory size and the productivity of the habitat defended, but because in some areas redwings forage extensively off territory, the relationship is likely to be loose. Upland habitats, such as grass or agricultural fields are generally less productive than marshes; accordingly, territory sizes in uplands are, on average, much larger than those in marshes. (Also, in regions where males regularly establish territories in uplands, most food is gathered on-territory (Orians 1980).) The average of average sizes determined for territories in upland habitats is 3058 m^2, and the average in marshes is 1658 m^2. If we exclude the exceptionally large marsh territories measured at one site in British Columbia (Picman 1987), the marsh average drops to only 937 m^2 (Table 8.2). Thus, upland territories are, on average, three times larger than ones in marshes. Some studies that compared redwing breeding characteristics in marsh and upland habitats in the same region have found upland territories to be larger (Case and Hewitt 1963, Allen 1977), although Ritschel (1985) found the opposite.

Within a habitat, territory sizes may be correlated with the proportion of food individuals gather on them, territories being smaller when a larger share of food is collected off-territory (Orians 1961). Conversely, on highly productive marshes, even a small territory could provide a large proportion of the diet, and the correlation could reverse – males might defend very small territories in the richest marsh areas *and* gather much of their food there.

Third, the local density of males competing for territories must influence average territory size (Nero 1956b). When population size is low and few high quality males are present, territories will be larger than when the population size is high and many high-quality males compete for space. In this way, redwing territories are elastic, expanding and compressing with the number of males able to establish themselves. There are, of course, limits to expansion and compression, with huge territories being economically indefensible and tiny territories being below the minimum size necessary for successful breeding.

The size of a male's holding can change frequently during a breeding season as males contest boundary locations, try to enclose within their borders nests under construction, respond to deaths, desertions and evictions of neighbors and insertions by new males, and as vegetation matures in spring, altering the shape and dimensions of a marsh. Juvenile Pocket territories at CNWR during the 1992 breeding season illustrate the size and shape changes of territories (Fig. 8.3). The number of males holding territories in the marsh varied from 17 on 14 March, to 19 on 15 April, to 16 on 12 May, to 15 on 12 June, each change in number having consequences for territory sizes and shapes. The territories changed for a variety of reasons. For example, Male 1, on territory in the eastern corner of the marsh early in the season, was evicted and replaced by Male 18, the individual who had owned a territory in that location in 1990 and 1991. Since territory sizes vary within seasons, it is not surprising that there is also broad variation between years. Juvenile Pocket, for example, contained at mid-season in 1990, 1991, and 1992, 17, 20, and 17 territories, respectively, with differing mean territory sizes (Fig. 8.4; Male 1 in Fig. 8.4 is the same individual as Male 18 in Fig. 8.3). Other illustrated examples of seasonal changes in redwing territories are provided by Nero (1956b), Orians and Collier (1963), Orians and Willson (1964), and Dickinson and Lein (1987).

Males with larger harems have, in general, better annual reproductive success (Weatherhead and Robertson 1977a, Orians and Beletsky 1989). Thus, any factor that influences male harem size should also influence breeding success. Logically,

Table 8.2. Mean territory sizes of male redwings in various habitats, and vegetation types

Mean size[1] (sq m)	Range (sq m)	*n*	Years	Location	Main vegetation	Reference
Upland habitats						
2180	120–4000	49	1960–61	New York	Alfalfa, timothy, 'weeds'	Case and Hewitt (1963)
3210	700–9600	10	1973	Iowa	Grass/forb	Blakley (1976)
5480	4240–8820	25	1971	New York	Ryegrass, timothy, alfalfa	Allen (1977)
1680–2670[2]	–	–	1972	Michigan	Grass/forb/ legume fields[3]	Albers (1978)
370 ± 140	–	13	1980–82	California	Mustards, wild radish	Ritschel (1985)
5816	2214–29235	–	mid-1980s	Ontario	Cattail	Eckert and Weatherhead (1987b)
Marsh habitats						
1435	297–3575	4	1934	Utah	Cattail	Linford (1935)
330	120–580	17	1948–53	Wisconsin	Cattail	Nero (1956b)
670–1270[2]	–	86	1959–60	California	Cattail	Orians (1961)
690	240–4500	51	1960–61	New York	Cattail/bulrush	Case and Hewitt (1963)
270	–	50	1965	Oklahoma	Cattail	Goddard and Board (1967)

332–419[2]	47–1607	46	1966–67	Turnbull NWR, Washington	Cattail	Haigh (1968)
561–603[2]	133–1249	43	1966–67	Turnbull NWR, Washington	Bulrush	Haigh (1968)
1810	487–4343	16	1966	Costa Rica	Cattail	Orians (1973)
1740–2300[2]	–	12	1963, 1965	Seattle, Washington	Cattail	Orians (1980)
1070	210–2380	53	1971	New York	Cattail	Allen (1977)
1045	153–2890	97	1974–75	Ontario	Cattail	Weatherhead and Robertson (1977a)
780	213–2078	11	1976?	Ontario	Cattail	Weatherhead and Robertson (1977b)
1410–1590[2]	–	34	1974–76	Indiana	Cattail/burreed	Patterson (1979)
–	100–1000	30	1974	Illinois	Cattail	Lenington (1980)
–	350–1400	37	1975–76	New Jersey	Cattail	Lenington (1980)
8548–9923[2]	1188–18450	95	1976–80	British Columbia	Cattail	Picman (1987)
810 ± 750	–	68	1980–82	California	Cattail	Ritschel (1985)
384	123–810	6	1977	Alberta	Cattail/bulrush	Dickinson and Lein (1987)
152 ± 90	10–770	495	1978–86	Washington, Columbia NWR	Cattail	L.D. Beletsky and G. H. Orians, unpublished data

[1]± SD, when provided by authors.
[2]Range of means, depending on month or year.
[3]Some wetland areas included.

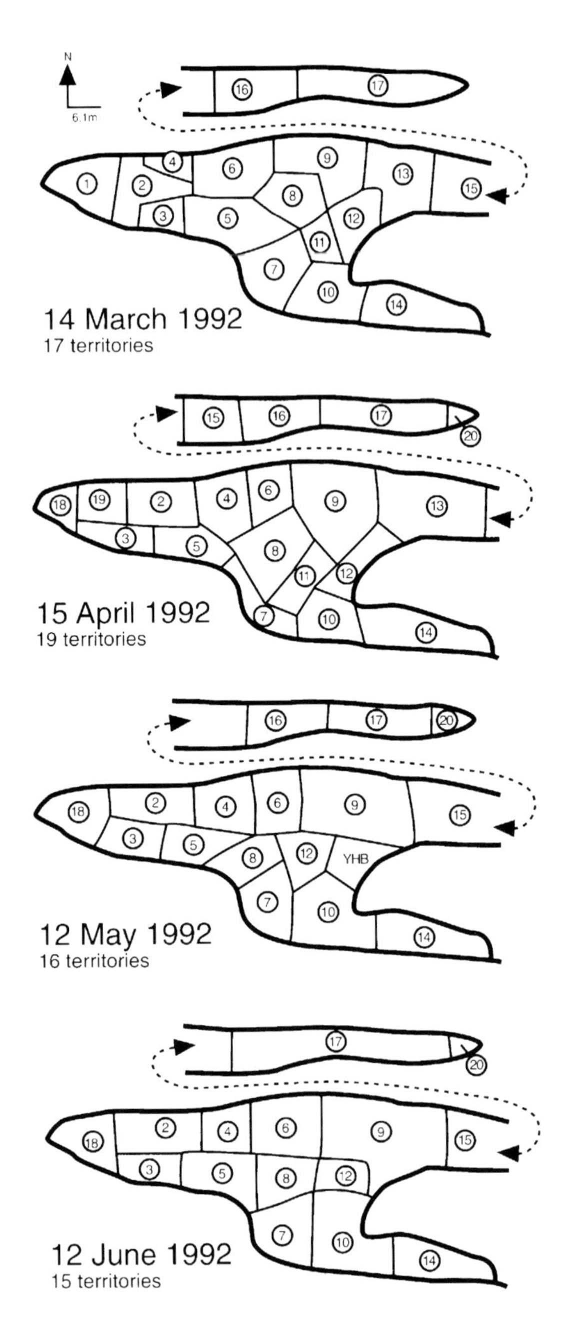

Fig. 8.3. Within-season changes in territory sizes, shapes, and numbers. Shown are mid-monthly territory maps of Columbia National Wildlife Refuge's Juvenile Pocket marsh during the 1992 breeding season. YHB = Yellow-headed Blackbird territory.

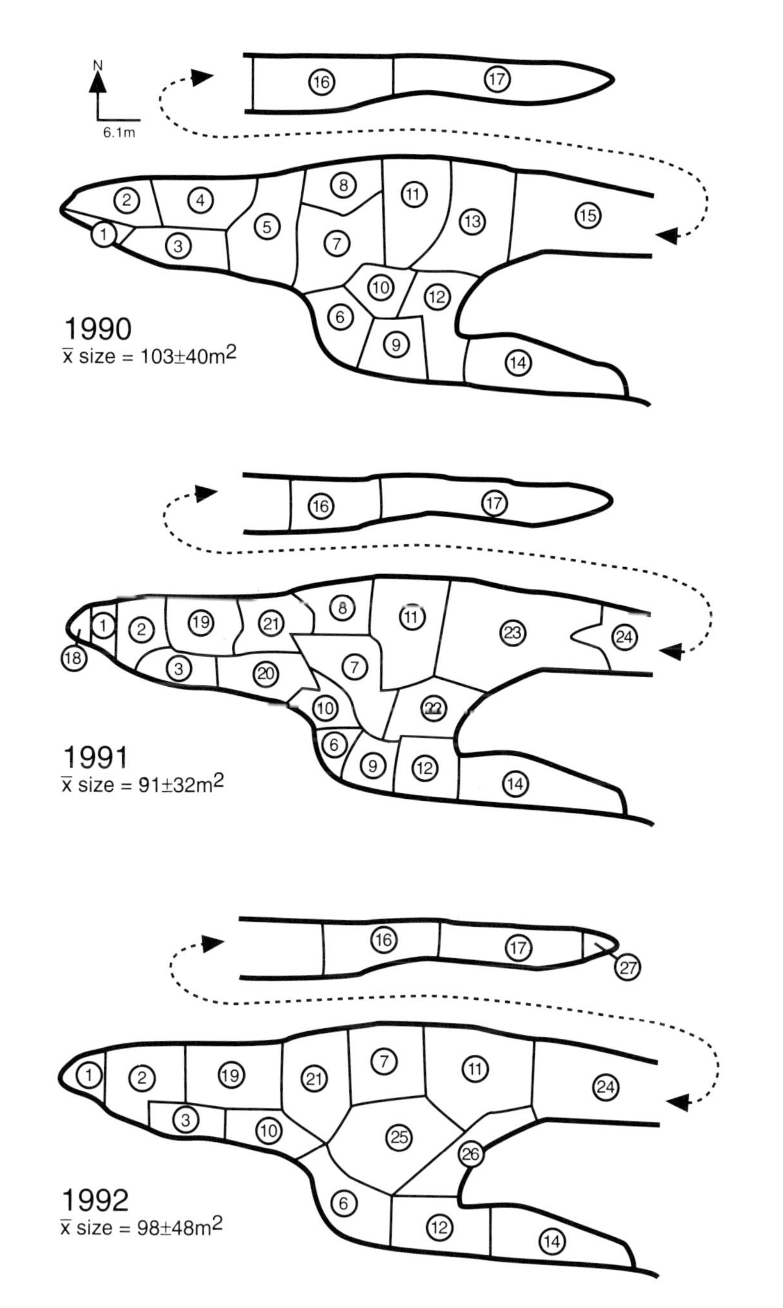

Fig. 8.4. Interyear changes in territory sizes, shapes, and numbers. Shown are territory maps of Columbia National Wildlife Refuge's Juvenile Pocket marsh at mid-breeding season of 1990, 1991, and 1992. Individual males retain the same number from year to year.

if larger territories provide more choices of nest sites and more food, territory size and harem size should be positively associated; but again, factors such as type of habitat involved and the proportion of on-territory feeding affect the relationship. Thus, there is no significant association between territory size and harem size in upland habitats in the Midwest (Blakley 1976, Dickinson and Lein 1987) or in marshes in Ontario (Weatherhead and Robertson 1977a). However, in the highly productive marshes of the West, territory size and harem size covary significantly, although not always positively. Haigh (1968), who studied breeding for two years in Washington, found a negative association: marshes that had smaller average territory sizes had larger average harem sizes. Our long-term information from CNWR shows that males with larger territories tended to have larger harems, the two being positively correlated (n = 849 male territory-years from 1977 to 1992, Spearman $r = 0.10$, $P = 0.003$), but annual male fledging success was not significantly correlated with territory size (Spearman $r = 0.06$, $P = 0.11$). A positive correlation between territory and harem sizes was also noted in a British Columbia population (Picman 1980b).

Interspecific defense

Redwing males defend their territories not only from competing conspecifics (individuals of the same species), but also from a few other species. Interspecific aggression can occur when two ecologically similar species overlap geographically and compete for the same resources. At CNWR and in other areas of the North-American West, Red-winged and Yellow-headed Blackbirds compete directly for the same marsh areas for breeding territories. Redwings have usually already occupied their territories in spring when the migratory male yellowheads arrive on the breeding grounds in early April.

The aggressive interactions between the two species are frequent, although usually not as intense as those between conspecifics (Orians and Willson 1964). Male redwings attempt to evict the settling yellowheads with their usual aggressive bill-up and song spread displays, while male yellowheads counter with their own distinctive vocal and visual displays (Orians and Christman 1968, Orians 1985). Outcomes of these disputes are rarely in doubt because yellowheads are so much larger and heavier than redwings – the average CNWR yellowhead male weighs 110 g, the average redwing male, 75 g. Still, yellowheads, even though they often arrive and settle in large numbers, do not occupy all marsh breeding habitat, nor do they exclude all male redwings. The usual outcome is that instead of redwings occupying an entire marsh as they did prior to the yellowheads' arrival, redwings are evicted or displaced over parts of the marshes. For example, Fig. 8.5 shows the eviction and displacement of redwings to the marsh's periphery following the arrival of yellowheads on Morgan Pocket marsh in 1988.

A male redwing, with a bill-up display, defends his territory from a male Yellow-headed Blackbird. The two species are interspecifically territorial. Yellowheads, larger and heavier than redwings, usually prevail.

The reasons that yellowheads do not establish territories in all marshes at CNWR and do not evict redwings from many areas are not entirely clear, but the territory requirements for the two species, although similar, are not identical. As stated previously, CNWR redwings prefer to establish territories among cattails, but yellowheads appear to prefer bulrushes. When yellowheads displace redwings on marshes that abut dry land, the yellowheads take only the deep-water areas, pushing redwings toward shore (Linford 1935, Orians and Willson 1964; Fig. 8.5). Moreover, yellowheads restrict their territorial sites to more productive lakes, and make little effort to occupy less-productive ones (Orians 1980). Finally, Orians noted an association in both British Columbia and Washington between the probability that a marsh was colonized by yellowheads and the height of surrounding structures. If the trees or cliffs surrounding a marsh created an angle of greater than 30° between the top of the marsh vegetation and the tree- or cliff-tops, yellowheads rejected the site (Orians 1980).

We can only speculate on the reasons for the interspecific difference in preferences, especially as the breeding biologies of the two species are so similar; yellowheads are also strongly polygynous, eat many of the same foods and feed insects to their young, and their nests are subject to many of the same predators. Choosing more productive marshes for their breeding territories is beneficial because their utilization means that yellowheads usually have adequate and close

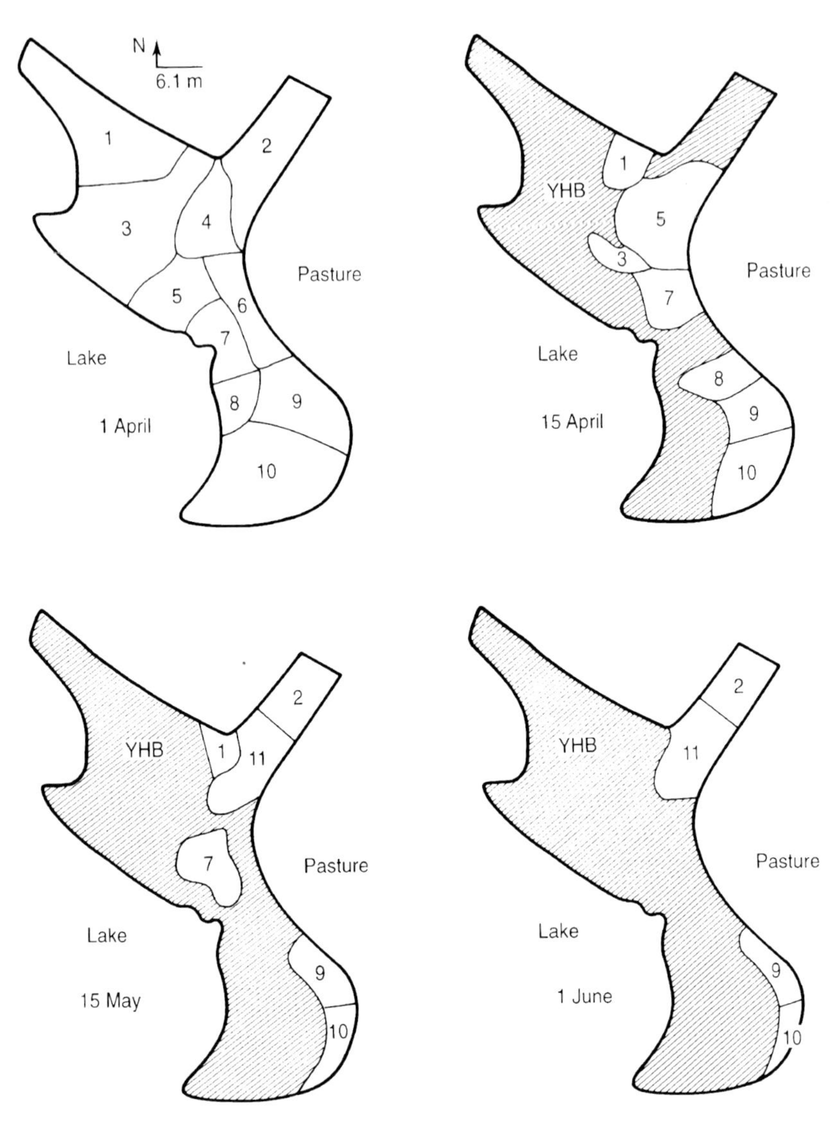

Fig. 8.5. Consequences of interspecific territoriality. Shown are the shapes, numbers, and positions of redwing territories on Columbia National Wildlife Refuge Morgan Pocket marsh at four times during the 1988 breeding season, as male Yellow-headed Blackbirds settled (shortly after 1 April) and commandeered most of the marsh for their own territories.

food supplies for their young; it is also logical, given that their size permits them to take the best sites from redwings. A preference for deep-water sites and those removed from shore is probably related to the enhanced safety from land-based nest predators afforded by these locations. The yellowheads' usual rejection of territories below cliffs or near trees might be related to the reduced visibility from the marshes caused by such obstructions, or perhaps it is just that yellowheads do not forage or move well in trees (and do not much discriminate between trees and other tall structures such as the cliffs surrounding many of the CNWR marshes). None of these speculations, however, quite explain why some adult male yellowheads become floaters rather than establishing territories in habitats that are good enough for redwings to breed quite successfully (Orians 1980).

Of the two species, the redwing is the more catholic in the types of sites acceptable for breeding territories. The redwing's success as the predominant marsh-breeding passerine over a wide range is probably partly due to that flexibility.

Redwing response to the annual yellowhead invasion is variable. The usual result of the interaction, again, is that male yellowheads evict male redwings, and territories of the two end up being mutually exclusive (Orians and Willson 1964). However, some redwings, instead of continually resisting until they are defeated and chased away, take a less aggressive stance, and persist on their territories by reducing interactions with the yellowheads, often by spending a good deal of their time adjacent to rather than on the territory. In this manner, territories of the two species sometimes overlap. Often, the redwings in these situations have nests underway (perhaps the reason they opt to remain in the vicinity), and striking variation in the males' motivation to defend nests and mates manifests. Although male yellowheads are not known to destroy redwing nests, they frequently chase and attack female redwings that are trying to tend nests. Sometimes the 110-g male yellowheads perch adjacent to 45-g redwing females as they incubate or brood young and peck their heads, but the stubborn females sometimes refuse to abandon their nests. We have caught female redwings with bloody heads, sometimes with skulls exposed from these attacks, and we suspect that some have died of their injuries. Some male redwings, even if by this time they are staying mostly at the periphery of the territory, will dive at and attack the male yellowhead each time the female is molested; others watch the interactions from a distance and do nothing to intercede.

Intriguingly, CNWR redwings appear to be aware that certain marsh areas are preferred yellowhead habitat and refrain from establishing territories there even before yellowheads arrive. Morgan Lake is a case in point. Morgan Pocket, on the lake's northeast corner, is a mixed redwing and yellowhead marsh (Fig. 8.5) and was one of our core marshes for monitoring breeding success. The fringe of bulrush and mixed cattail/bulrush marshes along the north and west sides of the lake are usually colonized only by yellowheads, some years at very high densities. During March, before yellowheads arrive in the study area, few if any redwing territories are established in these marshes. Occasionally, the marshes are used at this time of year for roosting and daytime loafing by groups of redwing floaters. The floaters act as if they know that territory ownership there would be short-lived and unproductive.

More curious is the yellowhead's frequent absence from Juvenile Pocket, CNWR's premier redwing marsh during the latter part of our long-term study. Redwings bred exceptionally well on Juvenile Pocket – territories were small, male turnover low, harems among the largest recorded anywhere, and annual breeding success per male and per female greater than the population means. Yellowheads apparently considered it unsuitable. They established territories there only occasionally, and when they did, redwings were usually not evicted. Typically five or more male yellowheads settled there, but after several days, only two or three remained on territories restricted to the area adjacent to the open water of Juvenile Lake. These males were able to attract females, who usually placed nests on the very outer fringe of emergent vegetation, abutting open water. Although redwings bred exceptionally well there, yellowheads may have considered the area to be suboptimal habitat because the vegetation is mostly cattail, which is not the yellowhead's first choice. Also, there are steep cliffs bordering the north side of the marsh, which apparently repel yellowheads. Finally, the dense concentration of redwings in Juvenile Pocket may have allowed them some success in defending their 'turf' from their larger cousins.

Redwings are also interspecifically territorial with the physically very similar Tricolored Blackbird, another closely-related marsh breeder. The two species interact throughout the tricolored's narrow range in central California and Oregon. Tricoloreds are dense colonial breeders with very small territories, large enough only for nest placement. Thousands of tricoloreds may settle simultaneously on a marsh that contains the territories of only several redwings. The previous residents try for a few days to defend their holdings, but their attacks are ineffective against the sheer numbers of invaders. Redwings either end up leaving or restricting their territories to the peripheries of a tricolored colony (Orians 1985). Redwing nests already underway when the tricoloreds arrive are abandoned. As with 'yellowhead marshes', redwings also appear to avoid placing territories in some areas that are regularly settled by tricoloreds (Orians and Collier 1963). Because redwings and tricoloreds look so much alike and diverged recently enough to retain some of the same behaviors, their interspecific aggressive displays are similar; for example, both species use bill-up displays (Orians and Christman 1968).

Redwings in certain areas are also strongly aggressive toward another marsh-breeding passerine, the Marsh Wren. Whereas yellowheads and tricoloreds compete with redwings for food and space, Marsh Wrens also may be chased because they are predators on the redwings' nests. Where the territories of the species overlap, the wrens often attack redwing nests, puncturing eggs (Orians and Willson 1964). In marshes in British Columbia and California, the wrens were the major predator on redwing nests (Picman 1980a, Ritschel 1985). In an experiment in which artificial redwing nests were placed in areas where redwings and wrens overlapped their territories, the wrens accounted for up to 88% of redwing nest destruction (Ritschel 1985). The wrens in California both punctured redwing eggs and ate the contents. British Columbia wrens were also responsible for death by pecking of redwing nestlings, particularly smaller ones (Picman 1980a), although this was not the case in the California study (Ritschel 1985). The breeding ranges of redwings and Marsh Wrens overlap to a large degree but the impact of the wrens on redwing breeding and the interactions between the two vary considerably. For example, from 1983 to 1992 at CNWR, Marsh Wrens bred in some of the core study marshes but their impact on redwing nesting was negligible.

Fidelity and movements

Between-year fidelity to territories is the rule among many passerine birds, including redwings. The main advantages of having a territory in the same place each year must be that long-term familiarity with the territory, its surroundings, and with neighboring conspecifics, lead to increased individual survival and enhanced reproductive success. Site familiarity means that males know the locations of food supplies, both on- and off-territory, the kinds and habits of local predators, and good hiding places. Familiarity with neighboring males reduces aggression during annual territory re-establishment and over boundaries. It may also lead to improved reproductive success because long-term neighbors are quicker or more efficient mobbers of nest predators (Beletsky and Orians 1989c). Also, territory fidelity allows males to avoid the time and energy costs associated with seeking and establishing new territories and also the risks involved with exploring unfamiliar areas. Males abandoning territories to search for new ones also forfeit site dominance on one territory before gaining equal power elsewhere. Site dominance is the almost automatic dominance over all male conspecifics that territory owners usually have while on their territories (see Chapter 10).

Given the benefits of territory fidelity, it is not surprising that redwings should practice it. Territory boundaries shift within and between years, so territories are rarely in *exactly* the same place from one year to the next; but usually there is a good amount of overlap (e.g. Fig. 8.4). At Turnbull NWR in Washington, 34 out of 39 males (87%) that returned to breed between years occupied about the same sites (Searcy 1979e). In British Columbia, the rate of fidelity was even higher at 94% (32 out of 34 returning males; Picman 1987). During an 8-year period at CNWR in which we mapped territories for more than 100 males that had, in total, 185 between-year opportunities to either show territory fidelity or to move, males returned to their same territories 87.6% of the time (Beletsky and Orians 1987a).

Although between-year site fidelity is the rule for breeding redwings, it is not absolute. Each year a small percentage of males move. The movers may be males trying to improve the quality of their territory. Lower set-up costs and higher survival associated with territory fidelity are of little value to males if their reproductive success is consistently poor. If breeding behavior has been shaped by natural selection to increase lifetime reproductive success, then males should have the ability to change their territory locations if initial choices prove to be poor producers of offspring. Because of the risks of forfeiting a present holding, males should attempt to move to new territories only under certain conditions. We made several predictions about territory movements that should be supported if they are indeed efforts by males to improve their lot (Beletsky and Orians 1987a).

We based our predictions on the premise that males should move to new territories in subsequent years only if they can collect comparative fledging success information on their own and on other territories to which they might move. If so, then (1) males should have below-average fledging success during the year before they move, and (2) they should significantly improve their success after moving; furthermore, for the males to decide to move, (3) the territories they moved to in year $x + 1$ should have had, on average, significantly higher fledging success in year x than they did on their own territories. Because males are absent from territories for only brief periods during breeding seasons, they can learn a great deal only about territories that are close to their own. Therefore, voluntary moves, if they are made as informed choices to improve reproductive success, should be over short distances. By implication, long-distance moves, if they occur, should be made in ignorance of past breeding success at the new sites and thus, generally should not lead to improved success.

Using data on the between-year territory movements of 30 males, we found statisitical support for predictions 1 and 2, above, and a supportive trend for prediction 3. The information on the distances males moved was also supportive. Eleven of the 30 males moved to territories directly adjacent to their previous year's site. The average distance moved was only 385 ± 642 m. The eight males that moved more than 200 m (range 500–2800 m) did not, on average, improve their seasonal fledging success after moving, but the others, those moving shorter distances, did (Beletsky and Orians 1987a). We also replicated this analysis using a subsequent 8-year data-set from CNWR, with essentially the same results (Beletsky and Orians 1996).

Thus, a small fraction of males each year apparently desert former territories and move to new but nearby ones, probably those that are vacant owing to the death of the former owners. Do we have evidence for this behavior aside from demographic comparisons? Usually, we do not witness territory shifts, but during experiments in which territorial males are removed from their territories (see Chapter 10), males with current territories on the same or on other, nearby, marshes, often move quickly to the newly-vacant sites. During one removal experiment a male switched from his current territory to a newly-vacant one about

200 m away, in the same marsh. The male did not make the switch at once, but continually flew between the two territories, which were separated by several intervening ones, and defended both from challengers. He contested boundaries and fought several times with a male that held a territory adjacent to the new site. Three hours after first intruding on the vacant territory, perhaps when he was confident that the previous owner of his new territory would not return and when challenges from the neighbor declined, he finalized the transition. This kind of rapid switch strongly supports the idea that males always have ready assessments of the quality of their own and surrounding territories and, given the opportunity, can act quickly on their assessments to improve their breeding success.

MATE-GUARDING AND EXTRA-PAIR COPULATIONS (EPCs)

After establishing first territories or re-establishing former ones in late winter or early spring, male redwings await the arrival and settlement of females. Males guard their mates during the females' fertile periods, that is they follow them and chase away other males to try to prevent EPCs (Beecher and Beecher 1979, Sherman 1989, Carroll 1993). David Westneat (1994) conducted two experiments in New York to investigate the effectiveness of mate-guarding and the conflicts it poses with other activities. In one experiment, he removed several males from their territories for 1 h during the fertilizable period of one of their females, observed behavior of the females and of neighboring males, and also determined paternity of young in the experimental females' nests by DNA fingerprinting. While the owner was absent, neighboring males approached and tried to court his females at a rate 100 times greater than on control, unmanipulated, territories; also, whereas Westneat rarely witnessed naturally occurring EPCs, he witnessed 36 in 29 h during the removal periods. Furthermore, the paternity information revealed that males who were removed when their females were about to lay (so that the resulting EPCs were most likely to be the last copulations the females had prior to fertilization) had very little chance of being the genetic father of their nestlings. In two out of 13 nests analyzed genetically, the removed males lost *all* paternity to their neighbors.

Westneat's first experimental result highlighted a critical conflict for males. Staying on territory to guard their mates was clearly advantageous (females in Westneat's population rarely leave territories), yet males left often, if only for several minutes, to forage elsewhere. In his second experiment, Westneat addressed the conflict between mate guarding and foraging by placing food on some territories so that the owners did not need to leave so often, and then compared their paternity rates with those of males on unmanipulated territories. Again, there were clear results: males with food placed on their territories made fewer off-territory trips and sired a significantly higher percentage of 'their' nestlings than did control males (Westneat 1994). These experiments convincingly demonstrate that males use their time on territory to guard their mates, that the guarding is effective at preventing some EPCs, and that there is a behavioral trade-off between mate-guarding and essential off-territory trips – males would stay to guard if they could. Male and female behavior varies in different regions, so comparative studies of mate-guarding would be interesting. For instance, females at CNWR, in contrast to those in Westneat's population, frequently solicit EPCs (Gray 1994) and regularly leave male territories on foraging trips. Also, with such large harems, males at CNWR should at times need to guard two or more females who have overlapping fertilizable periods.

PARENTAL CARE

Nest defense

Male redwings directly contribute in two ways to their offspring's development and survival; they defend nests and feed young. Females also provide these services, and thus males, who have competing interests, are often able to increase their reproductive success by reducing their parental care, particularly feeding nestlings. During a large part of the period when there are active nests on male territories that need to be defended and 'provisioned', unmated females are still choosing breeding sites. Males therefore can use their time and energy either to attract additional mates or invest in reproductive efforts already underway. Either tactic can increase seasonal breeding success and males are expected to pursue the tactic that, in their particular breeding environment, leads to greater success.

All male redwings defend their nests and do so vigorously and often. As Linford (1935) put it, 'Even if he is very slack in his other parental duties, it can never be said that the male redwing failed to defend his nest and young'. A change comes over territorial males during nesting, an alteration in personality with which researchers who study their breeding are all too familiar. Take the case of Paul and BA-RY. Paul, one of our field assistants, was a pleasant fellow who had never before worked with redwings, and BA-RY (blue-aluminum on his left leg, red-yellow on his right) was a 3-year-old male with a territory on McMannamon Lake. Before nesting began that year, Paul told me how he and BA-RY had become 'friends'. BA-RY would fly to a cliff-top to watch when Paul entered his territory; Paul would converse with the bird and later leave a few sunflower seeds on a rock that BA-RY would approach quickly to eat. But the friendship ended when eggs and nestlings appeared. A surprised Paul related to me that now whenever he set foot in the territory, BA-RY immediately approached and followed his every move,

Two male redwings and a female mob a Black-billed Magpie, a common predator on redwing eggs and nestlings.

landing on nearby cattails, giving alarm calls continuously. When Paul was near nests, BA-RY attacked with dives and even strikes. In fact, it got to the point that Paul had only to approach the general area of the marsh – not yet in the territory – and BA-RY came flying over to hover and swoop above Paul's head. The previous tolerance BA-RY had shown towards Paul's activities had given way to his parental responsibilities.

Nest defense, broadly defined, consists of watching the territory and its surroundings for nest predators, spending time physically near nests to guard them, giving signals to warn young and/or attract other defenders when potential danger is detected, and actively defending eggs and young by attacking predators. The alert calls males give while on their territories, which appear to function as a predator detection and surveillance system, were described in Chapter 5. An indirect indication of the beneficial effect of male vigilance on nest success was demonstrated in Wisconsin because nests that fledged successfully were, on average, closer to high, prominent perches that males used to maintain vigilance than were the nests lost to predators; investigators also found that females preferred to nest near prominent perches (Yasukawa *et al.* 1992a).

When a person nears an active redwing nest, both parents, if they are present, approach and mob the potential predator, as do adults from nearby territories. When researchers check nests for nestlings or handle them, the usual response of the parents is to hover close to and just behind or to the side of the person's head, the female at one ear, the male at the other (like earmuffs, we say). The male is usually the more aggressive, approaching more closely and striking the person more often. Knight and Temple (1988) demonstrated this with more natural predators, by placing models of a crow, a raccoon, or a hawk near active nests and recording male and female reactions. In response to a common nest predator, such as the crow, for example, males hovered over it about 1.5 times more often than did females, and dove at and struck the model seven times more often. This study confirmed previous work by D'Arms (1978), who tethered a live hawk or magpie inside redwing territories with active nests and found that males more often than females physically attacked the predators. Males, being larger and more aggressive, are the primary nest defenders.

Males dive at nest predators from various heights, up to at least 20 m (Searcy and Yasukawa 1995). They usually hit people from behind in the head, neck or back, apparently avoiding the danger to themselves of flying into the person's field of vision; they apparently attack real predators the same way, usually hitting birds in the rump or back of the neck (D'Arms 1978). Males usually come out of their dives and twist in the air to strike at people or predators with their feet, not bills (D'Arms 1978; personal observations). These strikes certainly get a person's attention, so there is little doubt that they would have some deterrent effect on the many kinds of nest predators that are much smaller and more vulnerable to injury from a 75-g bird than are humans. D'Arms noted that redwings diving and striking her tethered hawk or magpie sometimes surprised the predator and knocked it from its perch. Male redwings have been observed to attack many different animals with dives and strikes: crows, jays, magpies, herons, hawks, snakes, raccoons, minks, dogs, horses (Searcy and Yasukawa 1995). Several times at our CNWR site I watched as a frantic male redwing hovered over and repeatedly struck the enormous head of an unconcerned cow that was slowly munching its way through a clump of tall grass into which an unfortunate female redwing happened to have woven her nest.

Although all defend nests, males are not all equally effective. Individuals vary in the intensity with which they protect nests, and that variation apparently influences the probability of nesting success. Several investigators have scored territorial males for their aggressiveness toward potential nest predators. Searcy (1979e) monitored male responses in Washington when he entered their territories to check nests, and scored them from 0 to 7 (0 = male not seen; 1 = male present but no reaction; 2 = male gives low level alarm calls, stays perched; 3 = high level alarm calls, stays perched; 4 = circles over observer's head, greater than 3 m away; 5 = circles, less than 3 m away; 6 = attacks observer once or twice with dives or strikes; 7 = attacks repeatedly). The mean score for 43 males each tested several times was 3.3, with a range of 2.1–6.4. Male responses were quite variable to Knight and Temple's (1988) model presentations using a crow, a raccoon, and a hawk, and males were consistent in the responses among predators, i.e. if a male was timid toward the crow, he was also likely to be timid toward the hawk and raccoon. Male aggressiveness toward humans ranged from 2.0 to 5.1 on a scale of 0 to 7 in Ontario (Weatherhead 1990a).

Is this variation important? Nests in one study that eventually successfully fledged offspring had been, on average, defended significantly more vigorously than nests that eventually were lost to predators (Weatherhead 1990a). In another study, although the number of alarm calls given near an experimental nest by mobbing parents were significantly positively correlated with the eventual success of the nest, the total numbers of dives and strikes by the parents were not (Knight and Temple 1988). The effectiveness of male mobbing at deterring nest predation must vary with the kinds of predators involved and numbers of mobbers. Some predators, such as a raccoon or a large snake, obviously cannot be deterred by redwings regardless of their aggressiveness. A magpie, twice the size of a male redwing, intent on raiding a nest, probably will eventually be successful, unless a lucky blow causes serious injury. D'Arms (1978) tethered her live hawk or magpie in redwing breeding marshes for 15 min on about 50 separate occasions, and although they were attacked repeatedly, they never sustained injury.

Another aspect of male nest defense that has been considered in several studies, particularly because of its implications for female mate choice, is whether such behavior is 'shareable' among the females in a male's harem. Must females compete for this potentially valuable male parental care contribution, or do males defend all nests equally? If, for example, males defended only primary nests and the probability of nest success was enhanced by male defense efforts, then females should compete for primary status to secure this male resource. The two experimental studies that bear on this question yielded opposite results. Males in one study were significantly more aggressive when predator models were placed near nests of their primary females than near those of secondary females (Knight and Temple 1988); in the other, there was no apparent difference in defense intensity scores of males defending primary and secondary nests (Weatherhead 1990a). Therefore, this question remains an open one.

Feeding young

Feeding young is in some ways a more interesting sector of redwing paternal care because it varies so broadly. It is an exceedingly complex, finely-adjustable, facultative behavior. There are regional trends in the proportion of breeding males that participate in feeding young in the nest, variation within populations and, among those males that do feed, an array of possible adjustments according to nest rank and the number, size, age, and hunger level of the young. Greater proportions of males in all populations apparently feed their young after they fledge (Nero 1956a, Orians 1961, Wiens 1965, Beletsky and Orians 1990), but there is little detailed information because of the difficulty of monitoring the behavior of the mobile young, which are fed both on- and off-territory.

Research interest in male redwing feeding is strong. This stems primarily from an interest in a crucial behavioral trade-off in a polygynous system that provides males two stark choices for increasing annual reproductive success. They can either help mates care for current young by feeding them or they can advertise for additional mates (Trivers 1972). They cannot do both simultaneously, probably because high circulating testosterone levels that facilitate advertising and aggression are physiologically incompatible with feeding offspring (Wingfield *et al.* 1990). David Gori demonstrated this trade-off empirically with Yellow-headed Blackbirds, which are marsh-nesters with a breeding system similar to that of the redwing. He artificially elevated the circulating testosterone levels of territorial males (which usually fall dramatically when they begin feeding offspring) and instead of feeding at their first nests as they usually do, they increased their advertising rates and fed their young very infrequently. As a result, hormone-manipulated males fledged fewer offspring than controls (Beletsky *et al.* 1995).

All male redwings probably *know* how to feed nestlings but in some populations few do so. West of the Rocky Mountains, few males feed nestlings, whereas in the East and Midwest a much higher but variable proportion of males does so (Table 8.3). Productivity of breeding habitat, harem size, and breeding season length may all contribute to the pattern (Orians 1985, Muldal *et al.* 1986). In particular, the highly productive marshes of western North America permit females alone to raise their broods successfully, freeing males to exercise their options of prolonging advertisement and aggressively guarding their fertilizable females far into the breeding season.

Some natural 'experiments' reveal that males who do not usually feed young will do so under certain conditions, and vice versa. At CNWR, few males each year feed nestlings (Table 8.3), but in 1981, the year after the eruption of Mount Saint Helens covered the study area with 3–4 cm of volcanic ash, most males fed nestlings (Orians 1985). Survival of young in 1980 was probably very low. As a result, few yearling females, which usually arrive late, were present during the 1981 breeding season for males to attempt to attract, making feeding young the more attractive option that year. Also, with the ash-laden marshes probably suffering a temporary reduction in productivity, male feeding assistance in 1981 might have been essential for offspring survival (Beletsky and Orians 1990). Strehl and White (1986) observed that no males fed nestlings during a periodic cicada emergence year in Illinois, when food was super-abundant, but that at least a small proportion did so the next year.

Males feed nestlings only after they are 3 or 4 days old (Whittingham 1989,

Table 8.3. Percentage of male redwings that fed nestlings in various areas. Study sites were marshes unless otherwise noted

Location	Percentage of males feeding	Total number of males observed	Years	Reference
Utah	None	4	1934	Linford (1935)
Wisconsin	Rare	17–25/year	1948–53	Nero (1956a)
Central California	Rare[1]	>50	1957–60	Orians (1961)
New York, Ithaca[2]	Rare[1]	approx. 100/year	1960–61	Case and Hewitt (1963)
Central California	None	?	1961–65	Payne (1965)
Washington, Turnbull NWR	Rare	approx. 50/year	1965–67	Haigh (1968)
Wisconsin	None	9	1965–66	Snelling (1968)
Indiana	59	34	1975–77	Patterson (1979)
Iowa[3]	None	10	1973	Blakley (1976)
Washington, Turnbull NWR	Rare	approx. 15/year	1974–76	Searcy (1979e)
British Columbia	Rare	approx. 20/year	1976–79	J. Picman, personal communication
New York, Millbrook[4]	45	33	1981	Yasukawa and Searcy (1982)
Southern Coastal California	Rare	approx. 80/year	1980–82	Ritschel (1985)
Ontario	80	50	1983–84	Muldal *et al.* (1986)
Ontario	48	23	1984–85	Eckert and Weatherhead (1987a)
Washington, Columbia NWR	5	74	1983	Beletsky and Orians (1990)
	9	75	1984	
	5	75	1985	
	1	84	1986	
	8	77	1987	
	8	87	1988	
Wisconsin	71	139	1984–91	Yasukawa *et al.* (1993)

[1]One to three males reported to have fed nestlings.
[2]Each year, half the territories were located in uplands, half in marshes.
[3]Upland habitat.
[4]Marsh and old field habitats.

Beletsky and Orians 1990, Patterson 1991) and almost always feed only a single brood at a time, even when their territories contain two or more sets of nestlings. They will switch their feeding from a first nest to a second nest if the first is lost to predators (Muldal *et al.* 1986, Yasukawa *et al.* 1993). Generally, a male feeds at the most advanced nest on his territory. In some populations, male feeding regularly occurs at primary nests (Patterson 1991, Yasukawa and Searcy 1982) but in other areas there is broad variation in which nests receive assistance. In Ontario, 62% of primary, 74% of secondary, and 36% of tertiary nests received some male feeding (Muldal *et al.* 1986); in Wisconsin, 60% of primary, 75% of secondary, and 67% of later nests received male help (Whittingham 1989). Another Wisconsin study reported that 65% of primary and 26% of secondary nests were male-fed (Yasukawa *et al.* 1993) and during a Pennsylvania study, 47% of primary, 44% of secondary, and 43% of later nests received male help (Searcy and Yasukawa 1995). At CNWR, males were observed to feed at nests of all ranks (Beletsky and Orians 1990).

The flexibility of male feeding and how they decide to apportion it is also shown by several experiments. One investigator in Michigan enlarged redwing broods with the result that males that had not been feeding nestlings started to do so; when the brood size of feeding males was reduced, they stopped (Whittingham 1989). In Wisconsin, broods were exchanged between pairs of nests within territories; males (and females) responded by immediately adjusting their rates of feeding to the increased or decreased demands, based on nestling age and number (Yasukawa *et al.* 1993). Males also switched within 1 day to the original brood that they had been feeding, which was, on average, about 3 days older than the other. Finally, male feeding in Alberta was manipulated by either depriving nestlings of food, which increased their begging, or feeding them until they were sated (Whittingham and Robertson 1993). Males that previously had not been feeding young began feeding food-deprived nestlings unless they had a fertile mate to guard, and males that were feeding stopped when nestlings were sated and begging stopped.

These observational and experimental results show that: (1) males closely monitor the number, ages, and hunger levels of nestlings, and adjust their feeding accordingly; (2) males feed only when needed, i.e. with older nestlings and at larger-brood nests, and when females cannot do the job themselves; (3) male feeding is rare in the West, where high productivity marshes allow females to feed their broods adequately and, because of long breeding seasons, males benefit more by directing their reproductive energies to attracting additional mates and guarding their current ones.

Male feeding also attracts research interest from the viewpoint of female behavior because, if male feeding assistance enhances female fledging success and if it is apportioned unequally among females within harems, females should compete for it (see Chapter 7). Also, if males vary in their degree of competence in nestling feeding, and those traits are assessable prior to breeding, then females could select mates based on male feeding criteria. Several studies have confirmed that nests at which males assist their mates in feeding young have, on average, higher fledging success than nests at which only females feed (Muldal *et al.* 1986, Yasukawa *et al.* 1990, Beletsky and Orians 1990, Patterson 1991); this is because females do not lessen their own feeding rate when males assist (Muldal *et al.* 1986, Yasukawa *et al.* 1990, Patterson 1991). Thus, females clearly benefit from male feeding.

Is it possible that male feeding is the resource responsible for female–female aggression in this species, i.e. are primary females aggressive toward others trying to settle on their territories because they are defending their rank and thus the male's feeding assistance? Although on a day-to-day basis, male feeding, which is demonstrably beneficial, is not shareable or divisible among females, females do not appear to compete for it. Females are aggressive toward each other throughout the redwing's range, even in the West, where paternal care is much reduced. Although some studies have indeed indicated that males regularly feed primary nests, others also found males feeding later nests; in fact, in at least one study, males preferred to feed at their second rather than first nests (Muldal *et al.* 1986).

In summary, we have seen that for male redwings to achieve reproductive success, they must first establish and aggressively defend territories from competitors. Several major decisions are involved in territory acquisition and maintenance: when, where, and how to obtain a first territory (Chapter 10), how large an area to try to defend, how frequently to advertise the territory, how often to leave and how long to remain away, which species to defend against, how aggressively to defend the territory, and whether to stay between years or move. Once territories are secured, males advertise to attract mates, copulate with them as well as with females on adjacent territories, attempt to guard their own mates from neighboring males, defend nests and, in some cases, feed young. The breeding roles of the sexes overlap, except for copulation, in only two areas – defense of nests and young and, to a degree, feeding young. In the next chapter we combine male and female efforts and examine the result – annual and lifetime reproductive success.

Chapter 9

REPRODUCTIVE SUCCESS

INTRODUCTION

If an animal's life can be considered to have a purpose, it is reproduction. Most biologists more or less agree that, in general, an individual's behavior is the consequence of evolution by natural selection, as a result of the fact that some individuals in the past survived and reproduced better than others. The genetic traits of these more successful individuals have come to predominate in present populations. Further, an organism's essential being – its morphology, physiology, and behavior – is usually considered to be geared directly or indirectly to promote its own (and its kin's) reproductive success. Respiratory, nervous, and skeletal systems, plumage color and molt patterns, foraging patterns, social behavior, breeding schedules, etc., have all been shaped during evolution because some versions of these traits permitted their bearers to survive to reproductive age, become breeders, and excel in the race to produce offspring. Thus, we have an inescapably logical argument for life: differential reproduction in the past having produced the current 'crop' of highly-adapted individuals to compete in a reproductive race to leave more descendants to breed in the future.

Therefore, a chapter on reproductive success (RS) is a key one in any book that provides a general account of a species. Chapters on other aspects of biology – morphology and development and communication and breeding behavior – deservedly draw interest of their own but always their underlying agenda, stated or not, must be to prepare the reader to consider breeding success. That is, among the described and assessed attributes of the species that deliver a portion of a population each year to the threshold of breeding, which variations determine which individuals are more successful than others in producing young? Factors influencing variation in RS are particularly interesting because they are the forces to which a population is at that moment responding evolutionarily, i.e. the very forces that are shaping the species' future.

This chapter has two main objectives. One is to describe redwing RS from individual nests and cumulatively, over individuals' breeding seasons and lifetimes. The second is to examine environmental and other factors that influence individual variation in RS. Much of my own research has centered on assessing the constituents of lifetime RS (LRS) in redwings and therefore I include here a liberal amount of information from my work on the Columbia National Wildlife Refuge (CNWR) population. One caveat is that although ornithologists with an evolutionary view would like ideally to measure RS by the number of young that individuals succeed in rearing as future breeders, in practice that is rarely possible. For redwings and many other birds, fledging success must be used as an imperfect measure of RS because, owing to extensive dispersal after birth, most individuals disappear and are never seen again. Even marked juveniles can rarely be tracked to their breeding sites.

WHAT FACTORS DETERMINE NEST SUCCESS OR FAILURE?

By early spring on a redwing marsh, males have established and are vigorously defending their territories. They advertise almost incessantly for mates, secure a few, and copulation occurs. Females arrive on the marsh and choose male territories, build nests, mate, and produce clutches of jealously guarded eggs. What is the destiny of these nests? A half-century of monitoring tells a striking story of doomed efforts. During 27 studies of natural nesting conducted in 16 US states and Canadian provinces, only about 40% of more than 14 000 marsh-situated nests succeeded in fledging at least one nestling (Table 9.1). On average, each nest produced only 1.2 fledged young (ranging from 0.4/nest in one study in California, to 3.7/nest in Alaska); nests that were successful produced, on average, 2.7 fledglings. (To this dismal picture must be added fledgling mortality, which is bound to be high during the first few days outside the nest, but there is no way to quantify it.) This seemingly high rate of nest failure, about 60%, is actually not unusual for

Table 9.1. Average fledging success of redwing nests in various locations

Location	Year	Number of nests monitored	Percentage of nests that fledged ≥1	$\overline{X}$ no. of fledges/ nest[1]	$\overline{X}$ no. of fledges/ successful nest[2]	Reference
Marsh habitats						
Indiana	1927–28	24	–	1.96	–	Perkins (1928)
Pennsylvania	1936	37	43.2	0.95	2.19	Wood (1938)
Ohio	1939	67	–	1.57	–	Williams (1940)
Illinois	1941	356	–	1.90	–	Smith (1943)
California	1959–60	–	–	–	2.61	Orians (1961)
New York	1960–61	1112	31.8	0.81	2.54	Case and Hewitt (1963)
Maryland	1959–61	675	57.5	1.50	2.62	Meanley and Webb (1963)
Wisconsin	1959–60	518	29.0	–	–	Young (1963)
Wisconsin	1962–63	138	53.6	1.46	2.72	Wiens (1965)
Oklahoma	1965	243	26.7	0.81	3.02	Goddard and Board (1967)
Wisconsin	1964–66	186	38.7	1.02	2.64	Snelling (1968)
Ohio and Michigan	1964–65	157	28.0	0.82	2.91	Holcomb and Twiest (1968)
Washington	1966–67	307	–	0.79	–	Haigh (1968)
Connecticut	1968–70	738	53.0	1.44	2.71	Robertson (1972)
New Jersey	1973	164	–	1.19	–	Caccamise (1976)
Iowa	1972–74	111	–	1.05	–	Crawford (1977)
California	1976	150	34.0	0.66	1.94	Kundert (1977)
New York	1971	209	29.7	0.73	2.45	Allen (1977)
Ontario	1974–75	381	39.4	1.24	3.15	Weatherhead and Robertson (1977a)
New York	1977	25	56.0	1.40	2.48	Cronmiller and Thompson (1980)
British Columbia	1976–77	399	23.6	0.54	2.29	Picman (1980a)
	1976–82	1206	35.1	–	–	Picman et al. (1988)
California	1980–82	546	26.0	0.43	1.93	Ritschel (1985)
Ontario	1984	99	43.4	1.67	–	Muldal *et al.* (1986)
Illinois	1975–77	384	33.3	1.11	3.34	Strehl and White (1986)

Table 9.1.–*contd.*

Location	Year	Number of nests monitored	Percentage of nests that fledged ≥1	$\overline{X}$ no. of fledges/ nest[1]	$\overline{X}$ no. of fledges/ successful nest[2]	Reference
Alaska	1981–82	71	93.0	3.69	3.97	McGuire (1986)
New York	1988–90	173	35.7	–	–	Westneat (1992b)
Washington	1977–92	6208	33.4	0.86	2.57	Beletsky and Orians (1996)
Total or Grand mean		14684	40.2	1.23	2.67	
Upland habitat						
New York	1960–61	273	23.8	0.62	2.60	Case and Hewitt (1963)
Connecticut	1969–70	162	34.0	0.85	2.49	Robertson (1972)
Ohio	1969–73	211[3]	48.3	1.32	2.74	Francis (1975)
Ohio	1973–74	186	31.2	0.91	2.93	Dolbeer (1976)
California	1976	61	68.9	1.33	1.93	Kundert (1977)
New York	1971	49	53.1	1.35	2.54	Allen (1977)
California	1980–82	92	65.2	1.64	2.61	Ritschel (1985)
Total or Grand mean		1034	46.4	1.15	2.55	

[1]Total number of young fledged/total number of nests monitored.

[2]Total number of young fledged/total number of nests that fledged ≥1

[3]Includes 39 nests built in a marsh.

passerine birds using open-cup nests (Nolan 1978, Hotker 1989). The chief agents responsible for redwing nest failures and partial losses are identified and explored below.

Causes of nest failure

Predation

Predation of eggs and young is everywhere the single most important cause of redwing nest failure (Table 9.2). Nests are woven into concealing vegetation and adults are fierce in their defense roles, but nests are open and, compared with the 45–80 g adult redwings, the predators arrayed against them are formidable. Take, for example, our CNWR population. During the 16-year study, annual nest predation rates varied from about 35% to 75% (Beletsky and Orians 1996). The Black-billed Magpie, weighing about 200 g, was in most years the predominant predator

Table 9.2. Causes of failure of redwing nests in various regions

Location	Years	Total number of nests monitored	Percentage of nests terminated by					Reference
			Predation	Starvation	Desertion[1]	Weather[2]	Other	
Marsh habitat								
New York	1960–61	1112	33.9	–[10]	34.3	–	–	Case and Hewitt (1963)
Oklahoma	1965	243	53.0	–	–	–	–	Goddard and Board (1967)
Connecticut	1968–70	738	30.2[3]	–	–	–	–	Robertson (1972)
Costa Rica	1967	118	63.6[4]	–	–	–	–	Orians (1973)
Washington	1966–67	241	51.4	–	–	–	–	Holm (1973)
Iowa	1973	32	28.1[5]	–	12.5	21.9	9.4	Blakley (1976)
Ontario	1974–75	381	44.1	–	16.5	–	–	Weatherhead and Robertson (1977a)
New York	1971	209	39.2[6]	21.5	7.2	2.4	–	Allen (1977)
Washington	1975–76	558	44.0	–	–	–	–	Searcy (1977)
Illinois	1975–77	384	57.0[7]	–	6	–	5	Strehl and White (1986)
Ontario	1984–85	73	41.1	–	–	–	–	Eckert and Weatherhead (1987a)
British Columbia	1976–82	1206	44.2	–	–	4.8	15.9	Picman *et al.* (1988)
Washington	1985–88	1325	59.0[8]	–	–	–	–	Langston *et al.* (1990)
New York	1988–90	173	44.7[9]	0.0	1.7	5.6	4.5	Westneat (1992b)
Total or mean %		6793	45.2					

Upland habitat								
New York	1960–61	273	40.7	–	26.7	8.8	–	Case and Hewitt (1963)
Connecticut	1969–70	162	45.0	–	–	–	–	Robertson (1972)
New York	1971	49	12.2	26.5	4.1	4.1	–	Allen (1977)
Total or mean %		484	32.6					

[1]Some variation in this category is due to the fact that some authors considered nests that were never finished and/or finished nests that never received eggs to be 'deserted,' whereas others only included in this category nests that received at least 1 egg.

[2]Includes natural catastrophes (tide, fire, storm, wind) and, in uplands, field mowing.

[3]Major predator was raccoon.

[4]Minimum estimate.

[5]Major predator was raccoon.

[6]Major predators were marsh wren, raccoon.

[7]Major predators were raccoon, snakes.

[8]'Almost all (nest) failure seemed due to predation' (Langston *et al.* 1990).

[9]Major predators were American crow, raccoon.

[10]A dash signifies that authors of the referenced work did not mention a source of nest failure or that it was reported to be insignificant.

(Fig. 9.1). Several times I approached slowly one of our study marshes that lies beneath a 7-m high cliff and, until detected, was able to watch below me a single foraging magpie move systematically through the cattails, seeking redwing nests. The magpie occasionally directed its attention to the calling, hovering and diving redwing adults closely following its progress, but the mobbers scarcely impeded the predator's intensive search. Magpies at times terminated all active redwing nests on particular breeding marshes, and I was more than once approached by young field assistants in tears reporting that all of the 20 or 30 active nests that they had so carefully checked 3 days ago were now destroyed.

Sometimes a single predatory individual wreaks havoc on both redwing breeders and researchers. Raccoons occur throughout most of the redwing's breeding range and, almost invariably, when specific nest predators are enumerated in the literature, this 5–12 kg masked mammal resides at or near the top of the list (e.g. Table 9.2). Because raccoons are not plentiful at CNWR, their effect on redwing breeding during the study was patchy. However, one year a raccoon discovered that a main study marsh was a particularly good source of nutritious redwing eggs and young and raided it repeatedly. Unfortunately, that year a graduate student had planned to take blood samples there from redwing nestlings to perform a molecular paternity analysis. A spirited race developed between the student, who tried

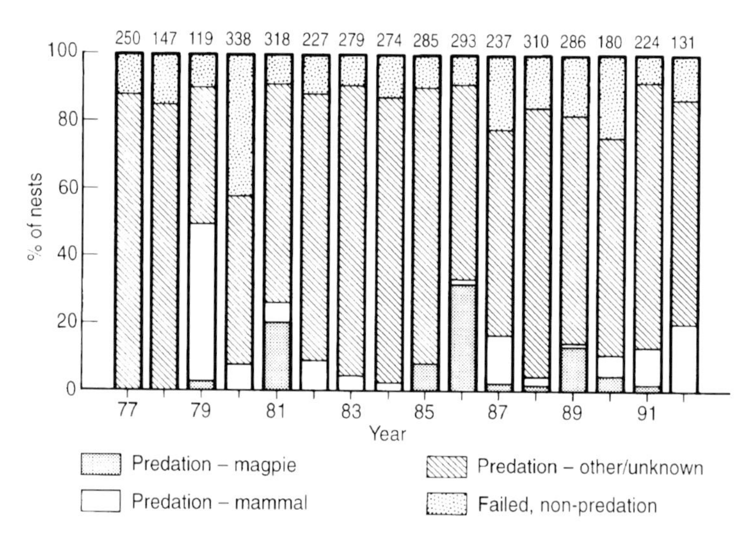

Fig. 9.1. Nest losses to predators of different types at the Columbia National Wildlife Refuge study site, 1977 to 1992. Values above the bars are the number of failed nests each year. Data include only nests that received at least one egg. Black-billed Magpies were credited with nest depredation only when we were sure they were responsible. However, we strongly suspect that magpies were also responsible for most of the losses credited to 'predation – other/unknown'.

to trap the raccoon while also bleeding nestlings at as young an age as possible (before the predator got to them), and the raccoon, which ignored the tastily-baited trap and continued its near-daily raids. The contest ended in a draw: the student obtained just enough blood for her work but the raccoon took most of the nests. CNWR redwing nests are also targets for a variety of snakes, including the venemous Western Rattlesnake, which is abundant there, and Gopher snakes. Many times making our rounds checking nests we met snakes at eye level in the marsh vegetation (not the most pleasant of experiences) and we, as well as many other researchers, have discovered them curled up inside empty nests that during previous checks had held eggs or nestlings. Other, less important, predators there were Marsh Wrens, Mink, Long-tailed Weasels, and Harvest Mice, which refurbish redwing nests by doming them to create shelters. Also, American Coots, for reasons unknown, occasionally were observed to leap up and, with their bills, pull down redwing nests.

Redwing nest predators noted by other researchers include Gray and Fox Squirrels, Groundhogs (Strehl and White 1986), such birds as American Crows, (Westneat 1992b), Ravens (Ritschel 1985), Blue Jays (Smith 1943), and Common Grackles (Bent 1958, Meanley and Webb 1963; but see Wiens 1965, Lenington and Scola 1982), and other snakes including Rat snakes, Cottonmouth, Water snakes, Black snakes and King snakes (Meanley 1971, Robertson 1972, Strehl and White 1986). Automatic cameras set out near artificial redwing nests in both marsh and upland habitats in Ontario captured on film a variety of visitors (Picman *et al.* 1993): Raccoon, Long-tailed Weasel, Red Squirrel, Blue Jay, American Crow, Gray Catbird, Virginia Rail, House Wren, Marsh Wren. In some populations, notably in British Columbia (Picman 1980a, Picman *et al.* 1988, 1993) and California (Ritschel 1985), Marsh Wrens are particularly destructive, but in others there is little if any interaction (Robertson 1972, Beletsky and Orians 1996). Costa Rican redwings, at the southern extreme of the range, in one study had very high nest predation rates by Boa constrictor and Indigo snakes as well as by Rice rats (Orians 1973; Table 9.2).

Redwing attacks on nest predators may regularly deter only birds up to a size only slightly larger than themselves, but the list of predators above, which includes many large birds, snakes, and mammals, and the high predation rates, argue that mobbing often must be ineffective. Mink and Raccoon, for example, continue their nest raids while being mobbed (Knight *et al.* 1985; personal observations). Also, some predators, especially mammals, operate at night, when redwings are asleep.

Redwings may employ other antipredator tactics. In addition to placing their nests in concealing vegetation and, at least sometimes, maintaining a degree of secrecy and stealth near nests, colonial breeding itself may have developed in this species as a response to high predation rates. Nesting in a dense group permits 'massive common defense' by colony members (Brown 1975). Indeed, we have found consistently better reproductive success per male and per female for redwings breeding on our CNWR pocket marshes, which are more densely-settled, with more adults per unit area available to mob (see Chapter 1; Beletsky and Orians 1989c, 1996), than strip marshes. Also, colonies of synchronous or near-synchronous nesters theoretically 'swamp' predators, so that although some nests are lost, the probability of any single nest being taken is lower than if they were highly dispersed in time or space (Robertson 1971). Such an enhancing effect of synchronous breeding was reported in a redwing population in New York. The probability that a female avoided nest predation increased with the number of other nests started on her marsh within 2 days, plus or minus, of her own (Westneat 1992b). Studies of many other species, which are reviewed by Wittenberger and Hunt (1985), also report a beneficial effect of synchronous nesting.

Other causes

Other causes of nest failure include abandonment of nests before or after laying of a clutch, infertile clutches, natural catastrophes in which the nest is physically emptied or destroyed (e.g. by wind, fire, uneven growth of anchoring vegetation, and even, at CNWR in 1980, by volcanic ash), interactions with Brown-headed Cowbirds or Yellow-headed Blackbirds, nestling starvation, and death of the tending adult. Many nests that are built never receive eggs. The proportion of these nests declines during the breeding season (Fig. 9.2). The pattern suggests that a main cause is worsening of weather early in the season after females decide to start nests, which interrupts their efforts and causes abandonment. After eggs are deposited, females probably abandon nests voluntarily only when environmental conditions deteriorate to the point that they must choose between their own survival and tending nests.

At CNWR, aggressive interference by male Yellow-headed Blackbirds caused some females to abandon nests, as did some interactions with Brown-headed Cowbirds, which are brood parasites (Fig. 9.3). Female redwings are cowbird egg acceptors (Rothstein 1975, Friedmann *et al.* 1977, Ortega and Cruz 1988), and they desert nests only rarely after single cowbird eggs are placed in their nests. However, when multiple cowbird eggs are deposited in a single nest (up to five at the CNWR site), the high degree of disturbance apparently causes some females to desert. Infertile eggs occur, but because many females participate in extra-pair copulations (EPCs), entire infertile clutches are rare. In fact, one reason female redwings may copulate with multiple males may be to reduce the chance that they will lay infertile eggs (Gray 1994). Finally, although starvation of entire broods occurs, it is not common (Table 9.2) – as with infertile eggs, starvation is much more often a cause of partial rather than total nest loss. At CNWR, entire broods rarely starved (Fig. 9.3); when they did, the reason was often suspected to be death of the mother.

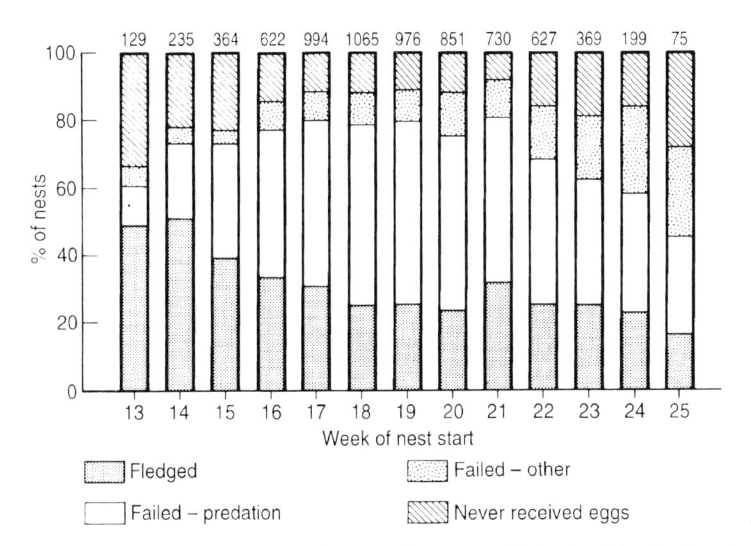

Fig. 9.2. Fates of all nests started during the 1977–92 Columbia National Wildlife Refuge study, by week of nest start. Week 13 is the last week of March; week 25 is the third week of June. Values above the bars are the total numbers of nests started each week.

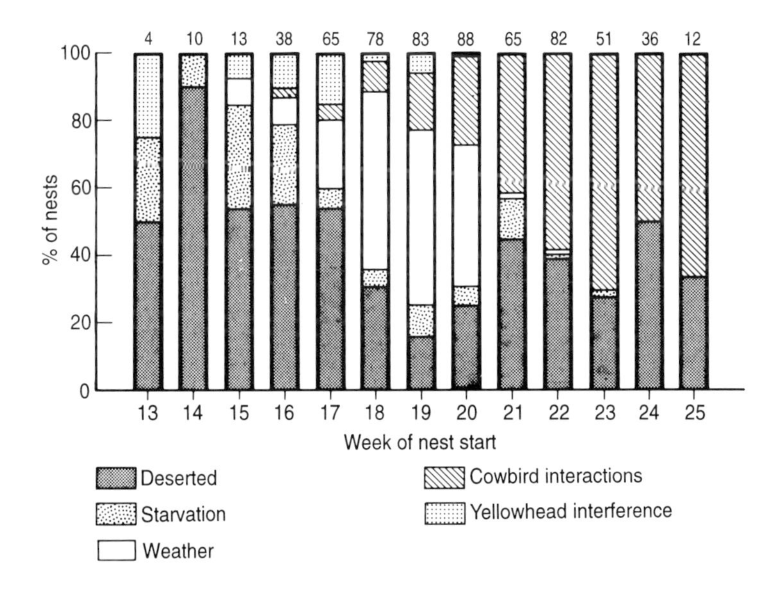

Fig. 9.3. Reasons for nest failures other than predation, by week of nest start, during the 1977–92 Columbia National Wildlife Refuge study. Values above bars are total numbers of nonpredation nest failures. Data include only nests that received at least one egg. Total numbers of nests started each week are given in Fig. 9.2.

Causes of partial loss

Many redwing nests that escape total nest failure fledge less than their full complement of eggs or hatched young. There are a number of causes for this partial reproductive loss, but predation, the leading cause of nest failure, usually is not one of them; most predators on redwing nests devour their entire contents.

Failure to hatch

A small percentage of eggs fail to hatch because they are infertile, because embryos die, or because they are accidentally damaged. This percentage ranged, in various studies: from 1.2% to 12.1% of eggs laid (1.2% of 83 eggs in 24 nests (Perkins 1928); 12.1% of 214 eggs in 67 nests (Williams 1940); 3.2% of 1140 eggs in 356 nests (Smith 1943); 3.0% of 1632 eggs in 797 nests (Young 1963); 7.1% of 2932 eggs in 900 nests (Robertson 1972); 7.6% of 1045 eggs in 324 nests (Caccamise 1978); 6.1% of 362 eggs in 173 nests (Westneat 1992a)). In Westneat's study, of 19 eggs that did not hatch, only four were infertile; the others had dead embryos or were dented. Thus, Westneat estimated an infertility rate of only about 1.1% (four out of 362 eggs). In the CNWR population, 88% of a large sample of unhatched eggs came from clutches with single sires (most clutches have more than one sire), providing evidence that a benefit to females of copulating with multiple males is assuring fertility of clutches (Gray 1994).

Starvation

Whereas starvation of whole redwing broods is infrequent, starvation is a major cause of partial brood reduction. Most redwings lay clutches of three to five eggs (Chapter 6) but, on average, they fledge only about 2.7 young per successful nest (Table 9.1). A small portion of the difference is due to egg loss, but many nestlings succumb, primarily to starvation, and secondarily to disease. The ratios of fledged to hatched young, for different clutch sizes, for the CNWR population are given in Table 9.3. The larger the brood, the more likely a nestling was to starve. Less than 50% of nests that hatched three or four young succeeded in fledging all of them. The overall ratio of fledging success to hatching success (total number fledged/total number hatched) for 3050 nests that fledged one or more young was 57.7%, a ratio consistent with that found in several other investigations: 45.9% in 231 successful nests (Young 1963); 60.9% in 103 successful nests (Wiens 1965); 59.6% in 65 successful nests (Goddard and Board 1967); 50.1% in 121 successful nests (Snelling 1968); and 62.6% in 615 successful nests (Robertson 1972).

Presumably, owing to individual and environmental variation, not all females each year can successfully feed all of their hatched young. If this is so, there should be broad variation among years in fledgling:hatching ratios within the same localities. That was what we found at CNWR; during the 16-year study of nesting

Table 9.3. Partial nest success: proportionate fledging success of nests at Columbia National Wildlife Refuge, 1977–92, that hatched one to six young

Number of young hatched	Number of young fledged						Number of nests
	0	1	2	3	4	5	
1	0.48	0.52	–	–	–	–	330
2	0.38	0.12	0.52	–	–	–	530
3	0.29	0.07	0.16	0.48	–	–	1007
4	0.29	0.04	0.11	0.22	0.34	–	1104
5	0.32	0.03	0.10	0.15	0.28	0.12	78
6	1.00	–	–	–	–	–	1
Total							3050

success, annual fledging:hatching ratios ranged from 41.5% in 1991 to 79.2% in 1979. In a perfect demonstration of the interaction of food availability and starvation of young, significantly fewer redwing nestlings starved during a period of peak cicada emergence in 1976 (one out of 112, 0.9%) than during the periods immediately prior to (10 out of 106, 9.4%) or after peak cicada emergence (19 out of 77, 24.7%; Strehl and White 1986). Fledging success per nest was also better during the period of peak emergence. Ewald and Rohwer (1982) placed feeders on some redwing territories to determine the effects of supplemental food on initiation of seasonal breeding. In contrast with an expected enhancing effect of the extra food on fledging success, they found that nest failure, measured by total or partial loss, was actually higher on experimental than on control territories, which they attributed to higher predation rates. Older, larger nestlings beg more and are more likely to survive periods when food abundance is low (Howe 1976, Clark and Wilson 1981, Teather 1992). Hatching of redwing eggs is partially asynchronous, clutches of three or four hatching over 2 and sometimes 3 days. The result is that last-to-hatch nestlings are smaller, get less food, grow slower, and are most at risk of starvation. This common drop in brood size, called adaptive brood reduction, may be why females begin incubation before the last egg is laid: when food is in short supply, asynchronous hatching leads quickly to a smaller brood needing to be fed. In a marsh in Illinois, among 41 successful nests, only 2.6% of 77 nestlings that hatched during the first day of hatching eventually starved in the nest or 'vanished', but 32.8% of 61 nestlings that hatched during the second or third day of hatching starved or vanished (Strehl 1978).

Brood parasitism

Brood parasitism by Brown-headed Cowbirds on redwing nests is common. However, its incidence rate varies from region to region, and its effect on nest success and, hence, individual RS, is apparently minor. The breeding ranges of redwings and cowbirds coincide over much of the continental USA and Canada. High incidences of cowbird parasitism are usually localized and are greatest in the central USA and Canada; cowbird parasitism on redwings in eastern North America is relatively infrequent (Friedmann *et al.* 1977). Parasitism rates – percent of redwing nests found containing one or more cowbird eggs – range from zero to more than 50%: 0% of 33 nests in Oklahoma (Wiens 1963); 5% of 99 nests in Michigan (Berger 1951); 35% of 382 nests in Manitoba (Weatherhead 1989); 42% of 258 nests in North Dakota (Linz and Bolin 1982); 54% of 59 nests in Nebraska (Hergenrader 1962). Within a locality parasitism rates can vary over small distances and between years. About 52% of redwing nests were parasitized on one breeding marsh in Kansas, but at another site 15 km away there was no parasitism (Facemire 1980). In a Wisconsin redwing population annual nest parasitism rate varied from 2% to 19% over 8 years (Searcy and Yasukawa 1995). During peak periods of parasitism, redwing nests in some areas accommodate an average of 1.4 to 1.5 cowbird eggs (range 1–5; Orians *et al.* 1989, Weatherhead 1989).

Much of the variation in parasitism rates results from differences in local densities of cowbirds and the availability of alternative host nests. The highest nest parasitism rates often occur near livestock feeding areas, where cowbirds thrive (Verner and Ritter 1983, Rothstein *et al.* 1980). The availability of alternative hosts may be especially important for redwings because owing to their superior size, redwings are poor hosts for cowbirds. CNWR redwing nests were frequently parasitized; in fact, the redwing is the most abundant host species for cowbirds in the area. This is probably not because cowbirds prefer to lay in redwing nests, but because there are more active redwing nests there during the cowbirds' laying period than nests of any other species (Orians *et al.* 1989). The first cowbird eggs usually appeared in nests in early May, 4 weeks after redwings started laying; the parasitism rate on new nests progressively increased through the remainder of the season until by mid-June, when the final nests of the season were started, 40% to 50% of them were parasitized (Fig. 9.4).

Because a female cowbird typically ejects a host egg from the nest when she lays her own (Friedmann *et al.* 1977, Røskaft *et al.* 1990) and because the parasitic nestling competes for food with the host nestlings, we might expect severely reduced redwing fledging success from parasitized nests. However, during a 7-year period, the average number of redwing young fledged from 55 parasitized and 654 unparasitized nests differed by only about one – 1.78 and 2.72, respectively (Røskaft *et al.* 1990). Hatching success was slightly lower for redwing eggs in parasitized versus unparasitized nests, perhaps owing to thicker-shelled cowbird eggs

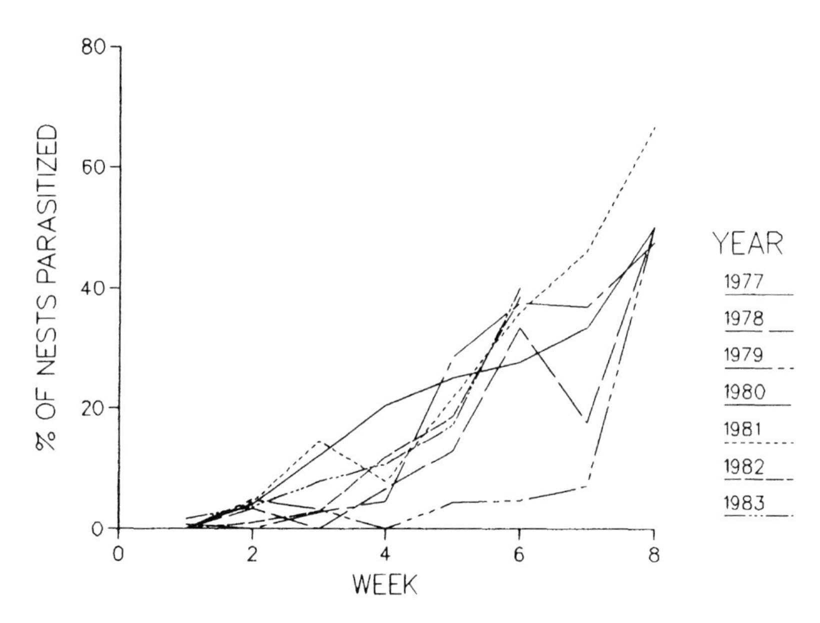

Fig. 9.4. Percentage of redwing nests parasitized by Brown-headed Cowbirds, by week, 1977–83. Week 1 is the fourth week of April; week 8 is the third week of June. (Reproduced with permission from Orians et al. 1989.)

bouncing against redwing eggs and denting them (Røskaft *et al.* 1990), but the main loss to the redwings from being parasitized was the egg ejected by the laying cowbird. Redwing nestlings, being larger than cowbird nestlings, presumably were unaffected by the parasitic nestling's presence in the nest. Also, very few cowbirds fledged from our redwing nests, suggesting that the redwing chicks, being larger than the cowbirds, effectively out-competed them for food brought to the nest. Redwings are therefore poor hosts for cowbirds in eastern Washington.

Our finding that redwing nesting success is only slightly reduced by accepting a cowbird egg and chick provides an explanation of why female redwings do not eject cowbird eggs, as some species have evolved the ability to do. When a female redwing returns to her nest to find a cowbird egg among her own, the main damage to her nest has already been inflicted – one of her eggs has been ejected and the female cowbird might have damaged others while attempting ejection. The cowbird chick will not compete well against the redwing's own chicks and so will not reduce the probability of their fledging, but trying to puncture and eject the cowbird egg (as a female redwing would have to do, because her bill is too small for her to grasp the egg; Rohwer and Spaw 1988, Rohwer *et al.* 1989) could cause damage to her own eggs (her bill ricocheting between eggs, etc.). Thus, her best tactic is to accept the damage that is already done, incubate the cowbird egg, and get on with her nest (Røskaft *et al.* 1990). Why she feeds the cowbird nestling is unresolved.

Brood parasitism by other redwings potentially could influence nest success, but an analysis of our CNWR data suggests that females very rarely lay eggs in other redwing nests. We found only 34 suspected cases in 7805 nests (Harms *et al.* 1991). Three separate studies of parental identity in redwings using DNA fingerprinting failed to find any of 749 nestlings to be the genetic offspring of females other than those attending the nests (Gibbs *et al.* 1990, Westneat 1993, Gray 1994). Thus, 'conspecific nest parasitism' must occur only in extraordinary circumstances, such as when nests are destroyed during a female's laying period and the next egg cannot be aborted or resorbed.

ANNUAL REPRODUCTIVE SUCCESS (RS)

Magnitude and variation

Because of differing reproductive strategies of the sexes that are ultimately tied to the relative sizes of eggs and sperm, the most successful males in a population theoretically should have greater RS than the most successful females (Trivers 1972, Emlen and Oring 1977). In polygynous species, where variation in individual male success is very high, the difference should be even more exaggerated. In fact, annual RS in redwings is characterized by differences and extremes. Male and female breeders sharply differ in average annual RS, in the magnitude of individual variation in RS, and in the major factors promoting variation. Males show extremes of RS that could only typify a nonmonogamous breeder.

There is a problem in assessing RS because although all young in a redwing nest are always the genetic offspring of the female tending that nest, the same cannot be said of the male who defends the territory in which the nest is located. Many offspring are sired during EPCs, but studies of redwing genetic paternity to date indicate that all or most males within breeding colonies participate in EPCs (Gibbs *et al.* 1990, Westneat 1993, Gray 1994). Thus, if I assume for males 100% paternity of their nests, as has been done in the past (under the rationale that what paternity a male loses through his harem's EPCs with neighboring males, he makes up with his own), I will not court a serious error. My assumption is supported by the CNWR population, where our information is that variation in male RS is about the same whether it is estimated by territorial fledging success or determined precisely by genetic analyses (Gray 1994). Of course, the gains and losses of paternity may not always balance (Gibbs *et al.* 1990); EPCs could bias results if some males are consistently much more successful than others in siring offspring in this way (Weatherhead and Boag, 1995).

In the various studies in which it was assessed, average annual RS for females varied from 0.23 to 3.15 fledged young/breeder, and for males, from 0.50 to 17.50 fledged young per breeder (Table 9.4). During our CNWR study, we searched each breeding marsh every 6 days to locate new nests, and we checked the contents of each nest every 3 days. Nestlings fledged usually at 11 or 12 days of age. We considered a particular nest to have fledged if (1) we saw fledglings near the nest at the appropriate time, (2) if the mother was observed feeding fledglings, or (3) if nestlings were known to have survived to eight days old, were absent during the next nest check (3 days later) and there was no physical evidence of predation, such as the nest or nearby vegetation being disturbed or damaged. Under these rules, we determined that annual RS per female in the CNWR population ranged from zero to eight fledged young, whereas for males, the range was zero to 33; average annual RS per female varied among years from 0.96 to 2.79, and for males, from 1.88 to 10.94 (Table 9.5). Thus, the most successful males have better RS than the most successful females. Variation in annual RS among breeders within a population was also considerably greater for males (Tables 9.4 and 9.5).

An array of ecological and individual factors influence variation among individuals in annual RS, and the factors that affect one sex are not necessarily the same as those influencing the other. The key factor driving the magnitude difference in RS between the sexes stems directly from the mating system: males have harems of reproducing mates but females do not.

Sources of variation in annual RS for females

Physical environment

Many aspects of the physical environment influence nesting success. For the most part they are attributes of the environment that protect nests from predators – concealing vegetation, high water levels, etc. How some of these influence nest success was discussed in Chapter 7. Here they are briefly explained, together with relevant information on breeding success.

Habitat type

We do not know why some redwing females nest in marshes whereas others nest in uplands. The choice could be dictated by imprinting on the birth habitat, by genetics, or by females actively choosing one or the other. Most investigators have studied redwing breeding success in marshes, as relatively little information exists for upland sites. The few comparative data available suggest that average annual RS/female is similar for both habitats (Table 9.4). This is interesting because it shows that although harems are generally larger in marshes (see Table 6.6), females distribute themselves such that each has access to approximately

Table 9.4. Annual fledging success among years and among individuals for male and female redwings

		Females			Males			
Location	Year	Mean	Variance[1]	n[2]	Mean	Variance	n	Reference
Marsh habitats								
New York	1960	1.62–3.15[3]	–	13–27[7]	3.50–6.00	–	6–14	Case and Hewitt (1963)
	1961	0.23–1.77	–	13–39	0.50–4.31	–	6–16	
Washington	1966	1.1	–	135[7]	3.24	–	46	Holm (1973)
	1967	0.7	–	111	2.14	–	43	
Iowa	1973–74	0.86[4]	–	35[8]	–	–	–	Crawford (1977)
		1.64[5]	–	35	–	–	–	
Ontario	1974–75	0.72–2.94[6]	–	272[7]	2.33–5.62	–	96	Weatherhead and Robertson (1977a)
Indiana	1974	1.8	–	42[8]	7.4	–	10	Patterson (1979)
	1975	2.0	–	38	6.8	–	11	
	1976	1.4	–	43	4.5	–	13	
Michigan	1975	1.86	3.01	28[8]	3.47	4.98	15	Payne (1979)
	1977	1.88	2.52	40	5.20	11.17	15	

California	1980	1.32	–	44[7]	4 .46	–	13	Ritschel (1985)
	1981	0.71	–	91	2.50	–	26	
	1982	0.89	–	120	3.24	–	33	
Ontario	1984	1.76	0.18	63[7]	4.74	3.61	42	Muldal *et al.* (1986)
Ontario	1986	–	–	–	4.0[9]	8.67	13	Gibbs *et al.* (1990)
New York	1988	–	–	–	2.7[9]	9.6	25	Westneat (1993)
	1989	–	–	–	3.0[9]	9.8	24	
Upland habitats								
Ohio	1973	1.5	–	68[8]	5.9	–	17	Dolbeer (1976)
	1974	1.1	–	64	5.0	–	14	
California	1980	2.0	–	35[7]	17.50	–	4	Ritschel (1985)
	1981	2.20	–	20	8.80	–	5	
	1982	1.35	–	26	8.75	–	4	

[1]Variances were provided in references or were calculated when possible from references.
[2]Number of individuals.
[3]Means and sample sizes (*n*) for two separate study marshes.
[4]Yearlings only.
[5]Nonyearlings only.
[6]Range of means from various marsh areas with differing female densities.
[7]Few if any banded females were used in this study. The number of females was estimated either by the maximum number of simultaneously active nests or by the minimum number of females that could account for all nests.
[8]Many females in this study were banded for individual identification.
[9]Based on DNA fingerprinting that revealed genetic paternity of all offspring.

Table 9.5. Mean annual fledging success for male and female redwings during the Columbia National Wildlife Refuge (CNWR) study

	Females			Males		
Year	Mean	Variance	*n*	Mean	Variance	*n*
1977	1.50	2.48	114	3.93	23.93	73
1978	2.46	4.11	201	8.06	50.30	77
1979	2.79	3.35	121	10.94	83.54	36
1980	1.11	2.65	245	4.97	32.21	68
1981	1.18	2.14	223	3.73	19.14	82
1982	1.06	1.97	132	3.87	17.26	69
1983	1.84	3.48	201	5.21	33.91	84
1984	1.55	2.74	225	5.38	34.07	82
1985	1.68	3.45	183	4.87	33.54	85
1986	0.96	2.35	157	2.33	15.65	84
1987	2.01	2.19	205	6.41	39.23	80
1988	1.16	1.82	211	3.18	16.48	92
1989	1.32	2.79	191	4.12	25.19	75
1990	1.21	1.62	161	3.73	16.65	59
1991	0.62	1.29	126	1.88	8.90	57
1992	1.13	2.27	130	3.58	21.19	52

equivalent breeding resources; however, this is not always the case, because annual success was generally higher at upland sites compared with marshes in one study (Ritschel 1985). If females breeding in uplands in some areas have slightly better RS than those in marshes, it may be due to lower nest predation rates (compare Allen 1977 in Table 9.2).

Vegetation type and nest placement

The vegetation in which females nest affects success because different plants provide different degrees of concealment, maximum possible nest heights, and degrees of openness around nests, each of which may influence the probability of nest detection by predators or the defense ability of redwings (Case and Hewitt 1963, Holm 1973, Weatherhead and Robertson 1977a, Picman 1980a). Water depth under nests often influences nest predation rates (Picman *et al.* 1993) and hence, RS (Weatherhead and Robertson 1977a). In Washington, nests in cattail, on average, fledged more offspring than nests in bulrush; however, annual RS per female was higher in cattail in only one of 2 years (Holm 1973). There is insufficient information available to say precisely which vegetation features and nest placement tactics consistently influence redwing RS, and in what ways; also, such relationships are bound to vary locally, depending on the identities of nest predators.

Food production per territory

Female redwings usually forage both on and off their mates' territories and, therefore, food production on territories is not expected to strongly influence RS. At CNWR, females with nests on marshes with high insect emergences had, in one study, about the same success rates as females on lakes with very low emergences (Orians 1980). In fact, fledging success was not enhanced in experimental over control sites in three separate studies during which supplemental food was placed on male territories (Ewald and Rohwer 1982, Ritschel 1985, Wimberger 1988). Where females rarely leave territories (Westneat 1994), food production on them should affect RS more strongly.

Proximity of nest to predators and sentinel perches

Where Marsh Wrens are the major predator on redwing nests, redwings were more successful if their nests were farther from wren nests (Picman 1980a, Ritschel 1985, Milks and Picman 1994). However, proximity to a major predator's nest does not always result in higher nest predation rates. For example, although Black-billed Magpies, the main CNWR nest predator, sometimes nested in shrubs directly adjacent to one of our redwing marshes, they always bypassed the marsh when departing their nest on foraging trips, almost as if they were inhibited from foraging directly below their own nest. We expected redwing breeding success on that marsh when magpies nested to be nil but that was not the case. Finally, nesting success may be higher when nests are placed near prominent perches on which males often spend long periods guarding nests (Yasukawa *et al.* 1992a).

Social environment

Social environments that females join when choosing breeding situations also may influence RS in a number of important ways.

Nest defense

Several lines of evidence suggest that nest defense by males and females affects RS because it is sometimes effective at deterring nest predation. First, occasional success is indicated by anecdotal observations of redwings mobbing and driving from their breeding colonies such predators as crows and jays. Second, at least two studies demonstrated a positive relationship between nest defense by parents and the probability of nest success (Knight and Temple 1988, Weatherhead 1990a). Third, indicating that nesting redwings successfully deter predators, other studies showed that artificial nests experimentally placed 1–5 m from active redwing nests had significantly higher survival rates than nests placed randomly or far from redwing nests (Ritschel 1985, Picman *et al.* 1988).

Success of nest defense is clearly tied to the number of active defenders and the vigor of their mobbing. Therefore, social and physical factors that increase the number or motivation of mobbers may influence nesting success. Results of several analyses suggest that denser aggregations of nests provide higher degrees of protection from predation (Picman 1980a, Orians 1980, Ritschel 1985, Picman *et al.* 1988). At CNWR, our marshes with smaller, more densely-packed territories experienced consistently better annual RS per male and per female than did our marshes that were more sparsely populated (Beletsky and Orians 1989c, 1996), i.e. where there were fewer potential mobbers nearby to defend nests. Also, neighboring males with whom females engaged in EPCs are more likely to help defend their nests (Gray 1994), and thus female copulatory behavior plays an indirect role in nest defense.

Nest density/synchrony

Closely allied with nest defense as an influence on RS is the density, in time and space, of active nests within a breeding colony. There may be an enhancing effect on RS of nest synchrony – a positive correlation between the number of simultaneously active nests in a given area and the probability of each one's success (Robertson 1973b, Westneat 1992b). Positive effects on RS of high-nesting density in time or space may result from either increased numbers of redwings within local areas, at about the same nesting phases, available to mob, or from predator swamping or dilution, or both. Under some conditions there may be the opposite effect, with higher nest or female densities leading to lower success per individual, because they more strongly attract predators (Weatherhead and Robertson 1977a).

Male feeding

Male assistance in feeding nestlings is valuable to females because it can significantly enhance their RS. Several studies in widely separated areas have demonstrated that male-assisted nests fledge, on average, more young than nests fed solely by the female. The fledging differential varied from 0.7 per nest in Ontario (Muldal *et al.* 1986) to 0.5 in Michigan (Whittingham 1989), 0.8 in Wisconsin (Yasukawa *et al.* 1990), 0.7 in Washington (Beletsky and Orians 1990), to 0.5 to 2.0 (depending on year) in Indiana (Patterson 1991).

Familiarity with mates, neighbors, and harem-mates

Long-term familiarity with conspecifics breeding on the same territory or on adjacent territories could influence a female's RS if familiarity reduced aggressive interactions or facilitated quicker or more vigorous group-mobbing of nest predators. The only evidence to date for a reproduction-enhancing effect of familiarity on redwing RS comes from our CNWR population. On three of our densely-settled marshes, the females on territories of males breeding for at least their second year, who had at least one adjacent male neighbor who had occupied the same position the previous year, fledged, on average, 0.3 more young annually than did females on territories without familiar neighbors (Beletsky and Orians 1989c). We speculated that their superior success might be due to males with long-term relations being more likely to assist each other in group mobbing or being quicker to do so; they could also, out of long practice, launch more coordinated attacks. Gray's (1994) work on EPCs (Chapter 7) suggests that females themselves influence the propensity of neighboring males to defend nests across territorial borders, but familiarity between males or between males and females could also play causative roles. Furthermore, females often have the same harem-mates in two or more years and it is possible that this type of familiarity plays a role in reducing female–female aggression (Beletsky and Orians 1996) and increasing group nest defense. The enhancing effects of social familiarity on RS that we observed in eastern Washington may not hold in all regions (Weatherhead 1995).

Individual attributes

The attributes of individual females also influence RS.

Age and experience

Age influences annual RS because the first time females breed they start later than older females, have smaller clutches, and are inexperienced parents. Yearling females usually arrive on the breeding grounds later than do mature females (those 2 years old or older) and therefore, on average, start nesting later. The time difference is significant because, often, the earliest nesters enjoy the highest probability of escaping nest predation and therefore the highest probability of nest success (Langston *et al.* 1990, Beletsky and Orians 1996). The difference in mean seasonal first-egg dates for yearlings versus mature females in different localities varies between 5 and 15 days: 15 days in Iowa, for 67 yearlings versus 41 matures (Crawford 1977); 5 days in Washington, 15 yearlings versus 85 matures (Langston *et al.* 1990); and about 10 days in Washington, 133 yearlings versus 1411 matures (Beletsky and Orians 1996).

Yearlings also have lower average annual fledging success because they lay smaller clutches than mature birds (Crawford 1977) and probably because they lack experience protecting nests and feeding nestlings. The latter possibility was tested by Yasukawa *et al.* (1990), who showed that the number of feeding trips/hour to the nest increased as females aged; the pattern suggests that with time foraging skills improved. During an 11-year period at CNWR, females we knew had bred during previous years fledged, on average, slightly more young annually than did yearlings (Table 9.6; Orians and Beletsky 1989); other researchers identified the same relationship (Crawford 1977). Surprisingly, however, females positively identified as yearlings for which we had breeding information for both their first and second years (n = 77) did not increase their RS from one year to the next: the mean annual RS for yearlings was 1.40 fledged young versus 1.39 for 2-year-olds (paired t = 0.06, one-tailed P = 0.24).

Rank

A female's nesting rank on her mate's territory, primary, secondary, or lower, may influence her RS because higher-ranked nesters, by starting earlier, are more likely to escape nest predation than lower-ranked females (Langston *et al.* 1990) or because, in those populations in which males regularly feed nestlings, the males apportion their assistance as a function of female rank. For example, males in Indiana preferentially fed at the nests of primary females (Patterson 1991), and in an Ontario study, they did so at secondary nests (Muldal *et al.* 1986). However, where males rarely feed young, a female's rank may not strongly affect her annual RS. At CNWR, where males rarely feed nestlings, there was no significant effect over an 11-year period (Beletsky and Orians 1996).

Table 9.6. Mean annual fledging success for female[1] and male[2] redwings of different ages and breeding experience during the Columbia National Wildlife Refuge (CNWR) study

	Mean	Variance	n
Females			
Yearlings	1.06	2.09	186
Mature females			
first year breeding as mature female[3]	1.45	2.49	605
second	1.64	2.98	376
third	1.60	3.21	242
fourth	1.42	3.27	161
fifth+	1.47	2.46	96
Males			
First year breeding	3.79	25.45	501
Second	5.40	33.74	242
Third	5.56	39.20	153
Fourth	5.15	32.38	106
Fifth+	5.68	36.06	76

[1]1982–92.

[2]1977–92.

[3]Includes information for females who had reddish epaulets when we first recorded them breeding, who were thus at least 2 years old, and information from the 'yearling' group who survived to breed the following year.

Rate of EPCs

Many offspring are sired via EPCs in all redwing populations where paternity has been analyzed. In some areas females apparently only acquiesce in EPCs but in others females actively solicit. Females that participated in EPCs (either observed or proved via genetic analyses of offspring) at CNWR fledged about 0.5 more young per nest than females that did not (Gray 1994).

Start date and genetic influences

Generally, field researchers have found strong negative relationships between nest start dates and nest success – the later the nest, the lower the probability of its success (Dolbeer 1976, Langston *et al.* 1990). At CNWR, we identified a significant relationship between a mature female's annual production of fledglings and the week she started breeding (Beletsky and Orians 1996). The main factors contributing to the pattern were that predation rates were lower earlier in the breeding season and that when early-nesting females' first nests failed, they had more time left in the season to renest.

The timing of when a female starts to nest can be influenced by a number of factors, both physical and social (see Chapter 7). Langston *et al* (1990) noted that larger females started earlier than smaller ones and, in our CNWR population, there was a significant positive correlation between years in individual female start dates, i.e. if a female started earlier than the population average in one year she tended to do the same the next year. Both of these results suggest some role of genetics that results in differences in morphology, physiology, or migratory and wintering habits.

Sources of variation in annual RS for males

An appropriate equation for determining an average male redwing's annual fledging success is harem size × average success per female. The product has to be corrected by an EPF factor, which would equal the average number of fledged offspring sired annually by breeding males through EPFs in the population minus the average paternity lost to neighboring males. Factors affecting female RS are described above. There are presently insufficient data to assess causes of variation among males in EPF rates. Therefore, in this section I will concentrate on harem size variation. There was a strong positive correlation in the CNWR population between harem size and annual RS per male ($r = 0.68$, $P < 0.0001$, $n = 1155$ male-territory-years); in fact, the number of young fledged per male territory rose almost linearly with harem size (Fig. 9.5). Physical and social environments as well as individual characteristics influence variation in this major constituent of RS.

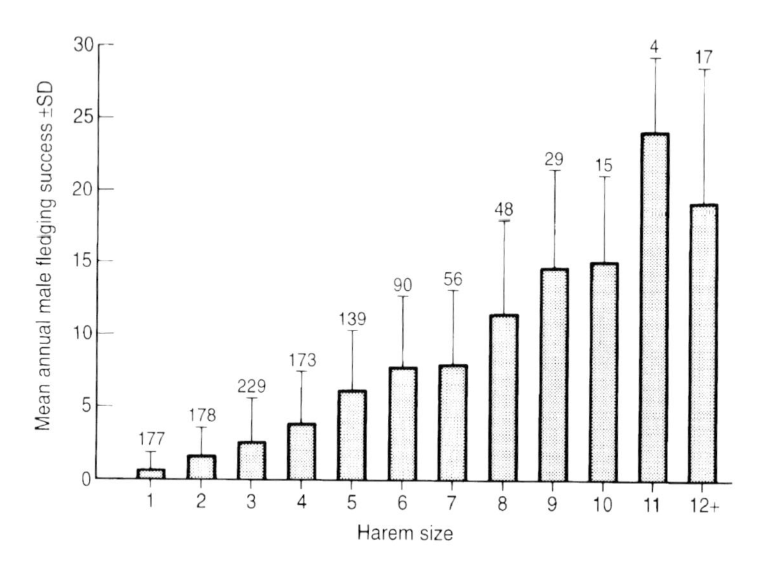

Fig. 9.5. Average fledging success for males as a function of harem size during the 1977–92 Columbia National Wildlife Refuge study. Values above the bars are the numbers of male breeding-years for each harem size.

Physical environment

Harems are generally larger in marsh than in upland territories in the same area (see Table 6.6). Within marshes, breeding colony shape and density of male territories may affect harem size. We found consistent differences in average harem sizes in our CNWR population between strip and pocket marshes. In strip marshes, which occurred in vegetation along the shores of lakes and waterways, territories were long and thin; territory density was usually low and each male had only one or two territorial neighbors. Pocket marsh was our designation for broad areas of marsh vegetation that contained many irregularly-shaped, usually smaller territories, where each male had up to five adjacent neighbors. Average harem sizes were significantly larger on the three pockets (4.6, 4.4, 5.5) during the 16-year study than on the five strips (3.6, 3.3, 3.2, 3.1, 3.4; Beletsky and Orians 1996). Because of these larger harems and also greater RS per female, males with territories on pockets had better annual RS than strip males in 15 of 16 years, often achieving a two- or three-fold advantage in average number of fledged offspring (Beletsky and Orians 1996).

Social environment

Males breeding next to others with whom they have long-term familiarity attract more females than males of equivalent experience in the same areas without familiar neighbors. Using 10 years of breeding data from the CNWR

A male redwing peers into the nest as his mate delivers insects to the young. When males help feed nestlings, the average number of young fledged per nest increases.

population, we determined that the average harem size for males with at least one familiar neighbor bordering their breeding territory (i.e. a male who occupied the same position during the previous breeding season), and their average annual fledging success rates, were significantly greater than those of males who had only new neighbors (average harem sizes, 5.0 ± 2.4 females, (n = 169), versus 3.9 ± 1.9 females (n = 148), respectively; annual RS, 6.6 ± 7.3 fledged young versus 5.0 ± 5.3 fledged young, respectively; Beletsky and Orians 1989c). To control for male and territory quality, we compared harem sizes for experienced individual males that held the same territories both in years in which they had familiar neighbors and in years they did not. In 62% of the cases (total n = 42) males had larger harems in the years they had familiar neighbors, 21% of the time they had smaller harems, and 17% of the time they had harems of the same size. Thus, breeding in groups in which members have long-term associations apparently has an enhancing effect on individual breeding success because these males attract larger harems.

Individual attributes

Some males may be more successful breeders than others due to inherited characteristics. For example, they may have inherited propensities for aggressiveness or dominance that permit them to defend larger territories in better areas, to be superior nest defenders, or to obtain more EPFs. At present there is no information for redwings on genetic influences on RS, principally because of the difficulty of comparing breeding histories of individuals within families and over generations. Young disperse over various distances making their breeding locations difficult to locate. Even for our intensively-monitored CNWR population, in which we banded all nestlings and monitored breeding over a wide area, we eventually identified only a handful of father–son pairs of breeders or sibling pairs that could be used to analyze genetic influences on breeding success.

Breeding experience could influence annual RS if first-time breeders attract relatively few females, are poor nest defenders, or if they feed young at lower rates than more experienced males. Average harem size increased slightly during each of the first 4 years that males bred in the CNWR population (Fig. 9.6); it increased significantly between first and second breeding years ($n = 202$, paired $t = -2.76$, one-tailed $P = 0.003$). At least part of the increase is due to males inheriting females when expanding their territories between years as neighbors die or leave (Yasukawa 1979). Annual RS for males increased significantly with breeding experience, between their first and second years on territory (Table 9.6; paired $t = -3.08$, one-tailed $P = 0.001$, $n = 242$ males for which we had breeding information for their first and second territorial years; first year RS = 4.14 ± 5.20 fledged young, second year RS = 5.40 ± 5.81 fledged young). We found no evidence of a negative effect on annual RS of senescence (Table 9.6); the noticeable decline in Fig. 9.6 in average harem sizes for males that held territories for more than 5 years is due to the fact that many males that bred that long had territories in lower-quality habitat, to which fewer females were attracted.

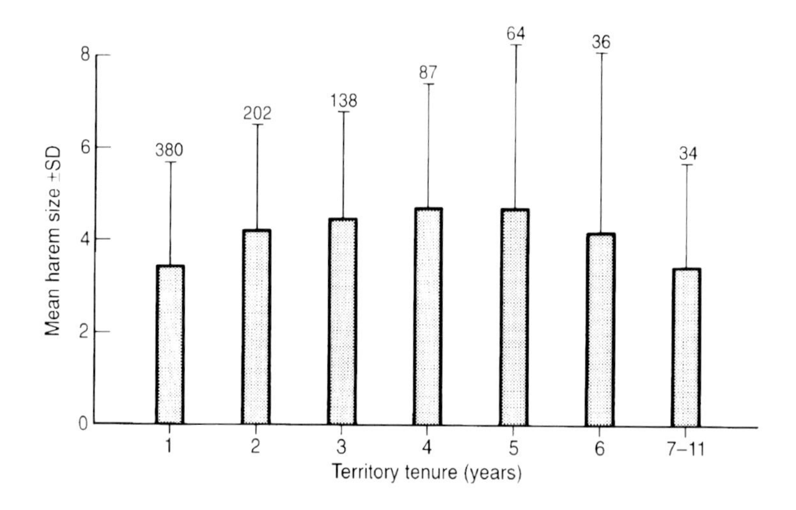

Fig. 9.6. Average harem sizes in the Columbia National Wildlife Refuge study for males with various numbers of years of territory ownership (territory tenure). Information is included for only those males who first owned territories in 1978–1990. Reproductive and harem-size monitoring ceased after the 1992 breeding season.

LIFETIME REPRODUCTIVE SUCCESS (LRS)

Ornithologists spend decades monitoring RS of populations of single bird species for the simple but essential reason that the assessment of LRS of marked individuals 'provides a better basis for estimating biological fitness than any other measure yet available' (Newton 1989, p. 441). The concept of biological fitness – the relative contributions of individuals with different genetic traits to future gene pools – is central to modern understanding of the origin of the diversity of species, of how the variation within species is maintained, and of how present biota changes. Yet until recently, when many long-term RS studies that were begun in the 1960s and 1970s 'matured,' good estimates of fitness were lacking for most long-lived vertebrates, such as mammals and birds. Many previous researchers had considered information on individual variation in RS during single breeding attempts or seasons to be indicative of relative fitness. However, there are major problems with such short-term measures of RS, among them that in variable breeding environments there is certain to be strong within- and between-season variation in nesting success (Perrins 1979). Also, an individual's RS is not guaranteed to accrue annually at the same rate, making snapshot views of relative RS among individuals potentially unrepresentative of their eventual relative successes. For example, in several cooperatively-breeding birds, young adults delay their own breeding for a year or more, but by waiting and initially helping their relatives, and only later breeding themselves, they often accrue substantial LRS (Stacey and Ligon 1987, Koenig and Mumme 1987); measuring their RS in a 'helping' year would give a very misleading estimate of their LRS.

LRS for birds can be calculated as the product of an individual's breeding lifespan and its fledgling production per year. Here I present information on LRS for the CNWR population, the best gathered to date for redwings, and explore the issue of breeding lifespans of males and females.

The average male (n = 461) at CNWR bred for 2.4 ± 1.8 years and had in that time 11.3 ± 15.7 young (range 0–176) fledge from his territory. The average female (n = 1311) bred for 2.4 ± 1.9 years and fledged 3.5 ± 4.0 young (range 0–25) over her lifetime. The distribution of the number of years breeding for individual male and female redwings during the study is shown in Fig. 9.7 and individual variation in LRS is shown in Fig. 9.8. In the latter, the peak at three fledglings produced by males and females during their lifetimes is a consequence of brood size – if a nest was successful, the mean number of young fledged was about 2.6 (Table 9.1). The information for males is precise: we knew exactly how many years they held territories and, approximately, how many young fledged from their nests. For females, however, our information is somewhat less precise because, although females attempt to breed each year, 13% of our banded females had gaps in their

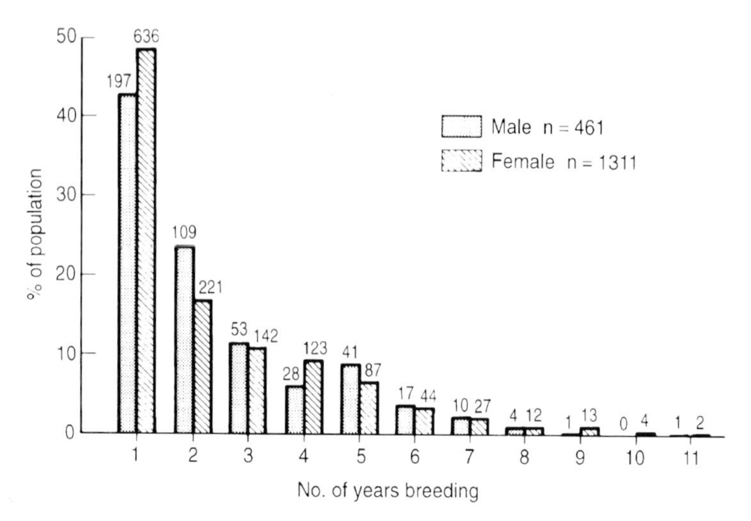

Fig. 9.7. The numbers of years breeding for individual male and female redwings during the 1977 to 1992 Columbia National Wildlife Refuge study.

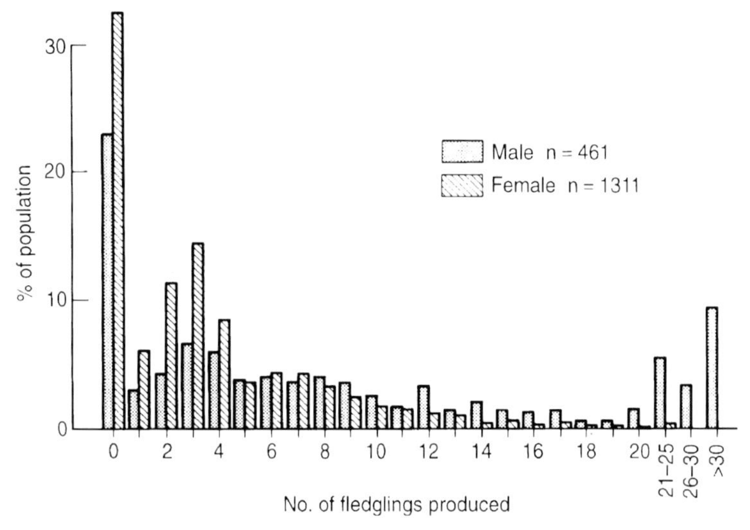

Fig. 9.8. Lifetime production of fledglings for male and female redwings during the 1977 to 1992 Columbia National Wildlife Refuge study (from Beletsky and Orians 1996).

breeding records of one or more years. Some may have temporarily moved out of the monitored area and then returned and others may have stayed but we missed in some years assigning them to nests. To avoid underestimating the LRS for these females, we credited them, in Figs 9.8 and 9.9, with the average number of fledglings produced by mature females during the years we missed them.

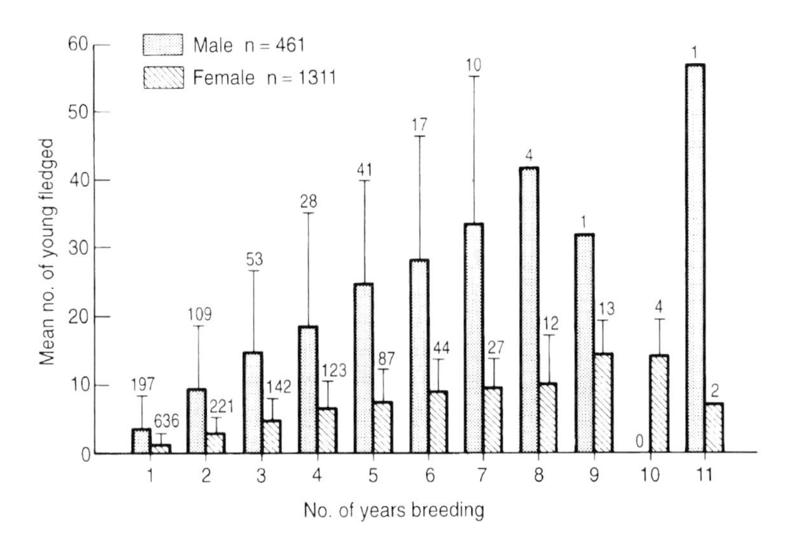

Fig. 9.9. Lifetime fledging success as a function of number of years breeding for male and female redwings during the 1977 to 1992 Columbia National Wildlife Refuge study. One male bred for 11 years, accumulating 176 fledged offspring (from Beletsky and Orians 1996).

Even with this correction, more female than male breeders produced no fledglings during their lives (Fig. 9.8; 23% of males, 33% of females). This is because rarely did all females in a harem fail to fledge any young, but individual females often fledged no young within a year. Of course, were *all* males included, a higher percentage of them would have no LRS because a variable fraction of the male population never acquires territories, and so has no chance to breed (see Chapter 10); conversely, all females attempt to breed each year that they are alive. Also, some males who had no young fledge from their territories may actually have sired young on adjacent territories, by EPFs. As is the case in many birds (Newton 1989), the distribution of LRS in the population is skewed, with many individuals of both sexes producing no or few fledged young and a relatively few individuals producing many. In the CNWR population, this phenomenon reached its zenith in one male who held a territory on our Morgan Lake marsh for 11 consecutive years and produced in that interval 176 fledged young, fully 4.2% of the monitored population's offspring from 1978 through 1988 (Fig. 9.9; Beletsky and Orians 1989b).

CONCLUSIONS AND IMPLICATIONS

On average, females at CNWR produced about 1.5 fledged young per year, whereas males produced about 5.5 (Table 9.6). To achieve at least the average LRS for the population, individuals had to breed for more than 1 year. The greater the number of breeding years, on average, the greater was an individual's LRS. In fact, the relationship, for both sexes, between number of years breeding and LRS was almost linear (Fig. 9.9). The main reproductive strategies of the sexes thus emerge, at least schematically. To maximize LRS, females need to select breeding situations in which the probability of nest predation is low and then continue breeding in as many years as possible. Males must obtain territories in areas where harems are large and nest predation low, and then retain their breeding status for as many years as they can. Except for a slightly lower average annual RS for first-time breeders of both sexes, there are no other age effects on RS (Table 9.6); a fifth breeding year is, on average, no more productive than a second.

Interestingly, for males, there is no relationship between the age at which they first gain territories and the total number of years they breed (Beletsky and Orians 1993). Males that first establish territories at age 2, 3 or 4 hold them for about the same average duration – about 2.5 years (the expected pattern if annual mortality is independent of age). Another intriguing observation is that not all breeding marshes supported equivalent average lifetime territorial lifespans. The average numbers of years that males held territories on our eight marshes ranged from 1.6 to 2.7. Therefore, *where* a male obtained a territory influenced his number of years breeding and, hence, his LRS (Beletsky and Orians 1996).

Patterns of LRS for our redwing population are consistent with those of many other passerine birds that have been studied. For example, between 20% and 40% of breeders in such species as Meadow Pipits, Black-billed Magpies, and Indigo Buntings produce no young during their lifetimes, and for a long list of monitored species, breeding lifespan is the chief contributor to individual variation in LRS (Newton 1989). Where redwing LRS patterns differ strongly from those of most other species is where we might expect – the magnitude of LRS and its variance in males is much greater than in females, as befitting a strongly polygynous breeder.

Female redwings, although saddled with the bulk of seasonal reproductive chores, have at least some advantages over males. They start breeding at age 1 year and nest more or less where and when they want. Any male should welcome any female into his harem and any harem should be pleased to accept a new member. Males, in contrast, have major impediments to overcome before reaching the threshold of breeding: they must first survive a nonbreeding, subadult, year, and, second, they must acquire a territory. Because in all populations there are apparently more potential male breeders than available territories, this last requirement is one that has a major effect on male LRS. As such, it is a subject that many researchers have spent considerable time thinking about and investigating, and it is the subject of the next chapter: how do males actually obtain territories and what determines which are successful and which are not?

Chapter 10

TERRITORY ACQUISITION AND TERRITORIAL DOMINANCE

INTRODUCTION: HOW DO BIRDS GET AND KEEP TERRITORIES?

An ancient Chinese proverb states that 'Two tigers cannot live on the same hill'. Thus, the ancients knew that animals had territories, but I did not discover it until I first began exploring the field of animal behavior. I presume that I had been dimly aware of territoriality before, but I had never applied the knowledge to everyday life. What it meant in practical terms was that the robin or jay that I noticed perched in my backyard maple on Tuesday was likely to be the same individual I had spotted there on Monday. The realization was hardly earth-shattering. Yet before, if I thought of it at all, I had assumed, I believe, that birds and other animals were usually transient and that one saw a robin in one's yard each day because robins, in ones or twos, continually moved through the area. Possession of a territory means that an individual, or a mated pair, or group, defends a particular parcel of land from other members of the species and more or less spends its days within the confines of the borders of the defended space. Thus, territories impose one kind of spatial order on animals and their activities, one that allows me, and others, to understand more of their biology.

Many kinds of animals establish territories and the practice is especially prevalent among birds. Much descriptive information exists about the different kinds of territories – those defended specifically for the food they contain, or a nest site, or all-purpose territories, in which birds essentially spend their lives during the breeding season. Large amounts of knowledge also exist on the advantages of territory owership, on the factors that determine territory size and quality, and on the behavior of birds once they own territories. But on two fundamental questions little was known, as I quickly discovered when I began studying territoriality. The first is a question that, after working for several years with redwings, I would ask myself frequently: how do individuals, amid what is plainly very strong competition, acquire their territories? That is, what determines which individuals get territories and which do not? Intuitively, it might appear that the biggest, strongest, or most aggressive individuals would get territories, but is that true? The second, closely-related, question is: given that some individuals do not have territories but keep trying to establish them, how are territories successfully maintained by their owners for periods long enough to breed for one or more years? The answers to the two questions might be identical, e.g. the best fighters are the ones able to acquire and to hold territories for long periods, or they might be different. For redwings and many other birds like them, these questions are crucial because, as we observed in Chapters 8 and 9, males need territories to enable them to reproduce.

If the number of available breeding territories in an area equals or exceeds the size of the local male population, there may be no competition; all potential breeders can have one (but competition might still occur for access to preferred sites). However, a good deal of evidence indicates that in most regions within their range, in most years, male redwings compete intensively for territories. Some males establish territories in poor-quality breeding habitat and attract no females. Many males that initially settle in an area are unable to establish permanent territories and are evicted by those that can. The best evidence for strong competition for territories is that a usually substantial fraction of the adult male population each breeding season fails to establish territories. These floaters are often observed before and during nesting trespassing in territories and challenging territorial males for possession of their holdings. Occasionally, they are successful in replacing territory owners that die or leave, or partly displacing owners to form a new territory (i.e., they insert) or, rarely, evicting a current owner and taking his territory.

There is no doubt that the floaters' main motivation in remaining near breeding areas, trespassing in territories, and challenging owners, is to acquire territories of their own. Floating is not an alternative reproductive strategy during which these males obtain a measure of reproductive success by, say, trespassing in territories and mating with females during extra-pair copulations (EPCs) or copulating with females in off-territory locations. Several studies have determined that almost all females engage in EPCs only with territory owners and never with floaters (see Chapter 7). Only by obtaining a territory does a male have a sure chance to breed. We also have strong experimental proof that not only are floaters motivated to establish and hold territories, they are physically, physiologically, behaviorally and 'psychologically' capable of it. Owners removed from their territories are quickly replaced by floaters, sometimes within minutes or hours and almost always within a day or two, as if floaters were resident in the area and watching for vacancies (Table 10.1; Orians 1961, Peek 1971, Rohwer 1982, Beletsky and Orians 1987b). Moreover, floater 'replacement owners' can be sequentially removed on the same site and replaced by additional adult floaters – up to nine or more times in regions where it has been tried (Orians 1961, Laux 1970, Shutler and Weatherhead 1991a). Eventually, during sequential removals, subadult males begin taking the

Table 10.1. Chronological history of removal experiments with redwings

Author	Location	Sequential removals at same site?	Rapid replacements?	Project to study
Beer and Tibbitts (1950)	Wisconsin	N	?	Breeding behavior
Orians (1961)	California	Y	Y[1]	Surplus breeders
Laux (1970)	New York	Y	Y[1]	Surplus breeders
Peek (1971)	Pennsylvania	N	?	Timing of breeding behavior
Yasukawa (1981a)	New York	N	Y[1]	Functions of songs
Rohwer (1982)	Washington	Y	Y	Territorial dominance
Eckert and Weatherhead (1987c,d)	Ontario	N	Y	Territorial dominance
Beletsky and Orians (1989d)	Washington	N	Y[1]	Territorial dominance
Shutler and Weatherhead (1991a, 1992)	Ontario	Y	Y[1]	Territorial dominance
Beletsky and Orians (1996)	Washington	N	N	Territory acquisition

[1]Some replacements reported to occur within an hour of owner removal.

territories (Orians 1961, Laux 1970), indicating that the supply of local adult floaters is probably exhausted and also indicating that, given the artificial situation of little or no competition from heavier, more aggressive adults, many subadult males would establish territories and breed. Adult floaters sometimes naturally become territory owners part-way through breeding seasons (Nero 1956b, Orians 1969, Rohwer 1982, Beletsky and Orians 1996).

Investigating how male redwings obtain territories involves studying floaters, which is a difficult task because of their itinerant nature. However, researchers have several ways to do so, both direct and indirect, and these methods will be explored here. Investigating how males maintain their territories in the face of constant threat and challenge by floaters involves studying the dominance relationships between owners and floaters, which is a somewhat easier job because the interactions occur on territories.

Before I discuss theories and tests of territory acquisition, two points need to be made. First, the size of a floater population determines competition levels and thus, is an important factor to assess in studies of territory acquisition. Unfortunately, because floaters do not reside in one place, assessing their population numbers is difficult; most investigators have not been able to do so. In general, at CNWR we operated under the premise that the adult male floater population in most years approximately equalled the size of the population of territory owners, and we had some evidence for this estimate (Beletsky and Orians 1996). However, the size of the floater population clearly varied. Some years large numbers of nonterritorial males were evident in the area but other years we trapped and saw few. Toward the end of our study, in 1993 and 1994, the floater population had fallen so low that when we removed territory owners, few replacements took the vacant territories (Beletsky and Orians 1996).

Second, in some species, all habitat within an area is divided into territories or all good feeding sites are included within defended space. The consequence for floaters is that they are relegated to live in a suboptimal habitat where opportunities to feed are limited, or they must exist as 'underworld' floaters, living covertly on territories of the dominant owners (Smith 1978, Arcese 1987). In either case, because of limited feeding opportunities, floaters may not be able to sustain the good physical condition necessary for them to challenge owners for territories. In contrast, redwings often inhabit areas where territories occupy only a small part of the landscape and spaces not defended offer superior feeding opportunities. At the Columbia National Wildlife Refuge (CNWR), territories occupied only the relatively small part of the refuge that consisted of marshes. Upland areas and lakes without marshes, which together constituted more than 95% of the terrain, were freely available to floaters. Furthermore, many of the uplands near lakes and marshes, as well as nearby agricultural fields, were excellent for foraging during most of the breeding season. Therefore, the undefended areas surrounding territories offered floaters space in which to feed unmolested and to maintain good physical condition, and could also serve as platforms from which floaters might observe territories and launch challenges.

GAME THEORY MODELS OF TERRITORIAL DOMINANCE AND HOW THEY CAN BE TESTED

Until recently there was little information available on the precise rules governing how birds establish and maintain territories. However, naturalists had long noted a perplexing and seemingly general phenomenon that characterized dominance interactions typical of territory owners and floaters. This phenomonon, or rule, simply stated is 'the resident always wins'. When birds are challenged for their territories, even during periods of high floater density and competition, the current owner, the resident, nearly always wins contests. Not only are residents always victorious, they usually win without real fights and thus, without having their fighting abilities tested. The contests usually involve visual and vocal threats by the resident and quick submission and retreat by the challenger. In fact, it sometimes appears as if the challenger quits as soon as the resident announces his presence (Rohwer 1982). What might explain this curious behavior?

The pervasive pattern of territorial dominance, otherwise known as the 'prior residency effect' on settling animal contests, was very puzzling particularly because a basic tenet of behavioral ecology is that when 'an individual's only chance of securing a vital resource, such as food or a mate or territory, depends on its winning a contest, then it should fight and risk the costs that this may entail' (Davies 1978). The implication is clear: if birds cannot reproduce without territories, they should fight for them. Because floaters usually do not fight, but acquiesce, other forces or strategies must be at work. Identifying those forces, the rules that underlie the prior residency effect, constitutes a major problem in explaining how birds get and hold territories.

Although biologists were aware of the linked phenomena of territorial dominance (territory owners are dominant to all conspecifics while on their territories) and challenger inhibition (challengers for territory ownership usually retreat before challenges escalate to physical combat), their causes were poorly understood until, in the 1970s, theoreticians such as John Maynard Smith and Geoff Parker began applying game theory models to animal contests. Game theory, developed in the field of economics, views interactions, or contests, in terms of the costs and benefits to each contestant of its own actions. The models assume that the behavior of one contestant depends on the behavior, or the predicted behavior, of the other.

In other words, the best strategy for obtaining the contested resource while avoiding debilitating injury depends on what competitors do. For example, such models try to predict the behavior of a challenger in a contest in which the challenged individual is prepared to escalate the contest to the level of physical combat; the probability that the challenger stays and fights, as opposed to capitulating and fleeing, might be related to factors such as the scarcity of the contested resource, the respective fighting abilities of the contestants, and the value of the resource to the contestants, which may differ depending on, among other things, amount of previous investment in the resource. Game theory models try to predict a behavioral strategy that results in maximum reproductive success for individuals that employ it, i.e. an 'evolutionarily stable strategy' (ESS). An ESS is a strategy that cannot be improved upon under current conditions and therefore is maintained in the population by natural selection.

A game theory model of challenger behavior might predict that challengers should always escalate a contest and fight if the prevailing conditions are that the contested resource is in limited supply, is valued equally by the contestants and the risk of injury during fights is low. An alternative model that predicts that challengers should refrain from engaging in escalated contests might hold if challengers are animals with inferior fighting abilities or value the resource less than the other contestant (and hence are less motivated to fight), or the resource under contention is in good supply. For the challenger, for the strategy in any of the last two cases to be beneficial (and to be an ESS), the possibility must exist that if they retreat now, they have some chance of acquiring the resource later, i.e. the cost of an immediate retreat must be offset by the chances of future acquisition of the resource. Such models are applicable to territory acquisition and retention in birds because the conflicts over territories between owners and floaters are clearly a special kind of animal contest.

Three general models encompass much of the game theory work on animal contests (Maynard Smith 1974, Parker 1974, Maynard Smith and Parker 1976, Parker and Rubinstein 1981). Although animals may not behave exactly according to the precepts of any one of the models – a combination of their effects is possible – they have been invaluable for our understanding of territorial dominance because they introduce a logical order to these dominance relationships and because their predictions have permitted rigorous testing in natural populations. The Resource Holding Potential (RHP) hypothesis asserts that residents acquire territories and hold them because they are bigger, stronger, or more aggressive than competitors, and thus, are better fighters. The Value Asymmetry (VA) hypothesis holds that residents win contests for their territories because they place higher value on them than do challengers and thus, expected 'payoffs', or benefits, to the current residents are greater than they would be to the challengers. The resident is therefore motivated to escalate contests beyond the level the challenger is willing to sustain. The third major hypothesis to explain animal contests is the Arbitrary Rule hypothesis, which states that contests are settled by an arbitrary rule that is respected by all contestants. Rules, such as 'first to attack, wins' or 'the individual that flies higher, wins', are likely to function only when there is a surplus of the contested resource. Therefore, they are unlikely to apply to most territorial contests in birds.

The RHP hypothesis is easier to test than the VA hypothesis because RHP makes the prediction that territory owners should differ measurably in their morphology, physiology, or behavior. Thus, RHP predicts that owners should be larger or heavier or more aggressive than floaters; this can be assessed in the field with moderate effort. Another prediction is that if individual RHP determines territory ownership, then the best males should acquire the highest-quality territories, with the consequence that there should be a positive correlation between male quality and territory quality (Eckert and Weatherhead 1987c). Testing the VA hypothesis is more difficult both because it predicts that owners and floaters *do not* differ morphologically or behaviorally, and because the relative value that owners and floaters place on territories (the 'currency' of the model) is less easily assessable.

We have found that the best method to test the VA hypothesis is the territory owner removal experiment. Removal of territorial individuals, often by shooting, has been for many years a productive research technique in ornithology. The method was used extensively during the 1960s and 1970s to test the hypothesis that territories place limits on the breeding densities of birds, and so limit population size (Orians 1961, Harris 1970, Krebs 1971, Knapton and Krebs 1974), and more recently, to investigate the dynamics of territorial breeding hierarchies (Hannon *et al.* 1985, Smith 1987). In our version of the removal experiment, territorial males are captured and held in cages for various durations away from their breeding marshes. After floaters take the vacant territories we release the original owners, who attempt to recapture their territories from their replacements. In effect, we orchestrate fights, escalated contests, between original and replacement owners. By varying the duration owners are held off the territories and how long the replacements own them, we attempt to manipulate how the contestants value the resource and, hence, outcomes of the contests. The main VA prediction is that successful territory retention by replacement owners should be time-dependent. That is, if initially floaters do not escalate challenges because the owners value the territories more highly than they do and so would escalate contests quickly to injurious levels, then with increasing time on a territory as a replacement, a point should be reached when a new territory has as much or more value to the replacement than it does to the original owner. At that point, the asymmetry in value is erased or balanced in the replacement's favor and he should be willing to escalate, fight, and win.

TERRITORY ACQUISITION AND DOMINANCE IN REDWINGS

Removal experiments and RHP comparisons

Redwings, for a number of reasons, are excellent subjects for removal experiments and other investigations of territory establishment and dominance. They are easy to catch and they do well in captivity. The openness of the habitats in which they establish their territories allows researchers to observe territorial contests closely and monitor their outcomes. Further, the presence of undefended areas directly adjacent to many territories allows the individuals that lose territorial contests the option of remaining near the contested territory and, from this base, conducting further forays and challenges.

Sievert Rohwer and Paul Ewald first used the removal method on redwings to test hypotheses of territorial dominance. They predicted that if fighting ability (RHP) determined territory ownership, then original owners should be significantly more aggressive than replacement owners (recently floaters) toward a redwing intruder (Rohwer 1982). They measured the aggressiveness of territory owners on small territories, which were physically isolated and whose owners had no adjacent neighbors, by placing a stuffed male redwing mount in each territory and scoring the owner's aggressive response for 10–15 min. They scored latency to attack, how close the males approached the mount, the amount of damage inflicted upon it, and other behaviors associated with strong aggression. After being tested, owners were removed and their territorial replacements were evaluated for fighting ability in the same way, generally within 1–72 h after assuming ownership. They also

Two male redwing territory owners in aerial combat, disputing the location of a common border. Such fights are relatively rare, probably because they are quite risky: severe injury is possible.

removed some of the first replacements and tested subsequent replacements. The result was that in 15 out of 20 cases, the original owners had higher aggression scores than did their first replacements, providing support for the RHP hypothesis. However, in five cases, the first replacements scored higher than the original owners, suggesting that some relatively poor fighters held territories. First and second replacements, on average, did not differ in their aggression scores, indicating that the floater pool was homogeneous with respect to fighting ability (Rohwer 1982).

An important observation that Rohwer made during his removals suggests one reason why fighting over territories in redwings (and in many other birds) is relatively rare. He confirmed that the redwing's strong feet and sharp beak were dangerous weapons:

> 'To my chagrin, presentation of a stuffed male for 10 to 15 minutes on a resident's territory often was sufficient for nearly total destruction of a beautiful new mount. Flight hits sometimes tore the entire back off a mount; powerful jabbing blows often chipped the glass eyes out of their sockets of hardened clay; similar jabbing blows at the back of the head would chip away the skull and its hardened clay filling; finally, attacking males often pounded into the back of these mounts in the area of the lungs where birds largely lack skeletal shields. These observations indicate that an escalated contest between two male redwings would be exceedingly risky even to the male that was the better fighter.' (Rohwer 1982)

Evidence that fighting is dangerous for all participants suggests a line of thought to explain challenger inhibition. Unless a floater detects a very weak male owner (RHP rules) or one that might be new on his territory and therefore does not yet value it sufficiently to defend it at all cost (VA rules), his most advantageous strategy may be to retreat in most of his challenges rather than to escalate and risk serious injury. He might be better off looking for an uncontested vacancy or an occupied-territory situation that balances more in his favor, such as one in which the current owner is sick or weak and so less able to defend himself.

In our removal experiments at CNWR we took males off their territories, held them for up to 7 days, released them to challenge their replacements for ownership, and monitored contest outcomes. We predicted that if RHP was the prime determinant of territory ownership, then original owners, that had attained and kept that status because of their superior RHP, should always be able to evict their replacements, who were, under the RHP hypothesis, floaters of low fighting ability. Even if a male's RHP declined slightly in captivity, when released he should be able to regain his condition and recover his territory. If VA considerations determined territorial dominance, then we predicted a negative correlation between the amount of time we held original owners and the probability that they could reclaim their holdings. In other words, the longer a replacement held an experimental territory, the greater should be its value to him and the less likely he should be to yield it to a returning original owner without engaging in an escalated contest that he was as likely or, because he was the current owner with a 'home field advantage', more likely to win than the original owner.

The typical behavior of released owners was to return quickly to their territories and fight their replacements for ownership. Males held for only 1 or 2 days usually returned to their territories within a few minutes of release. The latency to return for males held up to 1 week was more variable: some returned immediately but others apparently did not return to their home marshes until dawn of the day following release.

The important question is, who finally gets the territories? The original males, who owned the territories for months or, as in most cases, for years, or the replacements, who owned them for only 1–7 days? If the original owners always recovered their territories, we would have a difficult time discriminating between RHP and VA because both predict that the resident always wins. The RHP explanation would be that the original owners were high RHP males and the replacements were floaters of lower RHP, which is why they did not own territories before the experiment. The VA explanation would be that the original owners had so much more time and energy invested in the territories relative to the replacements that, given our maximum 7-day time-frame, they should always be more willing to fight and to persist, and thus they should always recover their territories. Perhaps if a pattern emerged of sometimes the original males winning and sometimes replacements winning, we could discriminate between the hypotheses.

We began with short-term removals, taking males from their territories for only 1 or 2 days. Replacements quickly took the vacancies. But because we removed only one or two males at a time, we created only small vacancies, some of which were taken by expanding neighbors; others were taken by 'new' males. In 91% of 55 cases, original males were able to recover their territories from their replacements (Beletsky and Orians 1987b). Because both VA and RHP hypotheses predict no reversal of dominance in the short term, this result was difficult to interpret. We next tried removals in which we held males for up to 7 days. In 1986 and 1987 we removed 25 males and held them 6–7 days. At the times of their releases, they faced replacements that had held the territories for 6–7 days. Only four of the 25 (16%) were able to evict their replacements and recover their territories. The others all tried to recover their territories, sometimes repeatedly, by engaging in escalated contests with replacements, but they could not do so. Two males held for just 4 days and who faced males that held their territories for 4 days, were able to recover their territories. During the 1986 and 1987 experiments we removed groups of owners simultaneously, clearing large expanses of marsh of territories, thus preventing neighbors from annexing empty territories and attracting more floater replacements.

The interesting result of our experiments was that males held 1–4 days generally were able to evict replacements and recover their territories, but males held 6–7 days could not. This pattern was most consistent with the VA hypothesis. RHP predicts that the original owners should always win, but that was not the case. VA predicts that who wins depends on how both contestants value the territory, with the victory going to the one who values it more, and that the balance can change over time. Our result indicated that replacements did not yet value their new holdings sufficiently after 1 or 2 days to defend them against a strong challenger, but after 6 or 7 days of ownership, they were prepared to defend them.

Even though our results pointed toward a VA explanation for territorial dominance, we still could not reject RHP as a significant contributor because the time the original owners spent in cages could have depressed their RHP levels. We needed to conduct another experiment, a 'double removal' experiment, to eliminate the possibility that captivity itself influenced the males' abilities to recover their territories. In 1988 we removed 11 males from a marsh simultaneously and held them for 7 days. We then removed the first set of replacements 5 days after we removed the original owners. A second set of replacements took the territories, so that when we released the original owners after 1 week, they faced replacements who had owned their territories for only 1 or 2 days. If 7 days in the cages debilitated males (depressed their RHP), then they should have been unable to regain their territories from the second replacements, even though, as we knew from our earlier removals, original owners usually can evict 1- or 2-day replacements. But the result was that eight out of 11 (73%) original owners evicted replacements and regained territorial status (Beletsky and Orians 1989d). This suggests that the reversal in dominance relationships that occurred somewhere between 4 days and 6–7 days was not a consequence of changing RHP levels but of shifting value asymmetries. In particular, the major change was in the replacements: after 1 or 2 days of ownership they were not yet motivated to defend against an escalated, potentially injurious challenge but after 6–7 days, they were.

Thus, our removals generally supported the VA hypothesis, whereas those of Rohwer (1982) supported the RHP hypothesis. How can we reconcile these conflicting results? Rohwer tested most of his replacements for fighting ability within 3 days of their assuming ownership, but our removals suggest that full territorial dominance does not develop in male redwings until after at least 5 or 6 days of ownership. Therefore, many of the replacements that Rohwer tested may not yet have attained full dominance behavior. Moreover, the aggression scores of Rohwer's replacements tended to increase with the number of hours that they owned territories prior to being scored; that trend is consistent with a VA but not an RHP interpretation of redwing removals.

Patrick Weatherhead and his students tested hypotheses of territorial dominance in redwings chiefly by evaluating the prediction that, if RHP is the prime determinant of territory ownership, owners should be measurably superior to floaters in morphological or behavioral attributes that contribute to competitive ability (RHP). For instance, in captive groups of male redwings, both wing length (a measure of body size) and epaulet length (a measure of epaulet size) are positively correlated with level of dominance (Searcy 1979d, Eckert and Weatherhead 1987b). Eckert and Weatherhead (1987d) compared wing and epaulet lengths of owners and their replacements during both experimental removal of territory owners and natural territory takeovers. In both cases, there were no significant differences in wing and epaulet lengths, on average, between owners and their replacements. For example, only seven out of 21 natural takeovers early during the breeding season were by new males that were larger than the males they replaced. Thus, Eckert and Weatherhead, rejecting RHP, concluded that the morphological characters that contribute to competitive ability do not directly determine who owns territories.

Eckert and Weatherhead (1987b) also tested the prediction that, under the RHP hypothesis, owners should distribute themselves so that the highest quality males also have the best territories (i.e. in an 'ideal dominance distribution'; Fretwell and Lucas 1970). They predicted that, in Ontario, where male redwings establish territories in both marsh and upland habitats, and where, because of their larger average harem sizes, marsh territories could be considered the higher-quality sites, marsh territory owners should be dominant to upland territory owners. They removed owners from both habitats, measured them, and placed them in the same aviaries to determine relative dominance. They concluded that, contrary to their prediction, upland males were, on average, significantly larger than and dominant to marsh males. Their interpretation of this test assumes that upland and marsh breeders come from the same population, that is, that given the chance, upland breeders would choose to move into marshes. Although this is by no means certain, given the high degree of genetic mixing of some redwing populations (see Chapter 2), it appears likely to be so. Another assumption of the test, that male reproductive success is higher on marsh than on upland territories, may not be generally true (see Chapter 9).

Shutler and Weatherhead (1991a) took another tack in removal experiments. They removed original owners and, sequentially, up to several floater replacements from the same territories and put them together in aviaries. Their prediction was that if RHP determines which males possess territories, then the same superior attributes that allowed original owners to acquire their territories should also allow them to dominate their replacements in the aviary. Relative RHP levels, in other words, should not be spatially dependent. In 15 aviary experiments, where each cage contained an original owner and his floater replacements, they failed to detect any difference in dominance between original owners and adult replacements – floaters were just as likely to dominate owners as vice versa. Thus, RHP as the explanation of territorial dominance was also rejected in that study.

Several investigations of redwing territoriality, therefore, have failed to find support for the RHP hypothesis, perhaps the more intuitively appealing of the available alternatives. However, one study determined that male redwings, even if they do not acquire their territories under RHP rules, at least pay attention to interactions that reveal individual RHPs and may use such information in territory boundary contests with neighbors. Freeman (1987) placed a stuffed male redwing mount inside territories and scored and ranked the aggressive responses of owners. He discovered that if a male failed to attack the 'intruder' vigorously, the adjacent neighbors, within a day, tended to enlarge their territories at the expense of the 'weak attackers'. Freeman's conclusion was that neighbor males watch the interactions between adjoining owners and floaters, continually assessing RHP levels, and use the information subsequently during boundary skirmishes. Such evaluations could explain some common observations of redwing behavior – that neighbors often intrude into adjacent territories to watch fights or to observe floaters that enter territories but are not quickly evicted.

RHP and physiology

Rohwer's (1982) mount presentations to territory owners and their replacements provide evidence that owners and floaters may differ in at least one measure of aggression – intensity of attacks on territorial intruders. The RHP presumption would be that the difference in aggressiveness is an important factor that separates owners from floaters. But precisely what aspect of aggression might be involved in territory acquisition? Simple aggressive motivation? Actual fighting ability? Males' self-perception of their RHP levels? Actual fighting ability among males is not thought to determine outcomes of territorial contests in redwings (Nero 1956b, Yasukawa 1979, Eckert and Weatherhead 1987d, Shutler and Weatherhead 1991a). In fact, during the CNWR study, four one-legged males held territories on our highly competitive core marshes, apparently with little trouble. They held territories for an average of 3.8 years (range 2–5), which exceeds the population average. If fighting ability alone determined territory ownership, these males should not have been able to win challenges and fights with intact males, which they must have done frequently to hold their territories.

What about motivation to fight? Rohwer's mount presentations showed that motivation to fight varied extensively among males. Other indicators of this variation are that males vary broadly in the intensity with which they defend nests (Searcy 1979e, Knight and Temple 1988) and battle Yellow-headed Blackbirds on their territories (see Chapter 8). Another situation that reveals tremendous variation in aggressiveness is how male redwings behave when held in the hand. I have held thousands for banding and for taking blood samples. The birds' responses to these indignities vary from quiet, immobile acceptance, to trembling fear, to biting, twisting resistance – so much so that in a few cases I have been unable to take blood samples. Such variation in aggressiveness may contribute to territory ownership and dominance. However, my 'in-the-hand' ratings are anecdotal and behavioral scoring all too often contains subjective judgements; an easily obtained and objective measure of aggressivness is needed for testing. A physiological measure may be appropriatc for this purpose.

Physiology underlies behavior and physiological measures are usually more objective than behavioral ones. For many years I worked with steroid hormones and their relationships to avian behavior. The sex steroid hormone testosterone influences aggressive behavior and dominance interactions in birds and other vertebrate animals. It is particularly involved in the expression of social dominance in reproductive contexts (Balthazart 1983, Wingfield *et al.* 1987), and thus may have strong effects during territorial contests. The precise nature of the relationship between testosterone and aggression is still not well understood but the hormone may allow aggressive behavior to be expressed at high intensity and frequency. Because acquiring and maintaining a redwing territory and engaging in escalated contests involve sustained and intense aggressive behavior, and if high circulating testosterone levels directly affect males' abilities to engage in such behavior, then a general RHP prediction is that territory owners should have higher circulating testosterone levels than floaters.

To test whether testosterone (T) is important for territory ownership in redwings, my colleagues and I determined the relationship between circulating T levels in the blood and territorial aggression and ownership with experiments in which, by implanting hormones or other substances into territory owners, we either artificially raised their T levels or decreased the effectiveness of their natural levels. The methods of these manipulations are fairly simple. Into experimental males we insert under the skin small pieces of surgical tubing 2.0–2.5 cm long (a minor operation that takes only a few minutes). The tubes contain crystalline testosterone, which moves out of the porous tubes and into the bloodstream, raising circulating T levels for about 4 weeks. We checked the effectiveness of this T- delivery system by catching males several days after they were implanted and taking blood samples for later analysis in the laboratory. The method results in significantly elevated circulating T levels, in fact, almost double the average levels of territory owners early during the breeding season (Beletsky *et al.* 1990). Control males got empty tube implants.

In an initial experiment, we implanted T into seven territory owners (locations of three of the seven are illustrated in Fig. 10.1a); we predicted that enhanced aggressiveness due to higher T levels would stimulate these males to expand their territories at the expense of their intact or control-implanted neighbors. However, only one of the seven had a major change in the size of his territory after the

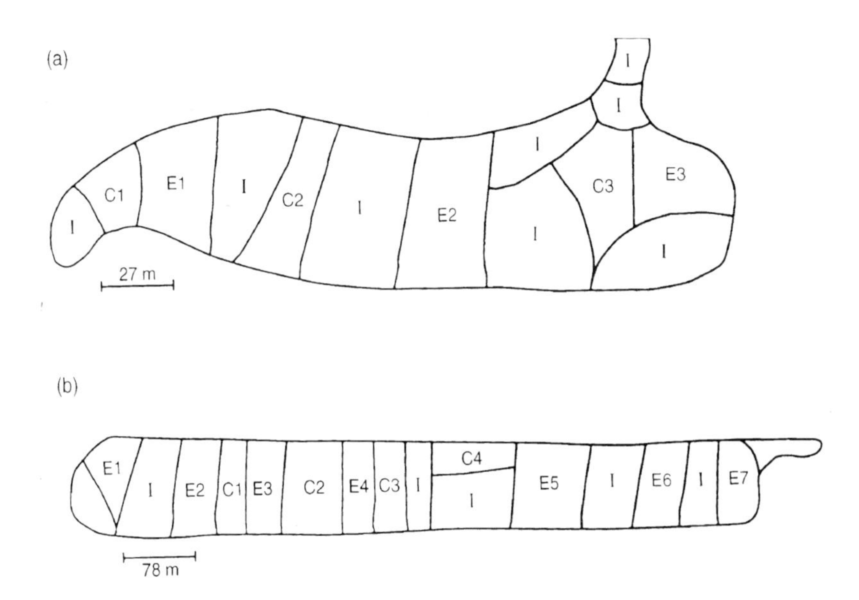

Fig. 10.1. Schematic diagrams of the territory positions and boundaries of male redwings used in implant experiments. (a) Herman Pond, the site of testosterone implants; (b) Marsh Unit 3, the site of flutamide/ATD implants (E = males that received experimental implants; C = control implant males; I = intact, unmanipulated males).

implants (male E2 in Fig. 10.1a). He lost half of his territory 4 days after his implant, but this was a special case because he lost it to a persistent floater who had owned that territory the year before. The other six T-implant owners sustained no changes in their territories during the 4 weeks after the implants (Beletsky *et al.* 1990). Thus, there was apparently no *territorial* effect on owners from having unusually high T levels. These males did engage in aggressive behavior at higher rates after their implants but this change in behavior did not alter territorial relationships.

In another experiment, we addressed the same question but from the opposite direction. We implanted into territory owners tubing containing two drugs that, together, effectively block the physiological action of testosterone. One drug, flutamide, binds competitively to T-receptor sites at target tissues, preventing access of T to its molecular receptors. The other drug, ATD (1,4,6-androstatriene-3, 17-dione) chemically blocks a biochemical pathway in which T is converted to the hormone estradiol, apparently another pathway along which T affects aggression (Walters and Harding 1988). In this study, seven territory owners on a marsh served as experimental birds, with flutamide/ATD implants, four males had control implants, and five were left intact (Fig. 10.1b). The result was dramatic. Within 10 days, one experimental male kept his territory, two lost 10–20%, one lost 30–40%, 2 lost 70–90%, and one lost 100% of his territory to neighbors and to floaters. Four weeks after the implants, we caught four of the experimentals and removed the tubes. Three of these four males then recovered all or part of the lost portions of their territories within a week. Therefore these drugs, which decrease the effectiveness of testosterone, certainly impaired site dominance and the ability of males to maintain their territories.

Our implants showed that although higher than normal T levels might not lead to 'super' territorial males, normal circulating levels were necessary for successful territory maintenance. The next question was, if that is the case, might males that establish territories have higher T levels than those that do not? To test this idea, we conducted a survey of circulating T levels during the entire 1987 CNWR breeding season. We took small blood samples (200–400 μl each) from a wing vein from all territory owners and floaters that we trapped. In all, we collected 204 samples from territory owners, 73 from adult floaters, and 77 from subadult floaters, distributed in time from late February to early June. In the laboratory we determined circulating T levels using radioimmunoassay techniques. The result of the survey is shown in Fig. 10.2. Owners generally had low T levels in late February and early March, when males occupied their territories for varying periods each day but before females arrived in the study area. T levels then rose during the third week of March, as first females arrived and settled on male territories. Owner T-levels rose sharply and significantly during the first week of April, as nesting began, and remained relatively high until the second week of May. Elevated T levels were associated in time with the period of maximum courtship, copulations, fertilizations, and mate-guarding (Beletsky *et al.* 1989). By mid-May to early June, T levels returned to those characteristic of the period before female settlement (Fig. 10.2a). The seasonal patterns of T levels in adult and subadult floaters were similar to that of owners, but the peak levels were lower (Fig. 10.2b, c). Territory owners had significantly higher T levels than adult floaters in March and May, and there was also a trend for higher T levels for owners in April. Subadult floaters had T levels even lower than those of adult floaters (Beletsky *et al.* 1989). Thus, our main finding was that during breeding, owners, on average, have significantly higher T levels than floaters.

This result is consistent with an RHP interpretation of territory ownership – territory owners differ physiologically from floaters and this difference at least partially determines which males can and cannot obtain territories. An alternative explanation also exists: circulating T levels in floaters rise to those characteristic of owners only after they acquire territories and are regularly challenged. Thus, the problem is one of separating cause from effect. If higher T levels are a major influence on a male's ability to establish a territory, then floaters with artificially high T levels should become territory owners. In another implant experiment we gave T implants to 12 resident floaters (these were males we trapped that had been color-banded during previous years and which our comprehensive census of territorial

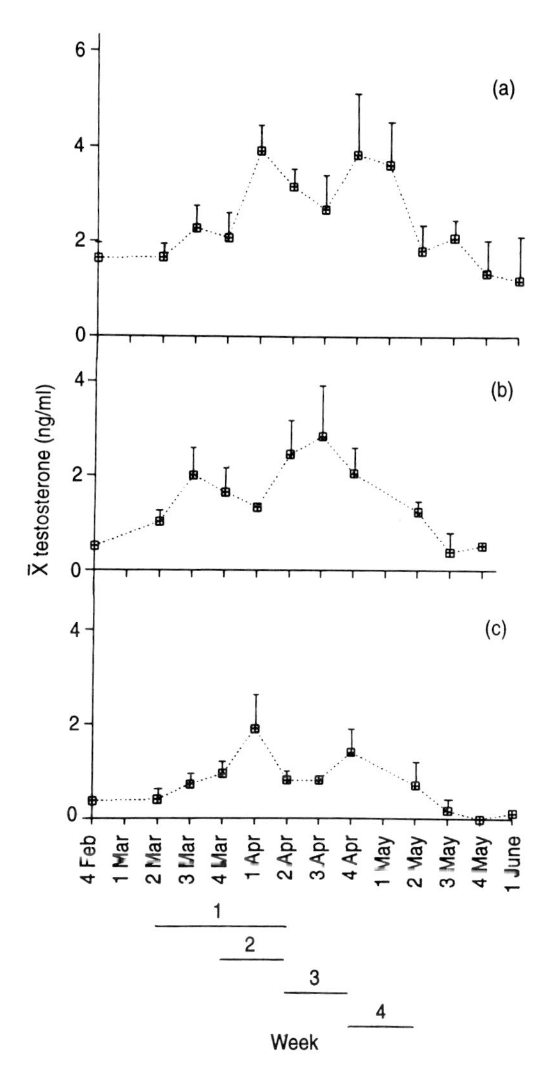

Fig. 10.2. Comparison in 1987 of weekly average circulating levels of testosterone in (a) adult territory owners, (b) adult floaters, and c) subadult floaters. Vertical bars show standard errors. Lines below the horizontal axis in (c) represent approximate reproductive stages on male territories: (1) female arrival; (2) first nests built; (3) first nestlings; and (4) first young fledged (from Beletsky et al., 1989).

males revealed not to be holding territories) early during the 1988 breeding season, raising their circulating T levels. We searched the study area repeatedly during the remainder of the breeding season to determine if they acquired territories (Beletsky *et al.* 1990). Only one of the 12 established a territory, a week after the implant, in the core study area, about 850 m from the trap where he was implanted. However, 12 unmanipulated floaters also established territories in the core study area during the 4 weeks after the implants. Thus, having artificially high T levels apparently did not enhance a floater's chances of acquiring a territory above a natural, background rate.

Our conclusion from these implant experiments is that although T is involved in territory maintenance in redwings, artificially elevating T levels in floaters is insufficient to cause significant changes in dominance relationships between owners and floaters – having high T levels, alone, cannot propel a floater into the breeding population. Although territory owners have, on average, higher circulating T than floaters, we rejected the RHP prediction that owners owe their territorial status to higher T levels. An explanation of higher T levels in owners that is gaining increasing support is the Challenge hypothesis (Wingfield *et al.* 1987, 1990), which suggests that T secretion is strongly stimulated when conspecific males challenge each other for territory ownership or access to mates. If the Challenge hypothesis is true, then only after floaters acquire territories do their circulating T levels rise to levels characteristic of owners, in response to challenges to their ownership from other males. Changing hormone levels in response to changes in social status, instead of vice versa, in this case is also the more logical order. After all, testosterone, a chemical derivative of cholesterol, is a 'cheap' substance for animals to manufacture. If territory ownership and, hence, advancement to breeding status, hinged solely on circulating levels of this sex steroid, floaters should simply produce more of it.

Site familiarity and floater tactics

What are the actual tactics male redwings use to acquire territories? Game-theory models can be applied to the tactics floaters use to search for territories. The RHP hypothesis predicts that, because the relative fighting abilities of a floater and the owners he may decide to challenge are of paramount importance, floaters should spend time evaluating owner RHP levels, either by challenging them directly or by observing their interactions with neighboring owners or trespassers. Only when a floater locates a relatively weak owner should he challenge seriously for ownership. To pursue this strategy, floaters might either monitor the RHP levels of a fairly small number of owners, waiting for one to sicken or be injured (a 'sit-and-wait' RHP strategy), or wander a larger area assessing RHPs at different sites. VA predicts that floaters should persist around a small set of territories to gather information not about the owners but about the territories themselves. Their strategy should be to learn enough about the territories so that their level of knowledge and investment rivals that of the owner, to a point where the value asymmetry no longer favors or only slightly favors the owner, at which time challenging is more likely to be successful. Also, should an owner die or leave, a VA floater, always nearby, should be the first to be aware of the vacancy and to claim it and, because of his previously-accumulated knowledge of the territory, should have a good chance of retaining it. Because the results of several studies of redwing territoriality suggested that chance may also play a role in determining which males get territories, a third major strategy has been proposed, the Lottery hypothesis (Searcy and Yasukawa 1995, Beletsky and Orians 1996). Lottery (L) floaters would search for territories over wide areas, looking only for vacancies. That is, they do not gather information on male RHPs or territory attributes; they look for suitable vacant habitat or territories whose owners are apparently missing, which they may do by simply observing territories or entering vacant areas and waiting for owners to appear.

There is little doubt that at least some male redwings obtain their territories as a result of chance events. In fact, all three major hypotheses, RHP, VA, and L, predict that, under conditions of strong competition, males without territories should immediately take any vacancies they encounter (Beletsky and Orians 1996), i.e. taking a vacancy is always preferable to engaging in risky challenges in which fighting and injury may occur. At CNWR, for example, where males are over-winter residents, many floaters probably detect and take vacancies during the period when males begin spending time on their territories at the breeding season's inception. Thus, males taking vacancies early during breeding seasons cannot be used to discriminate among the three hypotheses of territory acquisition tactics. Some males do not get territories early in the season but operate as adult floaters that continue searching for territorial opportunities. The three hypotheses make different, testable predictions about how they do so.

We used our long-term demographic information collected during the CNWR study to test a series of predictions, derived from the three hypotheses, as to how our redwing floaters search for territories (Beletsky and Orians 1996). As we might have predicted at the onset of testing, no single hypothesis garnered unqualified support, instead, each was at least partly supported. Our conclusion was that CNWR floaters follow a mixed searching strategy, perhaps incorporating elements of all three major hypotheses. A dominant theme coursing through all test outcomes, however, was positive support for an effect of site familiarity. Apparently, environmental information – knowledge of particular areas and/or inhabitants – is very important to redwing floaters. Indeed, all of our information indicates that most males spend their subadult breeding season near where they were born and, subsequently, as adults establish territories there.

Clues from the identities of territory owners

The identities of territory owners provide indications of searching strategies. What do we know of the origins of CNWR territory owners? First, as detailed in Chapter 2, our information on dispersal of banded nestlings suggests that most owners were born within 4–5 km of their territories. Furthermore, during a typical year,

20% or more of territory owners on our core study marshes had been born on those marshes and another 30%–40% of them were males banded initially as subadults in the study area (Beletsky and Orians 1993). Thus, most territory owners were born within a few kilometers of their eventual breeding sites and they spend at least a portion of their subadult year as floaters near their future territories. Clearly, most territories are obtained in areas with which males have deep, long-term familiarity, which is consistent with RHP or VA searching, but not with the Lottery hypothesis.

Sizes of floater 'beats'

Our extensive banding program at CNWR and annual censuses of all territory owners allow us to characterize the sizes of the areas that adult floaters search for territories, i.e. their searching 'beats' (Beletsky and Orians 1996). One large, multiple-capture trap (see illustration at chapter head) was operated in the north half of the study area, and one in the south half. In addition, many small traps were operated around the peripheries of our core marshes. We made nearly 20 000 male captures during the 16-year study. For each capture we recorded the male's identity, location, weight, and the date. Some males were trapped and banded before they established territories, some afterwards. A fraction of banded adult floaters never obtained territories.

Within a breeding season, about 50% of adult floaters that we caught at least twice had moved between traps located between 1.5 and 3.0 km apart. The other 50% of floaters caught at least twice had ranged over distances less than 1.5 km. When we compared the distribution of how far adult floaters eventually obtained territories from a central point they all frequented (one of our large traps) with a distribution of the number of territories available at increasing distances from the trap, territory acquisitions matched territory availability until about 2.5–3.5 km from the trap; beyond that distance, the number of acquisitions, but not number of territories available, dropped precipitously. Figure 10.3 shows the distribution of the number of territories available at increasing distances from our northern trap (the Ranch trap), in 500-m increments, and the distribution of adult floaters caught in that trap that later obtained territories. The pattern indicates that adult floaters searched for territories over a limited range of perhaps 3.5 km. Furthermore, many territory owners were trapped as adult floaters on the same marshes on which in later years they established territories; even more were trapped on marshes within 500 m of their eventual territories (Beletsky and Orians 1996). Data on floater beats, therefore, strongly indicates a relatively small area searched for territories, a pattern, again, more consistent with either a VA or RHP than with a Lottery strategy.

Clues from removal experiments

During our territorial male removal experiments, 74 floaters established territories on the experimental marshes. Their identities provide further information on

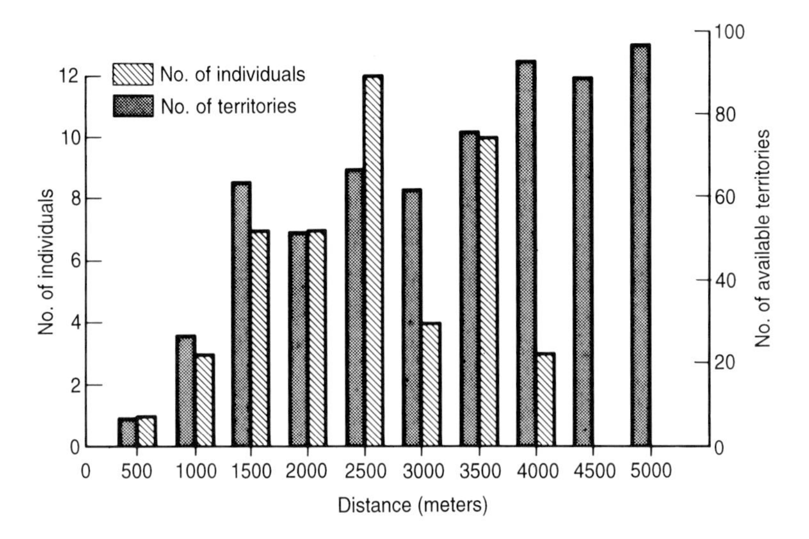

Fig. 10.3. Distances from the Columbia National Wildlife Refuge Ranch trap to the eventual territories of adult floaters caught at the trap that established territories in later years, 1985 through 1992. Also shown is the distribution of available redwing territories at 500-m intervals from the trap.

floater behavior. Forty-three per cent (32) of these replacements already wore bands. Eight wore nestling bands and seven had subadult bands, indicating that they were long-term residents of the area; 17 wore bands given to them as adults. Four of the 17 had never before owned territories, but the other 13 had – nine of them at the same marshes they now occupied as replacements. This last result is strongly supportive of the idea that males searching for territories, especially those that held them previously, restrict their searching to a small area. That males during removals obtain territories at marshes on which they previously held territories is indicative of a VA strategy – they had extensive knowledge of those sites and they remained near them, awaiting territorial opportunities. In cases when original owners lost their territories to replacements during removal experiments and then months or years later were able to regain territorial status, 70% of them did so on their original marshes or nearby (Beletsky and Orians 1996), further supporting VA.

Social stability and territory acquisition

Another method redwings and other birds may use to acquire territories is to seek out and exploit areas where territory ownership or territory boundaries are in flux.

This is the 'instability hypothesis' of territory establishment (Beletsky 1992). I decided to test this hypothesis because I had observed many instances in which floaters were strongly attracted to territorial situations that obviously reflected instability. The most consistent example for redwings occurs when territory owners are caught in decoy traps. The trap consists of a live 'decoy' male placed in a small cage with a spring-loaded net on top. It is placed on a pole within a target male's territory. The territory owner attacks the caged 'intruder,' eventually springing the net. It often takes 5–10 min of approaching, displaying at, and attacking the decoy before the target male hits the release lever and the net is sprung. During the interval, a common occurrence is for floaters to enter the territory, land within 2 m of the trap, and apparently closely observe the interaction between the decoy and the target male. These floaters, which are repeatedly chased away by the owner until he is trapped, may be assessing the situation for possible territorial openings.

Acquisitions can be divided into insertions, in which new owners partially displace one or more previous owners, replacements, in which the new owners replace dead or departed previous owners, and evictions, in which floaters challenge, defeat, and evict previous owners. Evictions are rare. I already knew that many redwings obtain territories by inserting between two established males. I hypothesized that males decide where to attempt insertion by watching for frequent disputes over territories or over boundary locations. It should be easier to insert at such a site than at others because ownership there may not be settled on each and every square meter of space – there could be areas that neither of the contesting males are quite sure about. A new male may be able to exploit such sites to establish a territorial foothold and later expand the holding. Also, locating areas with high male dispute rates may be beneficial to a floater because males in these areas may be tired and have low energy reserves, and so may be particularly vulnerable to persistent challenge. Territory boundaries also may be the best place to begin challenging because, being at the extreme periphery of a territory, they may be less strongly defended than more central parts.

I evaluated several predictions of the instability hypothesis, chiefly that floaters should be attracted to areas of social instability, such as areas with frequent or sustained boundary disputes, and that they should be more successful in gaining territories in unstable than in stable areas. In one test, I placed various numbers of male models on vacant territories for 30 min and recorded the attractiveness of the situation to floaters by noting numbers of overflights of and territorial trespasses near the models. In paired presentations I placed either one model in an alert posture with visible epaulets (intended to simulate a single male on his uncontested territory, i.e. a stable situation) or two models with visible epaulets, in bill-up, boundary dispute postures stationed 50 cm apart (an unstable situation). The models were attached with velcro straps high in the marsh vegetation (Fig. 10.4). In a second type of paired presentation, I placed either one model in the alert posture

Fig. 10.4. Models of male redwings in a simulated boundary dispute, placed in marsh vegetation.

or four models, all close to each other, in bill-up postures (simulating a particularly unstable, highly-disputed area). The instability hypothesis predicts more floater trespassing in areas with perceived boundary disputes, and that is what I observed. Floaters trespassed near all the models at fairly low levels, but significantly more of them approached groups of models in simulated boundary disputes than approached single models (Beletsky 1992). Other evidence that I accumulated in support of the prediction that floaters should be attracted to unstable areas included a high territory-owner turnover rate on a marsh on which we experimentally created socially unstable areas by cutting large swaths of cattail (thereby altering the quality of territories and removing boundary markers), suggesting that we attracted floaters who then obtained territories there. Also supporting the prediction was that experimentally blocking the action of testosterone in territorial implant males (described above) led to the attraction of many floaters to the experimental marsh and the creation of many new territories.

Another prediction of the instability hypothesis is that should many floaters for some reason be attracted to a breeding marsh, even a socially stable one, their presence on or near the marsh should increase instability and, hence, lead to new territory acquisitions. I tested this prediction by evaluating territory dynamics on a marsh that we used for a food addition experiment. The original purpose of the food addition was to examine the relationship between food availability and the onset of nesting. In mid-March, 1990, at Frog Lake, an isolated breeding marsh, we set eight seed trays, four within the marsh on territories and four about 30 m

Table 10.2. Changes in territory ownership on Frog Lake during a food addition experiment (from Beletsky 1992)

Date[1]	Number of territories on Frog Lake	Number of new territory owners
March 10	9	–
13[2]	9	–
19	11	2
20	11	1
21	10	–
22	11	1
24	13	2
26	13	–
27	12	–
31	13	1
April 2	15	1
16	14	1
20[3]	14	–
28	14	–
May 15	13	–

[1]Dates listed between 19 March and 16 April are those when changes in territory owners occurred, regardless of whether new males acquired territories.
[2]Food trays placed.
[3]Food trays removed.

upland from the marsh in undefended areas. Fresh seeds were added to the trays each day until 20 April, and territories were monitored each day. Frog Lake held nine territories when the trays were placed on 13 March. For 4 days all was normal; no changes in territory ownership occurred and few floaters were seen around the lake. The first groups of floaters, evidently attracted by the seed trays, were noted at the lake on 17 March, and a floater flock, usually with at least 12–15 male redwings, was present at the lake each day thereafter until we removed the trays. This permanent group of floaters apparently led to great territory instability on Frog Lake. Between 13 March and 20 April, the number of territory owners rose from nine to 14, but within that interval nine new males gained territories, some only temporarily (Table 10.2). Frog Lake in other years had been remarkably constant in the number of territories it held within a breeding season. Thus, the experiment supported the prediction that attracting floaters to a breeding area would generate territory instability, opening the way for settlement of new owners.

Another experiment that I performed to test the instability hypothesis prediction that boundaries should be preferred sites of territorial trespassing was to place male redwing models either in territory centers or on boundaries. The reason that floaters should observe and land on territory boundaries – in this region, usually lines of cattail plants – is that ownership of each part of a boundary may not be settled or remembered and therefore floaters may be able to persist there and begin territory insertions. The experiment was stimulated by observations of natural situations in which trespassing floaters that landed on boundaries were approached by owners but not attacked. In particular, I recall one incident in which an adult floater landed, by accident or design, at the exact intersection of three territories. The three owners approached the floater to approximately equal distances, about 1m, and directed their attention to him. However, none attacked the floater, as if none of the three knew which of them owned that particular site. The floater eventually flew off. The prediction in my experiment was that territory owners should quickly attack a model placed within their territories, but should not attack an intruder placed directly on their boundaries.

The typical response of 25 experimental males to a center-placed model was to approach it quickly and then hit it (24, 96% of males, did so). A typical response during a boundary presentation was to approach the model to 1–3 m, perch, and give bill-up displays. Often the two territory owners on either side of the boundary gave matching responses – approach and bill-up displays. Only four males (19%) during 21 presentations hit the model; in some of those cases I may have placed the model slightly off the boundary's position. Thus, redwing floaters landing on boundaries are less likely to be attacked than if they trespass within territories and, if they persist on a boundary, could eventually insert themselves as territory owners. Even if this is not a usual route to successful insertion, floaters landing on boundaries, if they are not chased immediately, have time to collect information about territories and owners that may be of use to them in later takeover attempts.

My simple experiments to test predictions of the instability hypothesis raise some interesting points concerning the behavior of both floaters and owners. For example, if it is true that floaters attracted to Frog Lake by the placement of food trays led to territory instability and the establishment of many new territories, and that floaters are partly attracted to long boundary disputes between neighbors because owners in these contests will be tired and thus susceptible to challenge and eventual insertion or replacement, then a logical strategy for floaters would be to prospect for territories in small groups. Territory owners easily evict single floaters from their territories, but evicting three or four would be more difficult and, if they persisted, would exhaust the owner. Redwing floaters often travel and trespass in small groups but they apparently do not obtain territories by cooperating to exhaust and evict single owners. A system of territory acquisition that worked that way, of course, would be highly unstable, because each group of floaters, in turn, would displace other males.

What are the implications of the instability hypothesis for the behavior of territory owners? New owners, who must engage in long boundary disputes with neighbors to establish and settle their borders and who are not yet secure in their territorial dominance, are perhaps at the highest risk of attracting floaters and losing all or parts of their territories to them. Thus, to minimize the attraction of floaters, new owners might be expected to act as if they were long-term owners, and to settle boundaries quickly. In fact, male redwings act like long-term owners within an hour of taking over a territory, immediately approaching and evicting intruders and giving high-intensity song spread displays (Hansen and Rohwer 1986).

All territory owners should also behave in ways to minimize instability or maximize the appearance of territory stability. One method would be to minimize the number and intensity of boundary disputes with neighbors. The phenomenon of territorial neighbors deciding a boundary location and respecting it in a 'gentleman's agreement' has been noted by many biologists, and even been given a name – the 'dear enemy' effect. Its underlying benefit is generally thought to be that it allows both neighbors to devote their energies to other activities, avoiding unnecessary time and energy spent unproductively arguing over borders (Wilson 1975). Another possible benefit of reduced frequency of boundary disputes is that it reduces behavior that may attract floaters. Such mutually beneficial avoidance of incidents that attract floaters might also explain why territory owners sometimes chase intruding floaters out of the territories of their neighbors when the neighbors are absent (Nero 1956b; personal observations). The instability hypothesis suggests that a stable group of long-term neighbors with few disputes over boundaries attracts relatively few floaters, to the benefit of all residents. Because new owners might attract floaters, they are a potential source of trouble for all other owners on a marsh, and thus owners may defend not just their own territories, but those of their neighbors.

CONCLUSIONS: HOW DO REDWINGS GET AND KEEP TERRITORIES?

We now have sufficient information in hand to attempt an overview of the strategies that male redwings employ to acquire breeding territories and also to address the sources of the dominant/subordinate relationships typical of owners and floaters. For few other songbirds is comparable information available. Male redwings in the CNWR population obtain territories within 4–5 km of their birth sites. They spend their preterritorial breeding seasons in the same area, possibly accumulating information on other males but, more likely, on potential territory locations. Because we trapped a significant proportion of eventual territory owners at or near their eventual territory sites, most males probably obtain territories on marshes with which they previously gained familiarity.

Floaters probably use a combination of VA-, RHP-, and Lottery-style searching strategies. A majority of males probably get their territories by assuming quick ownership of vacancies, encountered by luck (Lottery). But floater fidelity to particular areas strongly suggests that luck includes elements of a VA strategy. That is, floaters remain in areas with which they have good familiarity and locate within that area vacancies that rapidly accrue high value to them because of their previous knowledge of the sites, or their knowledge of particular marshes and territories facilitates successful challenges of already-established owners. Data on the distances from birth sites and from central subadult and adult trapping locations to eventual territory locations allow us to solidly reject a pure Lottery strategy (that floaters wander randomly until they find a vacancy). If a Lottery system for floaters exists at CNWR, it is restricted to wandering over a 2.5–4.0 km beat, which, during most years, encompasses about 200 to 300 territories. One study, conducted in Ontario, using methods somewhat different from ours, supported a Lottery system for redwings (Shutler and Weatherhead 1994); thus, it is possible that searching strategies differ between regions.

A substantial percentage of males obtain territories by inserting between established residents within and between breeding seasons; a few even manage to challenge and evict owners. It is logical to assume that males that do so are of high RHP and, because we have no reason to believe that any owner should easily relinquish any part of his territory, that superior abilities facilitated their acquisitions. Thus, redwings apparently achieve territorial success through some combination of luck, ability, and motivation – essentially the same mixture that confers success on people.

The most surprising conclusion of our inquiries into territory acquisition is that relative Resource Holding Potential (superior size, aggressiveness, or fighting ability) is not, by any measurements thus far, the primary determinant of territory ownership. Results of study after study that compare the morphologies of owners versus floaters, their relative fighting abilities, and hormonal correlates, fail to support the RHP hypothesis. However, even if a pure RHP strategy does not explain how most redwings acquire territories, relative fighting abilities probably take on increased importance in determining contest outcomes when owners and challengers are of radically different abilities.

Once a territory is acquired and site dominance attained, a male redwing exhibits strong fidelity to the site, within and between years, and even after losing the territory. Site dominance permits most territory owners to hold their territory for one or more breeding seasons and thereby achieve reproductive success. Such dominance apparently arises from a combination of value assymetries between owners and floaters – owners being more highly motivated to escalate and win contests over their valued holdings than are challengers – and a marked disinclination of floaters to engage in escalated battles. The precise sources of value asymmetries, whether owners are more motivated to escalate contests because of their superior time investment in a territory, or greater familiarity with local residents, or ignorance of alternative territory sites, etc., may be difficult to identify (Shutler and Weatherhead 1992), especially because a combination of factors is likely to be influential. Challenger inhibition may be directly related to the prevailing value asymmetries during territorial contests but one cause probably lies with the fact that fighting is dangerous for all participants, with the result that floaters that retreat to find lower-risk paths to ownership later have a higher likelihood of breeding successfully than those that choose to escalate.

Chapter 11

AGELAIUS PHOENICEUS AND HOMO SAPIENS

INTRODUCTION

The previous few chapters have concentrated on the details of redwing breeding and the small, cumulative decisions individuals make to increase their lifetime reproductive success. That these blackbirds are highly successful avian decision-makers and breeders is evident by their ubiquity and sheer numbers. Each year the cycle repeats: the annual spring dispersal to marshes and fields throughout much of subarctic North America, followed in autumn by a massive coming together to form huge flocks, sometimes mixed with other blackbirds and starlings. When such large numbers of individuals of almost any species convene, strong group properties are bound to emerge, particularly in their dealings with other organisms. This chapter details the continuing interactions of the Red-winged Blackbird with another dominant, pervasive species, Man.

I wish that I could state that the interactions of redwings and people are all positive or at least half-and-half; but I cannot. Because redwings and people compete in some areas for the same food, the dominant interaction between them is, unfortunately, negative. The main interaction is that between food producer and undesired consumer. Redwings, as agricultural crop pests, wield considerable economic impact. For instance, redwing consumption of and damage to field corn causes the greatest financial loss attributable to birds affecting any North American crop (Dolbeer 1990). This annual damage has engendered much research interest by state and federal agricultural agencies in redwings and their management. In this chapter I first discuss why redwings are so abundant and then, before describing agricultural damage and management efforts, tell of the beneficial effects redwings have for humankind. The chapter finishes with thoughts on the redwing's future.

ABUNDANCE

Recent estimates place the redwing as either the most abundant bird of North America or in the top 2 or 3 (Dolbeer and Stehn 1979, Dolbeer 1990). Not only are redwings astoundingly numerous today but, contrary to the downward population trends of so many of their songbird cousins, there are almost certainly far more alive now than there were 300 years ago. Why are there so many redwings, and why are their populations in some regions still growing?

Present-day redwing abundance is inextricably linked to the interaction between major land use changes by people and the bird's flexibility in adapting to human-altered environments (Dolbeer and Stehn 1983, Orians 1985). Redwings are specialized for marsh living (see Chapter 4) but they will also breed in a variety of habitats. For example, in California, Orians and Christman (1968) found redwings breeding in emergent marsh vegetation, ditchside, roadside, and fencerow vegetation, riparian areas, weed and brush patches, a variety of trees near water, occasionally in upland areas of mixed chaparral and grass, and in croplands. Contrast this broad habitat flexibility with the narrow requirements of an extreme specialist such as Kirtland's Warbler, which breeds only in stands of young jack pines, and one quickly understands why a species such as the redwing is superabundant, and Kirtland's Warbler is 'super-endangered'.

Other songbirds have benefited from people's alterations of the landscape (Brown-headed Cowbirds, Common Grackles, and Yellow-headed Blackbirds come quickly to mind) but none so much as the redwing. It is the redwing's ability to breed in and near croplands and its ability to exploit these sites for food that are key to its expansion and abundance. Since European colonization began, huge tracts of hardwood forest and dry grassland on the continent's eastern and mid-western sections, previously unsuitable for redwing breeding, have been cleared for agriculture. Redwings now breed over much of this area – in marshes and near agriculture-associated waterways. Redwings, for instance, are the most abundant avian inhabitants in the Midwest of grassed waterways that for decades have been used to prevent soil erosion in agricultural districts (Bryan and Best 1991). Redwings also breed increasingly in upland areas and in croplands themselves, such as in hayfields. Such upland breeding is now common in the East, Midwest, and parts of the Far West, particularly California. Redwing breeding success may be slightly lower in some upland habitats, owing to lower food densities than those normally found in marshes (but see Table 9.1), but the total amount of habitat available for breeding is much greater in uplands; as a consequence, the population of upland breeders now probably equals or even surpasses that of marshes. Redwing success also has been undoubtedly enhanced by a general increase in the productivity of their breeding marshes – a consequence of increasing fertilizer run-off from farmlands (Orians 1985).

Compounding the increase in redwing numbers resulting from expansion into new breeding habitats is enhanced overwinter survival due to agricultural food supplies. Whereas before dense human settlement redwings would have had to forage for natural, highly dispersed prey, redwings now find super-rich foraging patches (crops and crop residues and weed seeds in croplands) scattered throughout their winter ranges. Higher overwinter survival 'ripples' through populations, affecting many aspects of ecology and behavior. The mystery of why subadult males are reproductively mature yet only rarely hold territories and breed, for example, is solved if one imagines that in the not-too-distant past, adult overwinter survival was much lower, creating many more opportunities for subadult territory ownership (Rohwer *et al.* 1980).

The relationship between redwing populations and agriculture is even now not in equilibrium. Redwing numbers in some areas continue to rise as more and more land is brought under cultivation. For example, the St Lawrence Valley of Ontario was in the 1970s and 1980s probably the area with the fastest-growing redwing population as a result of the rapid increases in agricultural acreage; the population in southwestern Quebec more than doubled between the mid-1960s and mid-1980s (Dolbeer and Stehn 1983, Clark *et al.* 1986). Redwings apparently were first observed breeding in Alaska during the 1950s (Shepard 1962) but are now common breeders in a few areas of that state (Kessel and Gibson 1978). Of course, not every population increases. Redwing populations in western North America, on average, were stable during the past 26 years (DeSante and George 1994). Besser *et al.* (1984) reported a decline in breeding males in the northern plain states by about 40% from 1965 to 1981, which they attributed to increasing tillage of the redwing's breeding marshland for agricultural use.

One last attribute of redwings that contributes to their current abundance is that their personality includes a certain tenacity; if you will, the bird is stubborn. Some songbirds are shy or skittish near people and quickly desert areas altered or destroyed. Not so redwings. A common sight over large swaths of North America is redwings breeding in roadside ditches and in narrow cattail marshes contained within highway medians. The bird's strong tenacity to breeding sites, once they are established, borders on the bizarre. This streak is perhaps most tellingly illustrated by male response to the burning of their territories. The refuge personnel at CNWR periodically renew certain marsh areas by burning, in late winter, all standing emergent vegetation, sometimes draining marshes to do so (farmers also each year traditionally burn roadside and cropside vegetation in which redwings breed, as do many highway departments). Male redwing territory owners remained with their decimated territory, perching on whatever vegetation was spared – often on the one or two surviving, burnt, cattail stalks. But the strategy usually pays off; by mid-breeding season, some of the vegetation recovers, and these males are often successful at attracting mates to their territories.

The actual number of redwings alive at any one time is impossible to know, and, because of their huge numbers, all estimates are necessarily imprecise. Still, some researchers, with many years of relevant experience at their command, have offered their ideas. Meanley and Royall (1976) estimated that the USA population in the mid-1970s was 189 million individuals, and Weatherhead and Bider (1979) proposed a figure of 200 million. Richard Dolbeer (1990), who has monitored and researched redwing populations for more than 20 years and probably knows more about their numbers than anyone else, puts the number of redwings at the start of breeding each year at 165 million birds, which then doubles to more than 350 million by early July, when most young have fledged.

The time to count these birds is during autumn or winter, when they gather into large groups. But estimating numbers of individuals in flocks containing millions of birds is bound to be imprecise, regardless of the method used. Aggravating the situation for redwings is that they often feed and roost mixed with other blackbirds and with starlings. A huge southern roost observed each December during the late 1980s contained four species, each occurring in excess of a million individuals. In December 1987, 13 observers counted, roosting at the 2500 hectare Miller's Lake, Louisiana, 3 million starlings, 27.5 million Common Grackles, 53 million redwings, and 20 million cowbirds. Thus, added to the difficulty of establishing numbers is having to distinguish different species within huge groups. The estimate of 53 million redwings at that site was the all-time highest number of a single species ever recorded since 1900 during one Christmas bird count in the USA (Brugger *et al.* 1992). There are sufficient data to be able to state with some confidence that in recent years there were certainly, each April, at least 150 million redwings, and perhaps as many as 200 million – vast numbers. They combine with hundreds of millions of grackles, cowbirds, Tricolored, Rusty, and Brewers' Blackbirds, and starlings to total between 500 million and one billion North American 'blackbirds' that gather into large winter flocks. The locations of major blackbird roosts during the early 1960s are shown in Fig. 11.1; according to R. Dolbeer (personal communication) roosting patterns today are much the same.

Fig. 11.1. During the early 1960s, Meanley and Webb (1965) counted 63 roosts of at least a million or more blackbirds in the USA, mainly in rice-growing regions (map from Meanley 1971).

POSITIVE INTERACTIONS

Esthetics

Redwings have attained a certain conspicuousness in North American popular culture because they are visually striking birds, they are abundant over wide areas of the continent, and because they often breed in close proximity to people. As a result, they are familiar to many people who are not really bird-watchers. This familiarity manifests itself sometimes in strange ways. There is a professional ice hockey sports team, headquartered in the city of Detroit, Michigan, that calls itself the Redwings; nobody quite knows why, but it must have something to do with bird's fierce, athletic determination to defend territories and nests. Many people walking or jogging on suburban streets during spring are familiar with the sudden, surprise attacks of a male redwing swooping to defend one of his nests. Such attacks often rate small articles in local newspapers, usually including a short interview with the local college biologist, who assures the populace that they are not under siege by maddened birds. A few years ago a frequently-aired television commercial for coffee showed an expensively-outfitted young couple with binoculars, in the latest outdoors boutique attire, waiting near a marsh in the misty, gray dawn. As the sun rises, a single territorial song of a male redwing is heard. The man looks at the woman, whispers 'Red-winged Blackbird' in an awed tone, and then the two grab their thermos for some of that famous-brand coffee. Their morning, what with the rare sighting, was obviously successful, their coffee well-earned. And perhaps most notably, for a good many people all across the continent, the annual arrival of male redwings on their territories, and the the sound of their insistent 'conc-a-ree' songs, in February, March, or April, is a sure signal of approaching spring.

Contributions to research and agriculture

Contrary to the impression the reader may have to this point that scientific research on redwings must be limited to the species' ecology and reproductive behavior, the species is of interest to scientists in many fields. Owing to the bird's abundance, ease of capture, and high survival in captivity, redwings are utilized extensively in animal studies. A sample of investigations, culled from the recent literature, demonstrates this usefulness. In neurobiology Hill and DeVoogd (1991) used redwings to study the influence of changing daylength on brain anatomy in birds; Mason and Maruniak (1983) used them as subjects to study the effects of drugs on thermoregulation. In endocrinology redwings have a long history as study animals for hormonal research. Recently, Harding *et al.* (1988) used redwings to demonstrate that in birds, certain androgens, those that can be metabolized to estrogens, are necessary for the expression of male reproductive behavior. In captivity stress Burnet and Cyr (1990) studied redwings to determine how wild birds in aviaries respond physiologically to their captivity. In animal learning and social facilitation Gauthier and Cyr (1990) studied how light/dark cycles influence social facilitation in feeding and locomotory behavior. Mason and Reidinger (1981) and Mason *et al.* (1984) used redwings as subjects in a series of studies of observational learning, e.g. males had preferences for foods they first observed other males consuming. In optimal foraging Beauchamp *et al.* (1987) studied male redwings moving through mazes to examine food searching strategies. In mimicry and palatability studies several researchers, including Ritland (1991) and Evans and Waldbauer (1982) have used redwings as predators in investigations of how palatable insects such as butterflies and bees morphologically mimic similar species that are poisonous or unpalatable.

Redwings also have a long history as subjects in experiments conducted to evaluate potential environmental damage by agricultural pesticides and poisons. A usual protocol is to expose birds through their diet to various poisons at different dosages to determine LD_{50} (lethal dose 50), the dosage that is lethal to half the animals under test conditions. Redwings have been used to study such agricultural chemicals and environmental contaminants as PCBs (polychlorinated biphenyls;

Stickel *et al.* 1984a), organochlorines (DDT, DDE; Clark *et al.* 1988, Stickel *et al.* 1984b), and lead (Beyer *et al.* 1988). Redwings serve as biological indicators of environmental pollution because they are abundant visitors to agricultural fields, but also because they have been identified as highly sensitive to toxins. In fact, after testing 998 chemicals on 68 species of birds, investigators determined that the redwing was the most sensitive of all and they proposed the species as one upon which to base an avian chemical toxicity scale (Schafer *et al.* 1983).

Also, although their reputation is that of granivorous agricultural pest, destroyer of corn, rice and grain, redwings are also insectivores; they eat insects, which, historically, have been a farmer's worst enemy. Some agriculturalists regard redwings as beneficial to crops for their insectivorous ways, and also because they consume prodigious amounts of noxious weed seeds. Even during winter, redwing stomachs contain, in addition to purloined grain and weed seeds, significant insect pests such as armyworms, corn borers, and rootworm beetles (Robertson *et al.* 1978, Dolbeer *et al.* 1984, Bollinger and Caslick 1985a). Dolbeer (1990) approximated that during the breeding season, the estimated 8 million redwings in Ohio consume more than 5.4 million kg of insects, on average, about 53 kg/km^2. However, several investigators have done the arithmetic and found that although flocks of redwings in corn fields, for example, reduce insect damage to the crop, the reduction compensates for only 4% to 20% of the damage to the crop inflicted by the birds themselves (Weatherhead *et al.* 1981, Bollinger and Caslick 1985b). Therefore, on balance, the redwing, at least in some localities, is a pest.

NEGATIVE INTERACTIONS

Agricultural damage

Redwings have been considered agricultural pests ever since European settlement of the Americas; also, since colonial times, people have made efforts to reduce redwing damage by controlling their numbers. I describe how blackbirds impact agriculture by taking two important crops, rice and corn and, for each, examine four questions: (1) how do redwings cause damage, (2) when does damage occur, (3) what are the origins of the individuals responsible, and (4) what is the cumulative damage caused by redwings – is it major or minor? When assessing damage, one needs to bear in mind that redwings often damage crops in concert with other blackbirds (cowbirds, grackles) and starlings and, therefore the relative damage attributable to each can be difficult to determine. For much of the information below I rely heavily on the work of Brooke Meanley and Richard Dolbeer.

Rice growers, it seems, have only themselves to blame for redwings damaging their crops. What other group floods fields into marshes, the redwing's preferred

A flock of male redwings 'rolling' across a field. The rolling effect is created when individuals at the rear of the feeding group continually rise and fly over the other foraging birds, landing at the front of the ever-moving flock, on ground not yet searched.

habitat, and then stocks them with bird food? In colonial times, rice was cultivated along the East Coast in lowland areas in the Carolinas and Georgia, but by the late 1800s, most rice was grown in coastal Texas and Louisiana, areas that adjoin millions of acres of blackbird breeding marshes. Rice-growing spread in the early 1900s to eastern Arkansas (directly under the continent's largest blackbird flyway!), and then to nearby parts of Mississippi, Tennessee, and Missouri. Depending on location and variety, rice is planted from March through late June, and is harvested 100 to 170 days later. About 10–20 cm of water is maintained in ricefields to control weeds and to provide requisite moisture.

Damage to the rice crop by redwings was far greater before the advent of mechanized harvesting and drying. Harvested rice would be left in the field for up to 6 weeks to dry, during which it was exposed to blackbird foraging. Redwings now damage rice by attacking it as it sprouts and ripens, by pulling new sprouts, and by removing ripening grains – 'the bird pulls the rice plant out, places the grain crosswise in its bill and then exerts pressure and revolves the grain at the same time; the hull falls away . . .' (Meanley 1971). Redwings cause heavy damage to growing rice especially in southwestern Louisiana and eastern Texas. Redwings also feed later on the stubble of harvested fields. In Arkansas, redwing stomachs contained only 13% rice in May to July (during local breeding), but 67% rice in August to October, and 54% rice in November to April.

Rice losses occur in summer and autumn after redwing breeding is completed; local populations inflict the most damage (Meanley 1971, Brugger and Dolbeer 1990). Northern redwings, who do not begin their migrations until mid-September or October, do not reach the rice states in time to feed on ripening rice. In fact, only about 1% of the northern populations move into rice states before harvesting is completed. However, banding recoveries in one study showed that 84% of blackbirds in rice states in winter had migrated from more northerly areas, from Ohio to the Dakotas to Kansas, with most coming from the north-central area and a smaller portion originating in southern Canada (Meanley 1971). As harvesting ends in summer and autumn, millions of redwings are attracted to the last few unharvested fields, sometimes causing significant damage. From the end of harvesting until spring, redwings feed on waste rice and weed seeds left in the fields, performing, in effect, a clean-up service for rice growers.

Financial loss from blackbird foraging arises from a combination of decreased yield, reduced grade of damaged rice and, hence, reduced value, and increased costs for replanting (Brugger 1988, Brugger *et al.* 1992). Damage to rice by blackbirds in Arkansas was estimated 25 years ago as causing loss of about 4% of the crop – in 1971 the damage bill came to about $10 per acre for 426 000 acres. However, that amount was approximately equivalent to the amount lost as waste rice when the industry switched from hand to mechanized harvesting (Meanley 1971). Annual blackbird damage to Texas and Louisiana rice crops recently was estimated to be about $4 million per state (Brugger *et al.* 1992). The California rice industry still considers redwings to be their most damaging pest, spending, on average, over $30 per acre on control efforts (Dolbeer and Ickes 1994).

Redwings damage field corn (maize) by 'slitting husks and pecking out the contents of exposed kernels' (Dolbeer 1980). They damage corn throughout the period of kernel maturation (Dolbeer *et al.* 1984), but most damage occurs during early kernel ripening because once kernels harden, near harvesting, the birds have more difficulty handling and digesting it. Most of the damage in Ohio, for example, is caused by local breeders that, in late summer, congregate in large roosts, usually within about 60 km of their breeding territories (Dolbeer 1978). Corn is often planted in early May, with kernels ripening and becoming vulnerable to blackbirds about 75 days later. Most Ohio cornfields experiencing blackbird damage are attacked in August and September, and are located within 10 km of a major roost. Redwings may be attracted to cornfields initially by insect food, such as rootworm beetles (Dolbeer 1990). Migration starts in Ohio in late October and continues through November; few redwings winter in the state.

Most corn damage by redwings is apparently inflicted by males; females, perhaps because of their smaller bills, prefer other foods. In August and September, corn made up about 75% of the diet of male redwings in Ohio, but only about 6% of females' diet; similar percentages have been identified in other corn-growing regions (Dolbeer 1980).

Financial loss caused by blackbirds (redwings and grackles combined) foraging on field corn was estimated during the 1970s in Ohio to be $4 million to $6 million per year; less than 1% of the annual crop was lost. About 97% of cornfields had less than 5% damage attributable to blackbirds. Dolbeer (1990) estimated that continent-wide, redwings destroy less than 1% of the corn crop, a mere 2.5 to 3.0

million bushels (360 000 tons). Thus, the amount lost to redwings is a very small fraction of the total. The loss is much less than the amount of corn lost to insects, weeds, disease, and fungi (about 20% of the crop) and to mechanized harvest wastage (about 5%; Jugenheimer 1976).

Another crop attacked by redwings is sunflower, which is sometimes heavily damaged. (From many years of trapping redwings using sunflower seeds as bait, I can attest that they find the small, black seeds difficult to resist.) In North Dakota, 24% of the crop over a 70 km^2 area was lost to blackbirds in 1972; 16 fields in Ohio in 1973 suffered from 5% to 80% loss from blackbirds (Dolbeer 1975). However, most crop losses attributable to foraging blackbirds, including losses to millet, sorghum, oats, barley, flax, and soybean, are relatively minor. Even a huge roost that gathers annually in western Tennessee amid wheat, corn, and soybean fields inflicts, when compared with overall production, very minor losses (White *et al.* 1985), although the daily loss to all crops around that roost was estimated at $3000. Blackbirds also eat grain at livestock feedlots, causing some financial loss to farmers and ranchers but again, the amount taken is often relatively small (White *et al.* 1985). Redwings are not the prime suspects at feedlots. In fact, in one detailed study, at the west Tennessee roost, redwings preferred to forage in corn fields; grackles, cowbirds, and starlings were the main feedlot denizens.

Management and control

Decades of research on the economic impact of blackbirds on North American agriculture demonstrates that financial losses are, on average, and when considered as a fraction of total crop production, minor – a consistent finding for avian crop damage in general (Dyer and Ward 1977). The question then is, if less than 1% of the North American corn crop is damaged by blackbirds, why worry about it? Why is so much effort spent formulating and evaluating control agents? The answer is that, although overall damage might be relatively minor, in localized areas it can be heavy and losses to particular farmers can be significant. Although 97% of corn fields in Ohio experience less than 5% blackbird damage, 3% of fields sustain heavier damage. It is for these strongly-impacted areas that management is primarily aimed.

Efforts at reducing redwing crop damage go back at least to the 17th century, and perhaps further, before European settlement, when Native Americans may have posted people in fields to guard corn (Dolbeer 1980). Early control techniques consisted of guarding fields, scaring flocks, and shooting or burning roosting birds. One early Massachusetts law, aimed at reducing populations, required every man to kill six blackbirds before his marriage. Other laws demanded that landowners harvest an annual quota of blackbirds (Nero 1984). In recent times, nuisance blackbirds in roosts have occasionally been killed in huge numbers, up to a million at a time. Although in the USA redwings are a protected species under the Federal Migratory Bird Treaty Act, amendments permit farmers to kill blackbirds that are damaging or, by their proximity, liable to damage, their crops. Redwings are classified as agricultural pests in Canada and, as such, their killing is permitted.

Farmers and government agricultural specialists now have a spectrum of methods available to kill, repel, or otherwise interfere with blackbirds' ability to damage crops. Particular methods go in and out of style when research shows they are not effective, when chemicals are shown to be environmentally harmful, or when public unease grows over mass wildlife killing, and so new methods are continually evaluated. The general consensus, as one might expect, is that usually there is no single safe, effective, and economical means that provides adequate crop protection, and so the usual prescription is to combine several methods. Economic considerations, in particular, are of foremost concern to farmers – 5% blackbird damage to a corn crop must be balanced against the cost of sometimes expensive control efforts, such as aerial application of chemicals, the cost of which could equal or exceed that inflicted by the birds.

Alleviating crop damage by blackbirds can be approached in two ways – managing agriculture and habitat or controlling birds; the former is almost always easier, environmentally safer, and has the esthetic and political appeal of excluding harassing or killing wildlife. The various control strategies now in use or that were tested in the past are discussed below.

Crop timing

The dates at which fields are planted and mature influence the extent of blackbird damage. Therefore, farmers, knowing the stage of development that their crops are most susceptible and also the chronology of blackbird roosting in their localities, sometimes time planting to avoid most damage. Also, within susceptible localities, farmers strive not to have their fields mature unusually early or late because such fields concentrate birds and often experience the heaviest damage. A few days can sometimes make a significant difference. For instance, corn to be sold as fresh, harvested in one Ohio county had only 1% of ears damaged by blackbirds, but corn for canning, harvested 5 days later, had 26% damage (Stickley and Ingram 1977).

Hybrid varieties

Different varieties of the same crop can differ in their susceptibilities to blackbird damage. For example, various corn hybrids differ in the thickness and strength of their husks and in the length of the extension of husk beyond the ear tip. Tougher husks reduce the blackbirds' ability to slit them to expose kernels, and longer extension of the husk, in particular, apparently makes it much more difficult for birds to expose the ear. Thus, the advice to farmers: when all other factors (yield, maturation date, etc.) are equal, choose varieties with longer husks (Dolbeer 1980).

Alteration of habitat

Habitats can be altered or manipulated in a number of ways to reduce crop damage. For blackbird-resistant varieties of crops to be most effective there must be nearby alternative sources of food (the same applies for flock scaring strategies). Some farmers delay plowing under some of their harvested crops, leaving residues available for blackbird foraging. Sometimes fields are planted with cheap grains specifically to draw blackbirds and other wildlife away from main crops. Increasing habitat diversity near roosts and providing alternative food near roosts also can reduce blackbird damage in surrounding areas (Wiens and Dyer 1975). Another method to reduce blackbird numbers in particular areas is to clear brush and woods that border fields, because these areas attract birds because of the cover they provide.

Insect and weed management

Blackbirds in some cases are first attracted to agricultural fields by insect abundance. Redwings have been observed spending up to a week in corn fields, 'loafing' and feeding, particularly on rootworm beetles, before any corn damage occurs. In these cases, insecticides applied to fields just prior to the time the corn becomes susceptible to blackbirds should reduce damage. Together with clearing brush around fields and reducing weeds within fields with herbicides, this should also reduce the attractiveness of the field to birds, for the same reason, i.e. lack of cover.

Controlling the blackbirds themselves is sometimes successful in reducing local populations, but almost always there are associated negative effects on the birds, on the environment, or on crops in other areas, to which frightened or displaced flocks gravitate.

Control at the roost

Heavy agricultural damage by blackbirds usually occurs when crops are in close proximity to areas where large numbers of birds collect in roosts in late summer, autumn, and winter. However, the birds, by gathering in large numbers in predictable locations, leave themselves vulnerable to mass control efforts (Weatherhead 1981).

Simply scaring birds from roosts is rarely successful at dispersing them because they usually quickly return or move nearby *en masse*. In the 1800s, dry areas of roosts were set aflame at night, the flammability enhanced by strategically placed straw, and escaping birds were shot-gunned. Redwings then were also killed in large numbers at roosts to sell at market for human consumption as delicious 'reed-birds' (Neff 1942, Bent 1958).

In the 1950s, roosts were bombed to evaluate the effect of such drastic treatment on reducing blackbird numbers. One roost in Arkansas that contained an estimated 20 million blackbirds had 100 dynamite charges placed in it, set to explode simultaneously at night. About 200 000 birds were killed in the massive explosion. The practice, however, was deemed impractical for several reasons, which I leave to the reader's imagination.

A lethal method still occasionally implemented is the use of aircraft to spray inhabited roosts with detergent compounds, often PA-14 (Tergitol), an avian 'stressing' agent developed by the US Fish and Wildlife Service. These chemicals remove protective oils from feathers, lower surface tension of water on feathers, and so enhance wetting of the skin. If applied directly before or during cold, wet weather, birds die of hypothermia. Millions of birds have been killed using this method in Tennessee and Kentucky (Fig. 11.2). The method has its proponents because birds are killed, not simply moved elsewhere, because it kills only those

Fig. 11.2. A flock of blackbirds and starlings swarming into a huge roost near Milan, Tennessee in January 1977, shortly before a surfactant, or 'avian stressing' compound, was sprayed on the roost from a helicopter (from White et al., 1985).

birds at the roost, which usually are mainly the target species, and because it leaves little harmful residue. Still, the method has drawbacks. It can be used only in upland areas because of water pollution problems. It kills massive numbers of birds, thereby eliminating their useful properties, such as their weed-seed and insect eating. Most importantly, it usually proves to be only a short-term solution; even after up to 1 million birds are killed with surfactants, birds from surrounding areas soon move into the area, repopulating a roost (Dolbeer 1980). The abundance of birds in fields within 40 km of a huge roost in Tennessee returned to pre-application levels within 2 weeks after a surfactant applied there killed 1 million birds, more than 90% of its population (White *et al.* 1985).

Control in the field

Blackbirds can be killed while in agricultural fields, repelled from fields by being conditioned to avoid particular crops, or scared away; the last method enjoys the greatest popularity and success.

Some farmers apply lethal poisons to their fields. In the past, strychnine was used (Meanley 1971). Now, powerful insecticides such as parathion are used to treat bait corn or seeds, which is then spread near crops to be protected (Stone *et al.* 1984). Usually, relatively small numbers of birds are killed. Moreover, these poisons are nonselective, killing also nontarget species, and the poisons move up food chains, also killing animals that prey on blackbirds, such as falcons and hawks. Blackbirds are also shot while in fields or shot in conjunction with the operation of scaring devices. In addition, some farmers operate large blackbird traps adjacent to their crops. These traps, baited with food, water and live decoy birds, can catch thousands of individuals, which are subsequently killed. Twenty large traps operating in an Arkansas rice area during a growing season caught 40 000 blackbirds (Meanley 1971). At feed lots, a poison known as Starlicide, particularly toxic to starlings, is often spread about to kill that species.

Aversion learning to repel blackbirds has been tried. Methiocarb (Mesurol) is a broad-spectrum insecticide that may repel blackbirds from treated fields and render treated crops unpalatable (Rogers 1974); the birds, generalizing the unpalatability, should, in theory, avoid the same crop in other locations. Corn fields treated with methiocarb shortly before harvest have been shown to have significantly less damage than untreated fields. Methiocarb kills insects, which might also account for reduced blackbird activity in treated fields (Woronecki *et al.* 1981). One drawback is that in wet weather, methiocarb may have detrimental effects on plant crops.

Blackbirds can be prevented from entering fields or scared from them by the use of frightening devices. Over the years a number of types have been developed. Common are electronic noise-makers, propane exploders (gas-charged cannons that fire no projectiles but make loud noises at irregular intervals), broadcast taped distress calls, radio-controlled model airplanes, and all manner of visual frighteners: helium balloons tethered over fields, hawk-shaped kites, scarecrows and, relatively recently, reflective tapes and ribbons, metallic red or silver in color, strung over fields. An important tactic in using these devices is to scare the birds away quickly before they have a chance to adjust to a new location; the longer the birds are in a particular field, the harder it is to drive them away. Even with moving noise-makers or exploders around a field, or accompanying scare devices with occasional shooting into the flock, birds often quickly habituate to the noise and interruptions and refuse to leave.

Finally, chemicals have been pressed into service as insidious frightening agents. A compound, 4-aminopyridine (Avitrol) can be used to treat cracked corn, for instance, which is then mixed 1:100 with untreated corn and spread by airplane over a portion of a cropfield. Blackbirds that ingest even one treated corn particle die slowly, flying erratically and giving distress calls. The abnormal behavior often causes unaffected birds in the flock to leave (Dolbeer 1980). Although first reports were that the method showed great promise for reducing blackbird crop depredations, further research suggested that it was only partly effective. Also, like other chemical methods, Avitrol also killed nontarget animals.

New methods of control are continually researched. In keeping with modern human reliance on chemicals as cures, these include chemicals that may be applied to fields which repel blackbirds with their odors (Avery and Nelms 1990) and others that cause sterilization of individuals that consume treated baits (Lacomb *et al.* 1986); work also continues on improving blackbird resistance of new crop varieties. Again, the best contemporary advice to farmers with blackbird infestations is to take an integrated approach: plant pest-resistant varieties of crops, control insects, provide alternative food sources away from crops and, when all else fails, scare the birds away.

THE FUTURE FOR RED-WINGED BLACKBIRDS

What does the future hold for redwing populations during the next 50 or 100 years? If such a question were asked in a world empty of people or, at least, devoid of their massive environmental perturbations, the answer would be easy to formulate, almost certain to be correct, and it would be the same answer to the same question asked about any species that occupied a dominant position in animal abundance hierarchies, as does the redwing: subject to natural biotic and abiotic ecological processes – to competition and predation pressures, to habitat changes, to climatic shifts, etc. – redwings could be expected to continue with healthy populations, experience periodic population peaks and troughs, and eventually decline. But man's influence over habitats and animals prevents simple answers about 'average' expectancies. Often, the question, 'What is a particular species' future in the near term?', is equivalent to asking 'What is the species' future *vis-a-vis* people's current and expected influences on it?'

For many avian species, interactions with people have been no less than disastrous; extinctions of some birds and drastic population reductions in others are well-documented. For redwings, fortunately, their interactions with humans have, on balance, served only to increase their numbers. Thus, most prognosticators would agree on a generally bright redwing future. Consider these facts. Redwings are one of the most numerous of birds; most populations are stable or increasing. When and where other songbirds have drastically decreased in numbers from interactions with humans or other species, redwings have thrived. As more acreage is shifted from forest to farming (the rate of which has slowed but not stopped, particularly in southern Canada), and as damaged wetlands are restored (e.g. in the US Midwest), redwing populations should only benefit. Recent avian invasions and range expansions that have, in some cases, severely reduced other birds have not harmed redwings. In particular, redwings have not suffered noticeably from the invasion of European Starlings, of which 60 were released in New York in 1890 and which rapidly multiplied and spread to the present coast-to-coast population numbering in the hundreds of millions. Starlings, although competing often with redwings for winter forage, are cavity nesters and therefore do not compete with redwings for breeding habitat.

Also, although redwings serve as common hosts for the parasitic eggs of Brown-headed Cowbirds (as well as at least an occasional host for other, less studied cowbirds, such as the Bronzed Cowbird of the southwestern US, Mexico, and Central America), the parasitism does not significantly impact upon redwing breeding as it does upon that of many other songbirds. Cowbirds probably originated in the Great Plains of central North America but, with the continent-wide conversions of forests to farms, they have expanded, as open-country birds, into most North American regions below the Arctic – a process that still continues. Redwings do not eject cowbird eggs from their nests, but nonetheless, are relatively poor cowbird hosts because their young are larger than cowbird nestlings and can therefore compete well in the nest. Cowbird parasitism has not significantly reduced redwing populations.

Although people will doubtless continue efforts at blackbird control in agricultural areas, those efforts to date have had rather little effect on overall population numbers and they become increasingly unpopular. Most people frown on mass killings of wildlife. Politicians from farm states previously have supported control efforts such as roost killings, but environmental movements have arisen in all areas, and politicians are cognizant of environmentalist concerns. In particular, the best chances agriculturalists have of significantly reducing huge blackbird roosts reside with chemical agents; however, large-scale applications of chemicals, even ones claimed to be environmentally safe, become, in many areas, increasingly suspect and so unlikely. Furthermore, if control efforts in the past have successfully demonstrated anything, it is the incredible resilience of redwings and other blackbirds in flourishing despite these onslaughts.

Thus, conflict between farmer and redwing will continue, but the redwing should not suffer unduly for it. Farmers will come to rely on nonlethal, integrated blackbird management techniques (Dolbeer 1990), and redwings, as long as they maintain their open-country habits and as long as healthy, productive wetlands exist for their breeding, will survive and prosper. In summary, there is every reason to suppose that redwings, presently in an enviable position among the continent's avifauna, will continue to be a dominant, pervasive species for many generations, human and blackbird, to come. Still, given the capriciousness of nature, and the vicissitudes of people's interactions with their environment and with other species, their fads and fancies, one must exercise predictive restraint. After all, any ornithological prognosticator 200 years ago would surely have predicted a rosy future for the now-extinct Passenger Pigeon populations, which then comprised an estimated 1–3 billion birds.

The redwing's future in the longer term, millennia from now, is, of course, unknowable. However, the icterines are a rapidly evolving group (Orians 1985), and the genus *Agelaius* is actively evolving – witness the probable origin in what is now California of the Tricolored Blackbird from the redwing less than 10 000 years ago, and the various genetically diverse subspecies of redwings distributed about the continent (Gavin *et al.* 1991; see Chapter 2). Thus, we have some indication that changes for the redwing may be just around the evolutionary corner.

Chapter 12

CONCLUSIONS AND FUTURE DIRECTIONS

CONTRIBUTIONS OF REDWINGS TO ECOLOGICAL AND BEHAVIORAL RESEARCH

When at scientific conferences I meet fellow ecologists and ornithologists, after exchanging information on our interests, I am often challenged with the query, 'Why do you and others continue to work on redwings – don't we already know enough about them?' The short answer to the question, which I hope by this point in the narrative has been made abundantly clear to the reader, is no, we do not know enough; we have much more to learn. The large amount of scientific knowledge on redwings, summarized in this book, should not be viewed as an inhibitor of further research, but a stimulator. The reason is that the body of existing information on redwings is a tremendous boon to researchers.

The main advantage that such information provides is that new findings can be interpreted in the context of a known biological framework. A researcher journeying to a remote region of the Earth to study species never before scientifically monitored will spend years learning their basic biology before interesting questions can be formulated and tested, but with redwings, much basic biology is already in hand; we can proceed immediately to test new ideas and hypotheses and, importantly, we can interpret new results with respect to that which is already known. Two examples from my own work illustrate the point. My study of male redwings seeking territories by exploiting unstable boundaries between adjacent owners (Beletsky 1992) was stimulated by long observation and long-term information collected during the Columbia National Wildlife Refuge (CNWR) study and by other investigators that indicated that many floaters obtain territories by inserting themselves between established owners. Similarly, an analysis of male breeding success that demonstrated a reproductive advantage to males with territories adjacent to long-familiar neighbors (Beletsky and Orians 1989c) was possible and, indeed, pursued only because we had long-term information about territory histories and knowledge of the cooperative workings of redwing breeding colonies.

Although there are many aspects of redwing biology to explore further, research on the species to date has already contributed vastly to the sciences of ornithology and behavioral ecology. (I have always found it intriguing that so many major advances in our understanding of animal ecology and behavior accrue not from study of exotic or rare animals, but from ones so common as the redwing and, among birds, others such as European Robins, Great Tits, Pied Flycatchers, and Song Sparrows.) The chapters of this book reveal the major areas of the redwing's contributions. The most important may be on our understanding of polygynous mating systems. Why and how such systems developed, as well as mechanisms of mate choice and peculiarities of nesting, have been rendered far less obscure by research on redwings (see Chapter 6). One of the most comprehensive literatures on the effects of sexual selection on avian morphology – on size and ornamentation – and behavior is that on the redwing (see Chapters 3, 5; also Searcy and Yasukawa 1995). Our level of understanding of redwing communication is arguably unsurpassed among other birds; research on redwing signalling, to the degree that it generalizes to other species, has strongly augmented our understanding of animal communication. Other areas of knowledge to which redwing research has contributed, in addition to those mentioned in Chapter 11, are in our understanding of regional variation in breeding ecology and behavior and in comprehending the competitive versus cooperative forces at work that shape animal behavior, particularly in highly social or colonial aggregations; the last two areas are addressed below.

FUTURE DIRECTIONS

As I reviewed the literature on redwings during the preparation of this book, I found the regional differences that I encountered in their breeding ecology and behavior particularly compelling. The variation is significant, interesting from behavioral and evolutionary ecology viewpoints, and cries out for additional study; Table 12.1 lists some of the known major differences between eastern and western populations. Productivity differences in typical breeding habitats may be responsible for some of the variation. Marshes of the West are usually more productive of emergent insects, the food of redwing young, than eastern marshes. The chief reason may be that marshes of the arid West occur frequently on lakes with no outlets; these habitats concentrate nutrients and often have no fish to depress populations of aquatic insects.

Larger average territory sizes in eastern and midwestern populations versus those characteristic of the West (Table 12.1) probably reflect differences in habitat productivity, especially as most western redwing breeding is restricted to marshes, whereas many eastern redwings breed in less productive uplands. Differences in average harem sizes and proportions of breeding males that feed nestlings also may be related to habitat productivity. But why do females in the West actively solicit extra-pair copulations (EPCs) from territorial neighbors but females in the East do not solicit and even actively resist EPCs? One possibility is that, because males in the West rarely feed young they cannot 'retaliate' against females who 'cheat' on them by withholding that parental care; western females, in this view, are 'free' to pursue multiple sires for a single brood. Eastern females, however, when their 'cheating' is detected by their mates, may have to pay a price (Gray 1994, Weatherhead *et al.* 1994). Another potentially important difference is that western females very rarely, if ever, have young that result from EPCs with nonterritorial males (Gray, 1994), but a small percentage of redwing young in the East may be sired by such males (Weatherhead and Boag, 1995). The very different conclusions reached about how male floaters search for territories in our Washington state population (Beletsky and Orians 1996) and in an Ontario population (Shutler and Weatherhead 1994) may indicate real variation in searching strategies (males in Ontario employing a strategy of searching far and wide for vacancies, those in Washington remaining within quite restricted areas, gathering information, and eventually filling vacancies or inserting) or may be an artifact of how the information was collected during different research projects (see Chapter 10).

Table 12.1. Comparison of breeding attributes of eastern and western redwings

Attribute	Eastern and/or midwestern populations	Western populations
Breeding habitat	Marshes and some upland areas such as grasslands, pastures, and hayfields	Primarily marshes (some uplands in California)
Territory size (range of means from Table 8.2)	270 to 5816 m^2 (*n* = 13 studies)	152 to 2300 m^2 (*n* = 8 studies)
Harem size (grand means from Table 6.2)	2.4 (*n* = 17 studies)	4.2 (*n* = 7 studies)
Percentage of males that feed nestlings (ranges from Table 8.3)	0–80% (*n* = 9 studies)	0–12% (*n* = 8 studies)
Females solicit extra-pair copulations (EPCs)?	No (Westneat 1993)	Yes (Gray 1994)
Percentage of young sired by extra-territorial males	24% to 28% of nestlings (Gibbs *et al.* 1990, Westneat 1993)	35% of nestlings (Gray 1994)
Nesting success of females enhanced by EPCs?	No (Weatherhead *et al.* 1994)	Yes (Gray 1994)
Male territory acquisition by	Lottery strategy? (Shutler and Weatherhead 1994)	VA/RHP strategies (Beletsky and Orians 1996)

Prior to testing hypotheses that explain regional differences in breeding behavior, it would be illuminating to have even more comparative data on breeding ecology. For example, more data on the breeding behavior and characteristics of redwings in the northern parts of their range (Alaska and the Yukon) and also in the southern parts (Mexico and Central America) would be helpful for discerning geographic patterns. For instance, if the main strategies of male territory acquisition differ between Ontario and Washington, what might be the pattern in the tropical marshes of Costa Rica? That significant geographic variation in behavior has been already described for redwings in parts of their range hints that even more variation may be discovered over the species' broad range. Thus, the redwing offers a system in which variation in behavior within the same species can be used to make powerful tests of hypotheses of the evolution and function of behavior. Also, close icterine relatives, especially other marsh nesters that often breed alongside redwings, such as Yellow-headed and Tricolored Blackbirds, provide additional, between-species testing opportunities.

The presence of significant between-population differences in breeding ecology suggests the need for an admonition: redwing researchers need to be careful when interpreting and communicating results not to rush to generalize findings to the species as a whole unless the information warrants such treatment. For instance, that a female chooses a breeding situation primarily by factors other than the identity or quality of the male that owns the territory on which she will place her nest (see Chapter 7) is a robust, general result of several studies that were conducted in more than one region, and thus one that is generally true of the species. However, it has become clear that populations differ sufficiently in several aspects of breeding that results of individual studies should be interpreted to apply only to a local population. Results from single-population studies cannot be generalized even regionally: important female settlement patterns observed during one Washington study (Langston *et al.* 1990; see Chapter 7) were very different from those we found at the CNWR site – the two studies occurred at locations only 50 km apart!

Another fertile field of future research stems from advances in molecular determination of parentage. DNA fingerprinting to identify the genetic versus 'social' parents of each nestling within a brood is a powerful tool for illuminating an array of male and female breeding behaviors. Much of the behavior of breeding males and females may need to be reinterpreted in light of this new information. This field is in its infancy, but is beginning to provide new insights, for example, on female EPC behavior, on male mate-guarding and parental care, and on male nest defense (Gray 1994, Weatherhead *et al.* 1994, Westneat 1994, 1995).

The redwing will continue to be a strong contributor to research on the dynamics of group living, as biologists avail themselves of the many advantages, detailed in this book, of working with this flocking and colony-breeding species. There is much to be learned about the benefits of social familiarity and social cohesion. In particular, consideration of the social histories of banded individuals should provide further insights into competitive and cooperative aspects of colony behavior in general, and of redwing breeding biology in particular. Previously, most studies treated individual males and females as 'black boxes' – generic male or female birds without individual identities – but individuals have social histories, relationships, and bonds that affect their behavior, interactions with conspecifics and, ultimately, reproductive success. A case in point is the fact that a plethora of field investigations of the functions of female–female aggression in redwings, including my own, has failed to take previous social histories into account. If the reader will recall from Chapter 7, the question is why primary (first to settle and nest) females on male territories are often aggressive toward lower-ranked females when most studies show no detrimental affects of increased nesting densities, often no reproductive advantage to high rank, and, in some regions at least, females have no reason to compete for male parental care because most males do not help feed the young.

A secondary female with whom a primary female shared a territory during the previous breeding season, may elicit a milder response to a settling attempt than would a new, unfamiliar prospective harem-mate. Part of the reason could be simply the benefits of social familiarity – enhanced social facilitation in foraging, enhanced cooperation or group effectiveness in nest defense, etc. Also, incorporating unfamiliar individuals into a social group is costly, in terms of time and risk of injury during fights over dominance. Previous harem-mates may have dominance issues already settled and can devote their time to other activities that might improve their breeding success, such as additional foraging. The function of female–female aggression may also differ among regions. In the East and Midwest, where harems of two to three females are most common and males often feed young at the primary nest, most harems each year, because of the 40% annual mortality, must comprise females relatively unfamiliar with each other; the aggression that primary females direct at secondaries at these locations may function in defending their primary status and/or delaying nesting of secondary females. However, in the West, where harems are often much larger and males usually do not feed nestlings, the main function of female–female aggression may be to re-establish and maintain dominance hierarchies among females who have bred near each other during one or more previous breeding seasons (Beletsky and Orians 1996). It is even possible that harems or marsh groups of females winter together and each spring use aggression to reformulate their social hierarchies for another round of breeding.

As for males, there is anecdotal evidence that they form long-term territorial groups that may stay together throughout the year. A striking cohesiveness of territorial males at CNWR suggests that feeding flocks and roosts may have an internal social organization whose structure reflects that during the breeding season. In February and early March in eastern Washington, males sleep on their territories but, after a brief morning advertising period, they leave, usually in small groups, to forage in agricultural fields. My colleagues and I have observed repeatedly that when the males resident on a particular marsh return in late afternoon, they do so as a group. Typically, several males arrive in formation, each dropping down quietly to their respective territories as the small flock traverses the marsh. Returning *en masse* suggests that males depart feeding areas simultaneously, and therefore, that the group spends the day feeding together, perhaps as one component of a much larger flock. The broader implication is that there may be 'marsh units' of social organization – neighborhood groups of males that associate both during and outside of breeding, much as females may remain in harem or marsh groups.

A twist on this social cohesiveness is that males may maintain contact during the day and return together to their breeding marshes not only because of foraging or survival benefits of their association, but also out of mutual suspicion, each male concerned that portions of his territory could be usurped by others if he is not there. Neighbors, on occasion, can be cooperators, but they are always strong, direct competitors. Once, I was leading a novice field assistant to one of our main redwing marshes, explaining all the while that many of the males with territories on this particular marsh had long-term bonds because they had occupied adjacent territories for many years and thus, that we could expect little aggression over territorial boundaries. As we neared the marsh we heard thrashing sounds. We parted the vegetation to look. Two color-banded males who had held adjacent territories for 4 years were rolling around on the ground, locked in combat, presumably arguing over the position of their boundary.

That males resident on the same marsh early in the breeding season forage together far from their territories is demonstrated by the observation that color-banded CNWR males that we knew bred together showed up in the same trap at the same time. Because we placed color bands on unbanded birds in color-sequence orders, one unintended consequence of this social cohesiveness is that groups of males sometimes appeared later, resident on the same marsh wearing color combinations that we had issued consecutively to individuals that were caught in the same 'trapload'. For instance, three males wearing colorbands that were placed on unbanded males that were trapped together were later found to own adjacent territories on a marsh that was about 3 km from the trap where they received the bands. Either they all acquired territories there after being banded, which is possible but unlikely, or on that fateful day three unbanded males departed their territories and journeyed together to the trap. Even more striking, male marsh groups may persist for many years. On 24 February 1988, for instance, we trapped together at the core study site three males that had held territories for many years on the same marsh area located about 3 km away. Not only were they clearly long-term companions, two of the three had been initially banded as subadult floaters at the same trap on the same day 9 years before, on 19 April, 1979.

Marsh groups of males may remain together year round, even during migration, and in winter roosts. If so, then autumn and winter roosts may consist not of random aggregations of birds from many regions but structured groups of males that breed near each other. A roost could contain thousands of such breeding groups, aggregating temporarily for mutual benefit (see Chapter 4). The advantages of maintaining a social organization during nonbreeding periods might be similar to those postulated for breeding assemblages, namely the advantages of associating with familiar individuals, especially as they may relate to enhanced abilities to find food or good roost sites or detect and escape predators.

Maintaining the integrity of their breeding groups – having the same individuals breed together each year – may be important to males because, on average, males that breed next to long-term neighbors have greater success than males that breed next to new neighbors (Beletsky and Orians 1989c). This positive effect of social familiarity on reproductive success may provide an evolutionary basis for the development of cooperative behaviors, or ones that seem cooperative. The call alert system employed by the males in a breeding colony is one example (see Chapter 5). An intriguing observation that has been made by several researchers is that, when a male is temporarily absent from his territory, a neighboring male will enter the territory to drive off trespassing floaters; it is clearly in this good samaritan's interest to retain his current neighbors and preserve his breeding group, for preservation of the breeding enhancement effect and because establishing boundaries and relationships with new neighbors can be costly. Another curious observation was that during a few of our territorial male removals (see Chapter 10), after neighbors had expanded to take the removed males' territories, they relinquished the territories to the original owners immediately upon their return without a fight. There are two explanations. Either the 'expanding' neighbors did not value their new holdings, which appears unlikely, or the potential enhancing effect on their survival and breeding success of having back long-term neighbors (that is, maintaining the integrity of the breeding group) equalled or exceeded any benefit they could have accrued by fighting the returning owners for the territories. Thus, for many research purposes, individuals within breeding colonies (of redwings or other species) should not be considered interchangeable; rather, the lesson is that individual identities and social histories may significantly influence behavior.

I could continue to extol the virtues of redwing research, citing their contributions to ornithology and raising anecdotes and examples from the rich body of existing information that suggest productive and exciting future study, but this would simply belabor my main point: the Red-winged Blackbird has contributed mightily to several fields of research and should continue to do so for some time to come. In fact, my usual answer to those who ask me why we continue to study these well-researched birds is that, given what information is already available and what these birds have to offer, I am surprised that even more biologists do not study them.

Appendix

COMMON AND LATIN NAMES

Birds

American Coot	*Fulica americana*
American Crow	*Corvus brachyrynchos*
American Redstart	*Setophaga ruticilla*
American Robin	*Turdus migratorius*
Barn Swallow	*Hirundo rustica*
Black-billed Magpie	*Pica pica*
Blue Jay	*Cyanocitta cristata*
Bobolink	*Dolichonyx oryzivorus*
Brewer's Blackbird	*Euphagus cyanocephalus*
Bronzed Cowbird	*Molothrus aeneus*
Brown-headed Cowbird	*Molothrus ater*
Chestnut-capped Blackbird	*Agelaius ruficapillus*
Cliff Swallow	*Hirundo pyrrhonota*
Common Grackle	*Quiscalus quiscula*
Corn Bunting	*Miliaria calandra*
Eurasian Skylark	*Alauda arvensis*
European Starling	*Sturnus vulgaris*
Gray Catbird	*Dumetella caroninensis*
Great-horned Owl	*Bubo virginianus*
Great Reed Warbler	*Acrocephalus arundinaceus*
Great Tit	*Parus major*
House Finch	*Carpodacus mexicanus*
House Wren	*Troglodytes aedon*
Indigo Bunting	*Passerina cyanea*
Kirtland's Warbler	*Dendroica kirtlandii*
Long-tailed Widowbird	*Euplectes progne*
Marsh Wren	*Cistothorus palustris*
Meadow Pipit	*Anthus pratensis*
Montezuma's Oropendola	*Psarocolius montezuma*
Northern Mockingbird	*Mimus polyglottos*
Northern Oriole	*Icterus galbula*
Orchard Oriole	*Icterus spurius*
Pale-eyed Blackbird	*Agelaius xanthophthalmus*
Passenger Pigeon	*Ectopistes migratorius*
Pied Flycatcher	*Ficedula hypoleuca*
Prairie Warbler	*Dendroica discolor*
Raven	*Corvus corax*

Birds–contd.

Red-winged Blackbird	*Agelaius phoeniceus*
Rock Wren	*Salpinctes obsoletus*
Rusty Blackbird	*Euphagus carolinus*
Sharp-tailed Sparrow	*Ammodramus caudacutus*
Song Sparrow	*Melospiza melodia*
Tricolored Blackbird	*Agelaius tricolor*
Unicolored Blackbird	*Agelaius cyanopus*
Virginia Rail	*Rallus limicola*
White-crowned Sparrow	*Zonotrichia atricapilla*
Winter Wren	*Troglodytes troglodytes*
Yellow-headed Blackbird	*Xanthocephalus xanthocephalus*
Yellow-hooded Blackbird	*Agelaius icterocephalus*
Zebra Finch	*Poephila guttata*

Mammals

Domestic Dog	*Canis familiaris*
Fox Squirrel	*Sciurus niger*
Gray Squirrel	*Sciurus carolinensis*
Groundhog	*Marmota monax*
Harvest Mouse	*Reithrodontomys megalotis*
Horse	*Equus caballus*
Long-tailed Weasel	*Mustela frenata*
Mink	*Mustela vison*
Muskrat	*Ondatra zibethica*
Raccoon	*Procyon lotor*
Red Squirrel	*Tamiasciurus hudsonicus*
Rice Rat	*Oryzomys palustris*

Reptiles

Black Ratsnake	*Elaphe obselata*
Black Snake	*Coluber constrictor*
Boa Constrictor	*Boa constrictor*
Cottonmouth	*Agkistrodon piscivorus*
Gopher Snake	*Pituophis melanoleucus*
Indigo Snake	*Drymarchon corais*
King Snake	*Lampropeltis getulus*
Watersnakes	*Natrix* spp.
Western Rattlesnake	*Crotalus viridis*

Insects

Army Worms	*Spodoptera frugiperda*
Caddisfly	Order Trichoptera
Corn Borers	*Ostrinia nubilalis*
Damselflies	Order Odonata, suborder Zygoptera
Dragonflies	Order Odonata, suborder Anisoptera
Gnats	Order Diptera, suborder Nematocera
Midges	Order Diptera, suborder Nematocera
Mosquitoes	Order Diptera, suborder Nematocera
Periodic Cicada	*Magicicada* spp.
Rootworm Beetles	*Diabrotica longicornis*

Plants

Bulrush	*Scirpus* spp.
Bur-reed	*Sparganium eurycarpum*
Cattail	*Typha* spp.
Jack Pine	*Pinus banksiana*
Phragmites	*Phragmites communis*
Ponderosa Pine	*Pinus ponderosa*
Sagebrush	*Artemisia tridentata*

BIBLIOGRAPHY

Alatalo, R. V., Carlson, A., Lundberg, A. and Ulfstrand, S. (1981) The conflict between male polygamy and female monogamy: the case of the Pied Flycatcher *Ficedula hypoleuca*. *American Naturalist* **117:** 738–753.

Albers, P. H. (1978) Habitat selection by breeding Red-winged Blackbirds. *Wilson Bulletin* **90:** 619–634.

Allen, A. A. (1914) The Red-winged Blackbird: A study in the ecology of a cat-tail marsh. *Proceedings of the Linnaean Society of New York* **24–25:** 43–128.

Allen, H. R. (1977) Habitat structure, polygyny, and the evolution of upland nesting in Red-winged Blackbirds (*Agelaius phoeniceus*). Ph.D. thesis, Cornell University, Ithaca, New York.

Altmann, S. A., Wagner, S. F. and Lenington, S. (1977) Two models for the evolution of polygyny. *Behavioral Ecology and Sociobiology* **2:** 397–410.

American Ornithologists' Union (1957) *Check-list of North American Birds*, 6th edn. American Ornithologists' Union, Washington, D.C.

American Ornithologists' Union (1983) *Check-list of North American Birds*, 7th edn. American Ornithologists' Union, Washington, D.C.

Andersson, M. (1982) Female choice selects for extreme tail length in a widowbird. *Nature* **299**: 818–820.

Arcese, P. (1987) Age, intrusion pressure, and defence against floaters by territorial male Song Sparrows. *Animal Behaviour* **35:** 773–784.

Avery, M. L. and Nelms, C. O. (1990) Food avoidance by Red-Winged Blackbirds conditioned with a pyrazine odor. *Auk* **107:** 544–549.

Baker, R. R. (1978) *The Evolutionary Ecology of Animal Migration*. Hodder and Stoughton, London.

Ball, R. M., Freeman, S., James, F. C., Bermingham, E. and Avise, J. C. (1988) Phylogeographic population structure of Red-winged Blackbirds assessed by mitochondrial DNA. *Proceedings of the National Academy of Sciences, USA* **85:** 1558–1562.

Balthazart, J. (1983) Hormonal correlates of behavior. In Farner, D. S., King, J. R. and Parkes, K. C. (eds) *Avian Biology*, vol. 7, pp. 221–365. Academic Press, New York.

Beauchamp, G., Cyr, A. and Houle, C. (1987) Choice behaviour of Red-winged Blackbirds (*Agelaius phoeniceus*) searching for food: The role of certain variables in stay and shift strategies. *Behavioural Processes* **15:** 250–268.

Beecher, M. D. (1989) Signalling systems for individual recognition: An information theory approach. *Animal Behaviour* **38:** 248–261.

Beecher, M. D. and Beecher, I. M. (1979) Sociobiology of Bank Swallows: reproductive strategy of the male. *Science* **205:** 1282–1285.

Beecher, W. J. (1950) Convergent evolution in the American Orioles. *Wilson Bulletin* **62:** 51–86.

Beecher, W. J. (1951) Adaptations for food-getting in the American Blackbirds. *Auk* **68:** 411–440.
Beer, J. R. and Tibbitts, D. (1950) Nesting behavior of the Red-winged Blackbird. *Flicker* **22:** 61–77.
Beletsky, L. D. (1983a) Aggressive and pair-bond maintenance songs of female Red-winged Blackbirds (*Agelaius phoeniceus*). *Zeitschrift für Tierpsychologie* **62:** 47–54.
Beletsky, L. D. (1983b) Vocal mate recognition in male Red-winged Blackbirds. *Behaviour* **84:** 124–134.
Beletsky, L. D. (1983c) An investigation of individual recognition by voice in female Red-winged Blackbirds. *Animal Behaviour* **31:** 355–362.
Beletsky, L. D. (1985) Intersexual song-answering in Red-winged Blackbirds. *Canadian Journal of Zoology* **63:** 735–737.
Beletsky, L. D. (1989a) Communication and the cadence of birdsong. *American Midland Naturalist* **122:** 298–306.
Beletsky, L. D. (1989b) Alert calls of male Red-winged Blackbirds: Do females listen? *Behaviour* **111:** 1–12.
Beletsky, L. D. (1991) Alert calls of male Red-winged Blackbirds: Call rate and function. *Canadian Journal of Zoology* **69:** 2116–2120.
Beletsky, L. D. (1992) Social stability and territory acquisition in birds. *Behaviour* **123:** 290–313.
Beletsky, L. D. and Corral, M. G. (1983a) Song response by female Red-winged Blackbirds to male song. *Wilson Bulletin* **95:** 643–647.
Beletsky, L. D. and Corral, M. G. (1983b) Lack of vocal mate recognition in female Red-winged Blackbirds. *Journal of Field Ornithology* **54:** 200–202.
Beletsky, L. D. and Orians, G. H. (1985) Nest-associated vocalizations of female Red-winged Blackbirds. *Zeitschrift für Tierpsychologie* **69:** 329–339.
Beletsky, L. D. and Orians, G. H. (1987a) Territoriality among male Red-winged Blackbirds. I. Site fidelity and movement patterns. *Behavioral Ecology and Sociobiology* **20:** 21–34.
Beletsky, L. D. and Orians, G. H. (1987b) Territoriality among male Red-winged Blackbirds. II. Site dominance and removal experiments. *Behavioral Ecology and Sociobiology* **20:** 339–349.
Beletsky, L. D. and Orians, G. H. (1989a) Red bands and Red-winged Blackbirds. *Condor* **91:** 993–995.
Beletsky, L. D. and Orians, G. H. (1989b) A male Red-winged Blackbird breeds for 11 years. *Northwestern Naturalist* **70:** 10–12.
Beletsky, L. D. and Orians, G. H. (1989c) Familiar neighbors enhance breeding success in birds. *Proceedings of the National Academy of Sciences USA* **86:** 7933–7936.
Beletsky, L, D. and Orians, G. H. (1989d) Territoriality among male Red-winged Blackbirds. III. Testing hypotheses of territorial dominance. *Behavioral Ecology and Sociobiology* **24:** 333–339.
Beletsky, L. D. and Orians, G. H. (1990) Male parental care in a population of Red-winged Blackbirds, 1983–1988. *Canadian Journal of Zoology* **68:** 606–609.
Beletsky, L. D. and Orians, G. H. (1991) Effects of breeding experience and familiarity on site fidelity in female Red-winged Blackbirds. *Ecology* **72:** 787–796.
Beletsky, L. D. and Orians, G. H. (1993) Factors affecting which male Red-winged Blackbirds acquire territories. *Condor* **95:** 782–791.
Beletsky, L. D. and Orians, G. H. (1996) *Red-winged Blackbirds: Decision-making and Reproductive Success.* University of Chicago Press, Chicago, IL. (In preparation)
Beletsky, L. D., Chao, S. and Smith, D. G. (1980) An investigation of song-based species recognition in the Red-winged Blackbird (*Agelaius phoeniceus*). *Behaviour* **73:** 189–203.
Beletsky, L. D., Higgins, B. and Orians, G. H. (1986) Communication by changing signals: Call-switching in Red-winged Blackbirds. *Behavioral Ecology and Sociobiology* **18:** 221–229.

Beletsky, L. D., Orians, G. H. and Wingfield, J. C. (1989) Relationships of steroid hormones and polygyny to territorial status, breeding experience, and reproductive success in male Red-winged Blackbirds. *Auk* **106:** 107–117.
Beletsky, L. D., Orians, G. H. and Wingfield, J. C. (1990) Effects of exogenous androgen and antiandrogen on territorial and nonterritorial blackbirds. *Ethology* **85:** 58–72.
Beletsky, L. D., Gori, D. F., Freeman, S. and Wingfield, J. C. (1995) Testosterone and polygyny in birds. In Power, D. M. (ed.) *Current Ornithology*, vol. 12, pp. 1–41. Plenum Press, New York, NY.
Bendell, B. E. and Weatherhead, P. J. (1982) Prey characteristics of upland-breeding Red-winged Blackbirds, *Agelaius phoeniceus*. *Canadian Field-Naturalist* **96:** 265–271.
Bendire, C. E. (1895) Life histories of North American birds. *US National Museum Special Bulletin* **3:** 1–518.
Bensch, S. (1993) Costs, benefits and strategies for females in a polygynous mating system: a study on the Great Reed Warbler. Ph.D. thesis, Lund University, Lund, Sweden.
Bensch, S. and Hasselquist, D. (1992) Evidence for active female choice in a polygynous warbler. *Animal Behaviour* **44:** 301–311.
Bent, A. C. (1958) Life histories of North American blackbirds, orioles, tanagers, and allies. *US National Museum Bulletin* **211:** 1–549.
Berger, A. J. (1951) The cowbird and certain host species in Michigan. *Wilson Bulletin* **63:** 26–35.
Bernhardt, G. E., Van Allsburg, L. and Dolbeer, R. A. (1987) Blackbird and starling feeding behavior on ripening corn ears. *Ohio Journal of Science* **87:** 125–129.
Bertram, B. C. R. (1978) Living in groups: Predators and prey. In Krebs, J. R. and Davies, N. B. (eds) *Behavioral Ecology: An Evolutionary Approach*, pp. 64–96. Blackwell Scientific Publications, Oxford, UK.
Besser, J. F., DeGrazio, J. W., Guarino, J. L., Mott, D. G., Otis, D. L., Besser, B. R. and Knittle, C. E. (1984) Decline in breeding Red-winged Blackbirds in the Dakotas, 1965–1981. *Journal of Field Ornithology* **55:** 435–443.
Beyer, W. N., Spann, J. W., Sileo, L. and Franson, J. C. (1988) Lead poisoning in six captive avian species. *Archives of Environmental Contamination and Toxicology* **17:** 121–130.
Birkhead, T. R. and Møller, A. P. (1992) Sperm Competition in Birds: Evolutionary Causes and Consequences. Academic Press, Orlando, FL.
Birks, S. M. and Beletsky, L. D. (1987) Vocalizations of female Red-winged Blackbirds inhibit sexual harassment. *Wilson Bulletin* **99:** 706–707.
Blakley, N. R. (1976) Successive polygyny in upland nesting Red-winged Blackbirds. *Condor* **78:** 129–133.
Blank, J. L. and Nolan (1983) Parental manipulation of phenotypic sex ratio in Red-winged Blackbirds. *Biology of Reproduction* **26** (Suppl. 7): 123A.
Bollinger, E. K. and Caslick, J. W. (1985a) Red-winged Blackbird predation on northern corn rootworm beetles in field corn. *Journal of Applied Ecology* **22:** 39–48.
Bollinger, E. K. and Caslick, J. W. (1985b) Northern corn rootworm beetle densities near a Red-Winged Blackbird roost. *Canadian Journal of Zoology* **63:** 502–505.
Bond, J. (1961) *Birds of the West Indies.* Houghton Mifflin, Boston, MA.
Borror, D. J. (1959) Variation in songs of the Rufous-sided Towhee. *Wilson Bulletin* **71:** 54–72.
Borror, D. J. (1961) Intraspecific variation in passerine bird songs. *Wilson Bulletin* **73:** 57–78.
Bray, O. E., Kennelly, J. J. and Guarino, J. L. (1975) Fertility of eggs produced on territories of vasectomized Red-winged Blackbirds. *Wilson Bulletin* **87:** 187–195.
Brenowitz, E. A. (1981) 'Territorial song' as a flocking signal in Red-winged Blackbirds. *Animal Behaviour* **29:** 641–642.
Brenowitz, E. A. (1982a) The active space of Red-winged Blackbird song. *Journal of Comparative Physiology* **147:** 511–522.

Brenowitz, E. A. (1982b) Long-range communication of species identity by song in the Red-winged Blackbird. *Behavioral Ecology and Sociobiology* **10:** 29–38.
Brenowitz, E. A. (1982c) Aggressive response of Red-winged Blackbirds to mockingbird song imitation. *The Auk* **99:** 584–586.
Brenowitz, E. A. (1983) The contribution of temporal song cues to species recognition in the Red-winged Blackbird. *Animal Behaviour* **31:** 1116–1127.
Brown, C. R. (1986) Cliff Swallow colonies as information centers. *Science* **234:** 83–85.
Brown, J. L. (1975) *The Evolution of Behavior*. W. W. Norton, New York, NY.
Brugger, K. E. (1988) Bird damage to sprouting rice in Louisiana: Dynamics of the Millers Lake blackbird roost. *Proceedings of the Vertebrate Pest Conference* **13:** 281–286.
Brugger, K. E. and Dolbeer, R. A. (1990) Geographic origin of Red-winged Blackbirds relative to rice culture in south-central and southwestern Louisiana. *Journal of Field Ornithology* **61:** 90–97
Brugger, K. E., Labisky, R. F. and Daneke, D. E. (1992) Blackbird roost dynamics at Millers Lake, Louisiana: implications for damage control in rice. *Journal of Wildlife Management* **56:** 393–398.
Bryan, G. G. and Best, L. B. (1991) Bird abundance and species richness in grassed waterways in Iowa rowcrop fields. *American Midland Naturalist* **126:** 90–102.
Burley, N. (1981) Mate choice by multiple criteria in a monogamous species. *American Naturalist* **117:** 515–528.
Burley, N. (1985). Leg-band color and mortality patterns in captive breeding populations of Zebra Finches. *Auk* **102:** 647–651.
Burley, N. (1986) Sexual selection for aesthetic traits in species with biparental care. *American Naturalist* **127:** 415–445.
Burnet, R. and Cyr, A. (1990) Effect of handling stress on Red-winged Blackbirds (*Agelaius phoeniceus*). *Canadian Journal of Zoology* **68:** 1944–1950.
Busnel, R. G. (1968) Acoustic communication. In Sebeok, T. A. (ed.) *Animal Communication*, pp. 127–153. Indiana University Press, Bloomington, IN.
Caccamise, D. F. (1976) Nesting mortality in the Red-winged Blackbird. *Auk* **93:** 517–534.
Caccamise, D. F. (1978) Seasonal patterns of nesting mortality in the Red-winged Blackbird. *Condor* **80:** 290–294.
Caraco, T. (1979) Time budgeting and group size: A test of theory. *Ecology* **60:** 618–627.
Carey, M. and Nolan, V. (1979) Population dynamics of Indigo Buntings and the evolution of avian polygyny. *Evolution* **33:** 1180–1192.
Carroll, S. P. (1993) Divergence in male mating tactics between two populations of the soapberry bug. I. Guarding versus nonguarding. *Behavioral Ecology* **4:** 156–164.
Case, N. A. and Hewitt, O. H. (1963) Nesting and productivity of the Red-winged Blackbird in relation to habitat. *Living Bird* **2:** 7–20.
Catchpole, C. K. (1980) Sexual selection and the evolution of complex songs among European warblers of the genus *Acrocephalus*. *Behavior* **74:** 149–166.
Clark, A. B. and Wilson, D. S. (1981) Avian breeding adaptations, hatching asynchrony, brood reduction and nest failure. *Quarterly Review of Biology* **56:** 253–277.
Clark, D. R., Bagley, F. M. and Johnson, W. W. (1988) Northern Alabama colonies of the endangered grey bat *Myotis grisescens*: Organochlorine contamination and mortality. *Biological Conservation* **43:** 213–225.
Clark, R. G., Weatherhead, P. J., Greenwood, H. and Titman, R. D. (1986) Numerical responses of Red-winged Blackbird populations to changes in regional land-use patterns. *Canadian Journal of Zoology* **64:** 1944–1950.
Connell, C. E., Odum, E. P. and Kale, H. (1960) Fat-free weights of birds. *Auk* **77:** 1–9.
Corral, M. G. (1979) Intrapair communication in Red-winged Blackbirds, *Agelaius phoeniceus*: An analysis of female song. Master's thesis, State University of New York, Stony Brook, NY.

Cox, J. and James, F. C. (1984) Karyotypic uniformity in the Red-winged Blackbird. *Condor* **86:** 416–422.
Crawford, R. D. (1970) Mourning Dove and blackbird production in a Missouri pine planting. *Iowa Bird Life* **40:** 65–67.
Crawford, R. D. (1977) Breeding biology of year-old and older female Red-winged and Yellow-headed Blackbirds. *Wilson Bulletin* **89:** 73–80.
Cristol, D. A. (1995) Early arrival, initiation of nesting, and social status: An experimental study of breeding female Red-winged Blackbirds. *Behavioral Ecology* **6:** 87–93.
Cronmiller, J. R. and Thompson, C. F. (1980) Experimental manipulation of brood size in Red-winged Blackbirds. *Auk* **97:** 559–565.
D'Arms, E. H. H. (1978) Red-winged Blackbird mobbing behavior in the context of changing investment patterns. Master's thesis, University of Washington, Seattle, WA.
Darwin, C. (1859) *The Origin of Species*. John Murray, London, UK.
Darwin, C. (1871) *The Descent of Man and Selection in Relation to Sex*. John Murray, London, UK.
Davies, N. B. (1978) Ecological questions about territorial behaviour. In Krebs, J. R. and Davies, N. B. (eds) *Behavioural Ecology: An Evolutionary Approach*, pp. 317–350. Sinauer, Sunderland, MA.
DeSante, D. F. and George, T. L. (1994) Population trends in the landbirds of western North America. *Studies in Avian Biology* **15:** 173–190.
Dickerman, R. W. (1974) Review of Red-winged Blackbirds (*Agelaius phoeniceus*) of eastern, west-central, and southern Mexico and Central America. *American Museum Novitiates* **2538:** 1–18.
Dickinson, T. E. and Lein, M. R. (1987) Territory dynamics and patterns of female recruitment in Red-winged Blackbirds. *Canadian Journal of Zoology* **65:** 465–471.
Dolbeer, R. A. (1975) A comparison of two methods for estimating bird damage to sunflowers. *Journal of Wildlife Management* **39:** 802–806.
Dolbeer, R. A. (1976) Reproductive rate and temporal spacing of nesting of Red-winged Blackbirds in upland habitat. *Auk* **93:** 343–355.
Dolbeer, R. A. (1978) Movement and migration patterns of Red-winged Blackbirds: A continental overview. *Bird Banding* **49:** 17–34.
Dolbeer, R. A. (1980) Blackbirds and corn in Ohio. *US Fish and Wildlife Service, Resource Publication* **136:** 1–18
Dolbeer, R. A. (1982) Migration patterns for age and sex classes of blackbirds and starlings. *Journal of Field Ornithology* **53:** 28–46.
Dolbeer, R. A. (1990) Ornithology and integrated pest management: Red-winged Blackbirds *Agelaius phoeniceus* and corn. *Ibis* **132:** 309–322.
Dolbeer, R. A. and Ickes, S. K. (1994) Red-winged Blackbird feeding preferences and response to wild rice treated with portland cement or plaster. *Proceedings of the Vertebrate Pest Conference 16*.
Dolbeer, R. A. and Stehn, R. A. (1979) Population trends of blackbirds and starlings in North America, 1966–1976. *US Fish and Wildlife Service Special Scientific Reports – Wildlife* **214:** 1–99.
Dolbeer, R. A. and Stehn, R. A. (1983) Population status of blackbirds and starlings in North America, 1966–1981. *Proceedings of the Eastern Wildlife Damage Control Conference* **1:** 51–61.
Dolbeer, R. A., Woronecki, P. P. and Stehn, R. A. (1984) Blackbird (*Agelaius phoeniceus*) damage to maize: Crop phenology and hybrid resistance. *Protection Ecology* **7:** 43–63.
Dunson, W. A. (1965) Physiological aspects of the onset of molt in the Red-winged Blackbird. *Condor* **67:** 265–269.
Dwight, J. (1900) The sequence of plumages and moults of the passerine birds of New York. *Annals of the New York Academy of Science* **13:** 73–360.
Dyer, M. I. and Ward, P. (1977) Management of pest species. In Pinowski, J. and Kendeigh,

S. C. (eds) *Granivorous Birds in Ecosystems,* pp. 267–300. Cambridge University Press, Cambridge, UK.
Eckert, C. G. and Weatherhead, P. J. (1987a) Male characteristics, parental quality and the study of mate choice in the Red-winged Blackbird (*Agelaius phoeniceus*). *Behavioral Ecology and Sociobiology* **20:** 35–42.
Eckert, C. G. and Weatherhead, P. J. (1987b) Ideal dominance distributions: A test using Red-winged Blackbirds (*Agelaius phoeniceus*). *Behavioral Ecology and Sociobiology* **20:** 43–52.
Eckert, C. G. and Weatherhead, P. J. (1987c) Competition for territories in Red-winged Blackbirds: Is resource-holding potential realized? *Behavioral Ecology and Sociobiology* **20:** 369–375.
Eckert, C. G. and Wetherhead, P. J. (1987d) Owners, floaters and competitive asymmetries among territorial Red-winged Blackbirds. *Animal Behaviour* **35:** 1317–1323.
Emlen, S. T. and Oring, L. W. (1977) Ecology, sexual selection, and the evolution of mating systems. *Science* **197:** 215–223.
Erckmann, W. J., Beletsky, L. D., Orians, G. H., Johnsen, T., Sharbaugh, S. and D'Antonio, C. (1990) Old nests as cues for nest site selection: An experimental test with Red-winged Blackbirds. *Condor* **92:** 113–117.
Evans, D. L. and Waldbauer, G. P. (1982) Behaviour of adult and naive birds when presented with a bumble bee and its mimic. *Zeitschrift für Tierpsychologie* **59:** 247–259.
Ewald, P. W. and Rohwer, S. (1982) Effects of supplemental feeding on timing of breeding, clutch-size and polygyny in Red-winged Blackbirds, *Agelaius phoeniceus. Journal of Animal Ecology* **51:** 429–450.
Facemire, C. F. (1980) Cowbird parasitism of marsh-nesting Red-winged Blackbirds. *Condor* **82:** 347–348.
Fankhauser, D. P. (1971) Annual adult survival rates of blackbirds and starlings. *Bird Banding* **42:** 36–42.
Fiala, K. L. (1981) Reproductive cost and the sex ratio in Red-winged Blackbirds. In Alexander, R. D. and Tinkle, D. W. (eds) *Natural Selection and Social Behavior,* pp. 198–214. Chiron Press, New York.
Fiala, K. L. and Congdon, J. D. (1983) Energetic consequences of sexual size dimorphism in nestling Red-winged Blackbirds. *Ecology* **64:** 642–647.
Francis, W. J. (1975) Clutch size and nesting success in Red-winged Blackbirds. *Auk* **92:** 815–817.
Freeman, S. (1987) Male Red-winged Blackbirds (*Agelaius phoeniceus*) assess the RHP of neighbors by watching contests. *Behavioral Ecology and Sociobiology* **21:** 307–311.
Freeman, S. (1990) Molecular systematics and morphological evolution in the blackbirds. Ph.D. thesis, University of Washington, Seattle, WA.
Fretwell, S. D. and Lucas, H. L. (1970) On territorial behaviour and other factors influencing habitat destruction in birds. *Acta Biotheretica* **19:** 16–36.
Friedmann, H. (1929) *The cowbirds: A Study in the Biology of Social Parasitism.* C. C. Thomas, Springfield, MA.
Friedmann, H., Kiff, L. F. and Rothstein, S. I. (1977) A further contribution to knowledge of the host relations of the parasitic cowbirds. *Smithsonian Contributions to Zoology* **235:** 1–75.
Garrido, O. (1970) Variacion del genero *Agelaius* (Aves: Icteridae) en Cuba. *Poeyana* **68:** 1–18.
Garson, P. J., Pleszczynska, W. K. and Holm, C. H. (1981) The "polygyny threshold" model: A reassessment. *Canadian Journal of Zoology* **59:** 902–910.
Gartshore, R. G., Brooks, R. J., Somers, J. D. and Gilbert, F. F. (1982) Feeding ecology of the Red-winged Blackbird in field corn in Ontario. *Journal of Wildlife Management* **46:** 438–452.
Gauthier, S. and Cyr, A. (1990) Social facilitation in a light-dark cycle and under conditions

of constant luminosity in Red-winged Blackbirds (*Agelaius phoeniceus*). *Canadian Journal of Zoology* **68:** 451–456.

Gavin, T. A., Howard, R. A. and May, B. (1991) Allozyme variation among breeding populations of Red-winged Blackbirds: The California conundrum. *Auk* **108:** 602–611.

Gibbs, H. L., Weatherhead, P. J., Boag, P. T., White, B. N., Tabak, L. M. and Hoysak, D. (1990) Realized reproductive success of polygynous Red-winged Blackbirds revealed by DNA markers. *Science* **250:** 1394–1397.

Goddard, S. V. and Board, V. V. (1967) Reproductive success of Red-winged Blackbirds in north central Oklahoma. *Wilson Bulletin* **79:** 282–289.

Gori, D. F. (1984) The evolution of paternal care patterns and coloniality in Yellow-headed Blackbirds (*Xanthocephalus xanthocephalus*). Ph.D. thesis, University of Arizona, Tucson, AZ.

Gowaty, P. A. and Mock, D. W. (1985) Avian monogamy. *Ornithological Monographs* **37:** 1–121.

Grafen, A. (1988) On the uses of data on lifetime reproductive success. In Clutton-Brock, T. H. (ed.) *Reproductive Success*, pp. 454–471. University of Chicago Press, Chicago, IL.

Gray, E. M. (1994) The ecological and evolutionary significance of extra-pair copulations in the Red-winged Blackbird (*Agelaius phoeniceus*). Ph.D. thesis, University of Washington, Seattle, WA.

Greenwood, H. P., Weatherhead, P. J. and Titman, R. D. (1983) A new age- and sex-specific molt scheme for the Red-winged Blackbird. *Condor* **85:** 104–105.

Greenwood, P. J. and Harvey, P. H. (1982) The natal and breeding dispersal of birds. *Annual Review of Ecology and Systematics* **13:** 1–21.

Haigh, C. R. (1968) Sexual dimorphism, sex ratios, and polygyny in the Red-winged Blackbird. Ph.D. thesis, University of Washington, Seattle, WA.

Hamilton, T. H. (1961) On the functions and causes of sexual dimorphism in the breeding plumage characters of North American species of warblers and orioles. *American Naturalist* **95:** 121–123.

Hamilton, W. D. and Zuk, M. (1982) Heritable true fitness and bright birds: A role for parasites. *Science* **218:** 384–387

Hannon, S. J., Mumme, R. L., Koenig, W. D. and Pitelka, F. A. (1985) Replacement of breeders and within-group conflict in the cooperatively breeding Acorn Woodpecker. *Behavioral Ecology and Sociobiology* **17:** 303–312.

Hansen, A. J. and Rohwer, S. (1986). Coverable badges and resource defence in birds. *Animal Behaviour* **34:** 69–76.

Harding, C. F., Walters, M. J., Collado, D. and Sheridan, K. (1988) Hormonal specificity and activation of social behavior in male Red-winged Blackbirds. *Hormones and Behavior* **22:** 402–418.

Harms, K. E., Beletsky, L. D. and Orians, G. H. (1991) Conspecific nest parasitism in three species of New World Blackbirds. *Condor* **93:** 967–974.

Harris, M. P. (1970) Territory limiting the size of the breeding population of the oystercatcher (*Haematopus ostralegus*) – a removal experiment. *Journal of Animal Ecology* **39:** 707–713.

Hartley, I. R. and Shephard, M. (1994) Female reproductive success, provisioning of nestlings and polygyny in Corn Buntings. *Animal Behaviour* **48:** 717–725.

Hartshorne, C. (1956) The monotony threshold in singing birds. *Auk* **34:** 69–76.

Helms, C. W. (1962) Red-winged Blackbird killing a Sharp-tailed Sparrow. *Wilson Bulletin* **74:** 89–90.

Hergenrader, G. L. (1962) The incidence of nest parasitism by the Brown-headed Cowbird (*Molothrus ater*) on roadside nesting birds in Nebraska. *Auk* **79:** 85–88.

Hill, G. E. (1991) Plumage coloration is a sexually selected indicator of male quality. *Nature* **350:** 337–339.

Hill, K. M. and DeVoogd, T. J. (1991) Altered daylength affects dendritic structure in a song-related brain region in Red-winged Blackbirds. *Behavioral and Neural Biology* **56:** 240–250.

Hintz, J. V. and Dyer, M. I. (1970) Daily rhythm and seasonal change in the summer diet of adult Red-winged Blackbirds. *Journal of Wildlife Management* **34:** 789–799.

Holcomb, L. C. and Twiest, G. (1968) Ecological factors affecting nest-building in Red-winged Blackbirds. *Bird Banding* **39:** 14–22.

Holcomb, L. D. and Twiest, G. (1970) Growth rates and sex ratios of Red-winged Blackbird nestlings. *Wilson Bulletin* **82:** 294–303.

Holm, C. H. (1973) Breeding sex ratios, territoriality, and reproductive success in the Red-winged Blackbird (*Agelaius phoeniceus*). *Ecology* **54:** 356–365.

Horn, H. S. (1968) The adaptive significance of colonial nesting in the Brewer's Blackbird (*Euphagus cyanocephalus*). *Ecology* **49:** 682–694.

Hotker, H. (1989) Meadow Pipit. In Newton, I. (ed.) *Lifetime Reproduction in Birds,* pp. 119–133. Academic Press, New York, NY.

Howe, H. F. (1976) Egg size, hatching asynchrony, sex, and brood reduction in the Common Grackle. *Ecology* **57:** 1195–1207.

Hurly, T. A. and Robertson, R. J. (1984) Aggressive and territorial behaviour in female Red-winged Blackbirds. *Canadian Journal of Zoology* **62:** 148–153.

Irwin, R. E. (1990) Directional sexual selection cannot explain variation in song repertoire size in the new world blackbirds (Icterinae). *Ethology* **85:** 212–224.

Irwin, R. E. (1994) The evolution of plumage dichromatism in the New World Blackbirds: Social selection on female brightness? *American Naturalist* **144:** 890–907.

Jackson, J. J. (1971) Nesting ecology of the female Red-winged Blackbird. Ph.D. thesis, Ohio State University, Columbus, OH.

James, F. C. (1983) Environmental component of morphological differentiation in birds. *Science* **221:** 184–186.

James, F. C., Engstrom, R. T., Nesmith, C. and Laybourne, R. (1984) Inferences about population movements of Red-winged Blackbirds from morphological data. *American Midland Naturalist* **111:** 319–331.

Johnson, D. M., Stewart, G. L., Corley, M., Ghrist, R., Hagner, J., Ketterer, A., McDonnell, B., Newsom, W., Owen, E. and Samuels, P. (1980) Brown-headed Cowbird (*Molothrus ater*) mortality in an urban winter roost. *Auk* **97:** 299–320.

Jugenheimer, R. W. (1976) *Corn: Improvement, Seed Production, and Uses*. John Wiley and Sons, New York.

Kacelnik, A. (1979) Studies of foraging behaviour and time budgeting in Great Tits (*Parus major* L.). Ph.D. thesis, University of Oxford, Oxford, UK.

Kessel, B. and Gibson, D. D. (1978) Status and distribution of Alaska birds. *Studies in Avian Biology* **1:** 1–100.

Ketterson, E. D. and Nolan, V. (1976) Geographic variation and its climatic correlates in the sex ratio of eastern-wintering Dark-eyed Juncos (*Junco hyemalis hyemalis*). *Ecology* **57:** 679–693.

King, J. R. and Farner, D. S. (1961) Energy metabolism, thermoregulation, and body temperature. In Marshall, A. J. (ed.) *Biology and Comparative Physiology of Birds*, vol. 2, pp. 215–288. Academic Press, New York.

Kirkpatrick, M. (1982) Sexual selection and the evolution of female choice. *Evolution* **36:** 1–12.

Klomp, H. (1970) The determination of clutch-size in birds. *Ardea* **58:** 1–124.

Knapton, R. W. and Krebs, J. R. (1974) Settlement patterns, territory size, and breeding density in the Song Sparrow (*Melospiza melodia*). *Canadian Journal of Zoology* **52:** 1413–1419.

Knight, R. L. and Temple, S. A. (1988) Nest-defense behavior in the Red-winged Blackbird. *Condor* **90:** 193–200.

Knight, R. L., Kim, S. and Temple, S. A. (1985) Predation of Red-winged Blackbird nests by mink. *Condor* **87:** 304–305.

Knos, C. J. and Stickley, A. R. (1974) Breeding Red-winged Blackbirds in captivity. *Auk* **91:** 808–816.

Koenig, W. D. and Mumme, R. L. (1987) *Population Ecology of the Cooperatively Breeding Acorn Woodpecker*. Princeton University Press, Princeton, NJ.

Krebs, J. R. (1971) Territory and breeding density in the Great Tit, *Parus major*. *Ecology* **52:** 2–22.

Krebs, J. R. (1977) The significance of song repertoires: The Beau Geste hypothesis. *Animal Behaviour* **25:** 475–478.

Krebs, J. R. and Dawkins, R. (1984) Animal signals: Mind reading and manipulation. In Krebs, J. R and Davies, N. B. (eds) *Behavioral Ecology: An Evolutionary Approach,* pp. 380–402. Sinaur Associates, Sunderland, MA.

Krebs, J. R., Avery, M., and Cowie, R. J. (1981) Effect of removal of mate on the singing behaviour of Great Tits. *Animal Behaviour* **29:** 635–637.

Kroodsma, D. E. (1980) Winter Wren singing behavior: A pinnacle of song complexity. *Condor* **82:** 357–365.

Kroodsma, D. E. and James, F. C. (1994) Song variation within and among populations of Red-winged Blackbirds. *Wilson Bulletin* **106:** 156–162.

Kundert, K. F. (1977) Reproductive success of the Red-winged Blackbird (*Agelaius phoeniceus*) with relation to nesting habitat selection in Goleta, California. Senior thesis, University of California, Santa Barbara, CA.

Lack, D. (1954) *The Natural Regulation of Animal Numbers*. Clarendon Press, Oxford, UK.

Lack, D. (1966) *Population Studies of Birds*. Oxford University Press, Oxford, UK.

Lack, D. (1968) *Ecological Adaptations to Breeding in Birds*. Methuen, London, UK.

Lacomb, D., Cyr, A. and Bergeron, J. M. (1986) Effects of the chemosterilant Ornitrol on the nesting success of Red-winged Blackbirds. *Journal of Applied Ecology* **23:** 773–779.

Lande, R. (1980) Sexual dimorphism, sexual selection, and adaptation in polygenic characters. *Evolution* **34:** 292–305.

Lande, R. (1981) Models of speciation by sexual selection on polygenic characters. *Proceedings of the National Academy of Sciences USA* **78:** 3721–3725.

Langston, N. E., Freeman, S., Rohwer, S., and Gori, D. (1990) The evolution of female body size in Red-winged Blackbirds: the effects of timing of breeding, social competition, and reproductive energetics. *Evolution* **44:** 1764–1779.

Lanyon, S. M. (1994) Polyphyly of the blackbird genus *Agelaius* and the importance of assumptions of monophyly in comparative studies. *Evolution* **48:** 679–693.

LaPrade, H. R. and Graves, H. B. (1982) Polygyny and female–female aggression in Red-winged Blackbirds (*Agelaius phoeniceus*). *American Naturalist* **120:** 135–138.

Laux, L. J. (1970) Nonbreeding surplus and population structure of the Red-winged Blackbird (*Agelaius phoeniceus*). Ph.D. thesis, University of Michigan, Ann Arbor, MI.

Lenington, S. (1977) Evolution of polygyny in Red-winged Blackbirds. Ph. D. thesis, University of Chicago, Chicago, IL.

Lenigton, S. (1980) Female choice and polygyny in Red-winged Blackbirds. *Animal Behaviour* **28:** 347–361.

Lenigton, S. and Scola, R. (1982) Competition between Red-winged Blackbirds and Common Grackles. *Wilson Bulletin* **94:** 90–93.

Lightbody, J. P. and Weatherhead, P. J. (1988) Female settling patterns and polygyny: tests of a neutral-mate choice hypothesis. *American Naturalist* **132:** 20–33.

Linford, J. H. (1935) The life history of the Thick-billed Red-winged Blackbird, *Agelaius phoeniceus fortis* Ridgeway in Utah. Master's thesis, University of Utah, Salt Lake City, UT.

Linz, G. M. and Bolin, S. B. (1982) Incidence of Brown-headed Cowbird parasitism on Red-winged Blackbirds. *Wilson Bulletin* **94:** 93–95.

Linz, G. M., Vakoch, D. L., Cassel, J. F. and Carlson, R. B. (1984) Food of Red-winged Blackbirds, *Agelaius phoeniceus*, in sunflower fields and corn fields. *Canadian Field-Naturalist* **98:** 38–44.

Lorenz, K. Z. (1949) Über die Beziehungen zwischen Kopfform und Zirkelbewegung bei Sturniden und Ikteriden. In Mayr, E. (ed.) *Ornithologie als biologische Wissenschaft.* Carl Winter, Heidelberg, Germany.

Lucas, A. M. and Stettenheim, P. R. (1972) Avian anatomy/integument. *Agricultural Handbook 362.* US Department of Agriculture, Washington, D.C.

Marler, P. and Isaac, D. (1960) Song variation in a population of Brown Towhees. *Condor* **62:** 272–283.

Marler, P. and Tamura, M. (1962) Song 'dialects' in three populations of White-crowned Sparrows. *Condor* **64:** 368–377.

Marler, P., Mundinger, P., Waser, M. S. and Lutjen, A. (1972) Effects of acoustical stimulation and deprivation on song development in Red-winged Blackbirds (*Agelaius phoeniceus*). *Animal Behaviour* **20:** 586–606.

Mason, J. R. and Maruniak, J. A. (1983) Behavioural and physiological effects of capsaicin in Red-winged Blackbirds. *Pharmacology, Biochemistry, and Behaviour* **19:** 857–862.

Mason, J. R. and Reidinger, R. F. (1981) Effects of social facilitation and observational learning on feeding behavior of the Red-winged Blackbird (*Agelaius phoeniceus*). *Auk* **98:** 778–784.

Mason, J. R., Arzt, A. H. and Reidinger, R. F. (1984) Comparative assessment of food preferences and aversions acquired by blackbirds via observational learning. *Auk* **101:** 796–803.

Maynard Smith, J. (1974) The theory of games and the evolution of animal conflicts. *Journal of Theoretical Biology* **47:** 209–221.

Maynard Smith, J. and Parker, G. A. (1976) The logic of asymmetric contests. *Animal Behaviour* **24:** 159–175.

McDonald, M. V. and Greenberg, R. (1991) Nest departure calls in female songbirds. *The Condor* **93:** 365–373.

McGuire, A. D. (1986) Some aspects of the breeding biology of Red-winged Blackbirds in Alaska. *Wilson Bulletin* **98:** 257–266.

McNicol, D. K., Robertson, R. J. and Weatherhead, P. J. (1982) Seasonal, habitat, and sex-specific food habits of Red-winged Blackbirds: Implications for agriculture. *Canadian Journal of Zoology* **60:** 3282–3289.

Meanley, B. (1965) The roosting behavior of the Red-winged Blackbird in the southern United States. *Wilson Bulletin* **77:** 217–228.

Meanley, B. (1971) Blackbirds and the southern rice crop. US Fish and Wildlife Service, Bureau of Sport Fisheries and Wildlife, *Resource Publication* **100:** 1–64

Meanley, B. and Royall, W. C. (1976) Nationwide estimates of blackbirds and starlings. *Proceedings of Bird Control Seminars,* Bowling Green State University, Bowling Green, Ohio **7:** 39–40.

Meanley, B. and Webb, J. S. (1963) Nesting ecology and reproductive rate of the Red-winged Blackbird in the tidal marshes of the Upper Chesapeake Bay region. *Chesapeake Science* **4:** 90–100.

Meanley, B. and Webb, J. S. (1965) Nationwide population estimates of blackbirds and starlings. *Atlantic Naturalist* **20:** 189–191.

Metz, K. J. and Weatherhead, P. J. (1991) Color bands function as secondary sexual traits in male Red-winged Blackbirds. *Behavioral Ecology and Sociobiology* **28:** 23–27.

Metz, K. J. and Weatherhead, P. J. (1993) An experimental test of the contrasting-color hypothesis of red-band effects in Red-winged Blackbirds, *Condor* **95:** 395–400.

Milks, M. L. and Picman, J. (1994) Which characteristics might selection favour as cues of female choice in Red-winged Blackbirds? *Canadian Journal of Zoology* **72:** 1616–1624.

Miskimen, M. (1980) Red-winged blackbirds: I. Age-related epaulet color changes in captive females. *Ohio Journal of Science* **80:** 232–235.

Møller, A. P. (1988) Female choice selects for male sexual tail ornaments in a monogamous swallow. *Nature* **32:** 640–642.

Møller, A. P. (1994) *Sexual Selection and the Barn Swallow*. Oxford University Press, Oxford, UK.

Monnett, C., Rotterman, L. M., Worlein, C. and Halupka, K. (1984) Copulation patterns of Red-winged Blackbirds (*Agelaius phoeniceus*). *American Naturalist* **124:** 757–764.

Moore, W. S. and Dolbeer, R. A. (1989) The use of banding recovery data to estimate dispersal rates and gene flow in avian species: Case studies in the Red-winged Blackbird and Common Grackle. *Condor* **91:** 242–253.

Morris, L. (1975) Effect of blackened epaulets on the territorial behavior and breeding success of male Red-winged Blackbirds, *Agelaius phoeniceus*. *Ohio Journal of Science* **75:** 168–176.

Morse, D. (1970) Territorial and courtship songs of birds. *Nature* **226:** 659–661.

Morton, E. S. (1973) On the evolutionary advantages and disadvantages of fruit eating in tropical birds. *American Naturalist* **107:** 8–22.

Morton, E. S. (1975) Ecological sources of selection on avian sounds. *American Naturalist* **109:** 17–34.

Morton, E. S. (1983) Geographic and structural stability in song dialects of Californian Red-winged Blackbirds: A functional view. Address, 101st Stated Meeting of the American Ornithologists' Union, New York City.

Muldal, A. M., Moffatt, J. D. and Robertson, R. J. (1986) Parental care of nestlings by male Red-winged Blackbirds. *Behavioral Ecology and Sociobiology* **19:** 105–114.

Muma, K. E. and Ankney. C. D. (1987) Variation in weight and composition of Red-winged Blackbird eggs. *Canadian Journal of Zoology* **65:** 605–607.

Muma, K. E. and Weatherhead, P.J. (1989) Male traits expressed in females: Direct or indirect sexual selection? *Behavioral Ecology and Sociobiology* **25:** 23–31.

Muma, K. E. and Weatherhead, P. J. (1991) Plumage variation and dominance in captive female Red-winged Blackbirds. *Canadian Journal of Zoology* **69:** 49–54.

Murton, R. K. (1971) Why do some birds feed in flocks? *Ibis* **113:** 534–536.

Neff, J. A. (1942) Migration of the Tricolored Redwing in central California. *Condor* **44:** 45–53.

Nero, R. W. (1954) Plumage aberrations in the Redwing (*Agelaius phoeniceus*). *Auk* **71:** 137–155.

Nero, R. W. (1956a) A behavior study of the Red-winged Blackbird. I. Mating and nesting activities. *Wilson Bulletin* **68:** 5–37.

Nero, R. W. (1956b) A behavior study of the Red-winged Blackbird. II. Territoriality. *Wilson Bulletin* **68:** 129–150.

Nero, R. W. (1984) *Redwings*. Smithsonian Institute Press, Washington, D.C.

Nero, R. W. and Emlen, J. T. (1951) An experimental study of territorial behavior in breeding Red-winged Blackbirds. *Condor* **53:** 105–116.

Newton, I. (1989) Synthesis. In Newton, I. (ed.) *Lifetime Reproduction in Birds*, pp. 441–469. Academic Press, New York.

Nolan, V. (1978) The ecology and behavior of the Prairie Warbler *Dendroica discolor*. *Ornithological Monographs* **26:** 1–595.

Olson, S. (1985) Weights of some Cuban birds. *Bulletin of the British Ornithological Club* **105:** 68–69.

Orians, G. H. (1961) The ecology of blackbird (*Agelaius*) social systems. *Ecological Monographs* **31:** 285–312.

Orians, G. H. (1966) Food of nestling Yellow-headed Blackbirds, Cariboo Parklands, British Columbia. *Condor* **68:** 321–337.

Orians, G. H. (1969) On the evolution of mating systems in birds and mammals. *American Naturalist* **103:** 589–603.

Orians, G. H. (1972) The adaptive significance of mating systems in the Icteridae. *Proceedings of the XV International Ornithological Congress*, 389–398.

Orians, G. H. (1973) The Red-winged Blackbird in tropical marshes. *Condor* **75:** 28–42.

Orians, G. H. (1980) *Some Adaptations of Marsh-nesting Blackbirds*. Princeton University Press, Princeton, NJ.

Orians, G. H. (1985) *Blackbirds of the Americas*. University of Washington Press, Seattle, WA.

Orians, G. H. and Beletsky, L. D. (1989) Red-winged Blackbird. In Newton, I. (ed) *Lifetime Reproduction in Birds*, pp.183–197. Academic Press, New York, NY.

Orians, G. H. and Christman, G. M. (1968) A comparative study of the behavior of Red-winged, Tricolored, and Yellow-headed Blackbirds. *University of California Publications in Zoology* **84:** 1–81.

Orians, G. H. and Collier, G. (1963) Competition and blackbird social systems. *Evolution* **17:** 449–459.

Orians, G. H. and Horn, H. S. (1969) Overlap in foods and foraging of four species of blackbirds in the Potholes of Central Washington. *Ecology* **50:** 930–938.

Orians, G. H. and Wittenberger, J. F. (1991) Spatial and temporal scales in habitat selection. *American Naturalist* **137:** S29–S49.

Orians, G. H. and Willson, M. F. (1964) Interspecific territories of birds. *Ecology* **45:** 736–745.

Orians, G. H., Røskaft, E. and Beletsky, L. D. (1989) Do Brown-headed Cowbirds distribute their eggs at random in the nests of potential hosts? *Wilson Bulletin* **101:** 599–605.

Ortega, C. P. and Cruz, A. (1988) Mechanisms of egg acceptance by marsh-dwelling blackbirds. *Condor* **90:** 349–358.

Otis, D. L., Knittle, C. E. and Linz, G. M. (1986) A method for estimating turnover in spring blackbird roosts. *Journal of Wildlife Management* **50:** 567–571.

Owen, D. F. (1954) The winter weights of titmice. *Ibis* **96:** 299–309.

Parker, G. A. (1974) Assessment strategy and the evolution of fighting behaviour. *Journal of Theoretical Biology* **47:** 223–243.

Parker, G. A. and Rubinstein, D. I. (1981) Role assessment, reserve strategy and acquisition of information in asymmetric animal conflicts. *Animal Behaviour* **29:** 221–240.

Patterson, C. B. (1979) Relative parental investment in the Red-winged Blackbird. Ph.D. thesis, Indiana University, Bloomington, IN.

Patterson, C. B. (1991) Relative parental investment in the Red-winged Blackbird. *Journal of Field Ornithology* **62:** 1–18.

Patterson, C. B., Erckmann, W. J., and Orians, G. H. (1980) An experimental study of parental investment and polygyny in male blackbirds. *American Naturalist* **116:** 757–769.

Payne, R. B. (1965) The breeding seasons and reproductive physiology of Tricolored Blackbirds and Red-winged Blackbirds. Ph.D. thesis, University of California, Berkeley.

Payne, R. B. (1969) Breeding seasons and reproductive physiology of Tricolored Blackbirds and Red-winged Blackbirds. *University of California Publications in Zoology* **90:** 1–137.

Payne, R. B. (1979) Sexual selection and intersexual differences in variance of breeding success. *American Naturalist* **114:** 447–452.

Payne, R. B. (1984) Sexual selection, lek and arena behavior, and sexual size dimorphism in birds. *Ornithological Monographs* **33:** 1–52.

Peek, F. W. (1971) Seasonal change in the breeding behavior of the male Red-winged Blackbird. *Wilson Bulletin* **83:** 383–395.

Peek, F. W. (1972) An experimental study of the territorial function of vocal and visual display in the male Red-winged Blackbird (*Agelaius phoeniceus*). *Animal Behaviour* **20:** 112–118.

Perkins, S. E. (1928) City park nests of Red-winged Blackbirds. *Bird Lore* **30:** 393–394.

Perrins, C. M. (1979) *British Tits*. New Naturalist, Collins, London, UK.

Picman, J. (1980a) Impact of marsh wrens on reproductive strategy of Red-winged Blackbirds. *Candian Journal of Zoology* **58:** 337–350.

Picman, J. (1980b) Behavioral interactions between Red-winged Blackbirds and Long-billed Marsh Wrens and their role in the evolution of the Redwing polygynous mating system. Ph.D. thesis, University of British Columbia, Vancouver, Canada.

Picman, J. (1981) The adaptive value of polygyny in marsh-nesting Red-winged Blackbirds; renesting, territory tenacity, and mate fidelity of females. *Canadian Journal of Zoology* **59:** 2284–2296.

Picman, J. (1987) Territory establishment, size, and tenacity by male Red-winged Blackbirds. *Auk* **104:** 405–412.

Picman, J., Leonard, M. and Horn, A. (1988) Antipredation role of clumped nesting by marsh-nesting Red-winged Blackbirds. *Behavioral Ecology and Sociobiology* **22:** 9–15.

Picman, J. Milks, M. L. and Leptich, M. (1993) Patterns of predation on passerine nests in marshes: Effects of water depth and distance from edge. *Auk* **110:** 89–94.

Pitelka, F. A. (1958) Timing of molt in Steller Jays of the Queen Charlotte Islands, British Columbia. *Condor* **60:** 38–49.

Power, D. M. (1970) Geographic variation in the surface/volume ratio of the bill of Red-winged Blackbirds in relation to certain geographic and climatic factors. *Condor* **72:** 299–304.

Raikow, R. J. (1978) Appendicular myology and relationships of the New World Nine-primaried Oscines (Aves: Passeriformes). *Bulletin of the Carnegie Museum of Natural History* **7:** 1–43.

Ridgway, R. (1902) The birds of North and Middle America. Part II. *Bulletin of the US National Museum* **50:** 1–834.

Ritland, D. B. (1991) Unpalatability of viceroy butterflies (*Limenitis archippus*) and their purported mimicy models, Florida queens (*Danaus gilippus*). *Oecologia* **88:** 102–108.

Ritschel, S. E. (1985) Breeding ecology of the Red-winged Blackbird (*Agelaius phoeniceus*); tests of models of polygyny. Ph.D. thesis, University of California, Irvine, CA.

Roberts, L. B. and Searcy, W. A. (1988) Dominance relationships in harems of female Red-winged Blackbirds. *Auk* **105:** 89–96.

Roberts, T. A. and Kennelley, J. J. (1980) Variation in promiscuity among Red-winged Blackbirds. *Wilson Bulletin* **92:** 110–112.

Robertson, R. J. (1971) Optimal niche space of the Red-winged Blackbirds. Ph.D. thesis, Yale University, New Haven, CT.

Robertson, R. J. (1972) Optimal niche space of the Red-winged Blackbird (*Agelaius phoeniceus*). I. Nesting success in marsh and upland habitat. *Canadian Journal of Zoology* **50:** 247–263.

Robertson, R. J. (1973a) Optimal niche space of the Red-winged Blackbird. III. Growth rate and food of nestlings in marsh and upland habitat. *Wilson Bulletin* **85:** 209–222.

Robertson, R. J. (1973b) Optimal niche space of the Red-winged Blackbird: Spatial and temporal patterns of nesting activity and success. *Ecology* **54:** 1085–1093.

Robertson, R. J., Weatherhead, P. J., Phelan, F. J. S., Holroyd, G. L. and Lester, N. (1978) On assessing the economic and ecological impact of winter blackbird flocks. *Journal of Wildlife Management* **42:** 53–60

Rogers, J. G. (1974) Responses of caged Red-winged Blackbirds to two types of repellents. *Journal of Wildlife Management* **38:** 418–424.

Rohwer, S. (1978) Passerine subadult plumages and the deceptive acquisition of resources: A test of critical assumption. *Condor* **80:** 173–179.

Rohwer, S. (1982) The evolution of reliable and unreliable badges of fighting ability. *American Zoologist* **22:** 531–546.

Rohwer, S. (1983) Testing the female mimicry hypothesis of delayed plumage maturation: A comment on Procter-Gray and Holmes. *Evolution* **37:** 421–423.

Rohwer, S. (1986) A previously unknown plumage of first-year Indigo Buntings and theories of delayed plumage maturation. *Auk* **103:** 281–292.

Rohwer, S. and Butcher, G. S. (1988) Winter versus summer explanations of delayed plumage maturation in temperate passerine birds. *American Naturalist* **131:** 556–572.

Rohwer, S. and Spaw, C. D. (1988) Evolutionary lag versus bill-size constraints: A comparative study of the acceptance of cowbird eggs by old hosts. *Evolutionary Ecology* **2:** 27–36.

Rohwer, S., Fretwell, S. D. and Niles, D. M. (1980) Delayed maturation in passerine plumages and the deceptive acquisition of resources. *American Naturalist* **115:** 400–437.
Rohwer, S., Spaw, C. D. and Røskaft, E. (1989) Costs to Northern Orioles of puncture-ejecting parasitic cowbird eggs from their nests. *Auk* **106:** 734–738.
Røskaft, E. and Rohwer, S. (1987) An experimental study of the function of the red epaulettes and the black body colour of male Red-winged Blackbirds. *Animal Behaviour* **35:** 1070–1077.
Røskaft, E., Orians, G. H. and Beletsky, L. D. (1990) Why do Red-winged Blackbirds accept eggs of Brown-headed Cowbirds? *Evolutionary Ecology* **4:** 35–42.
Rothstein, S. I. (1975) An experimental and teleonomic investigation of avian brood parasitism. *Condor* **77:** 250–271.
Rothstein, S. I. (1977) Cowbird parisitism and egg recognition of the Northern Oriole. *Wilson Bullletin* **89:** 21–32.
Rothstein, S. I., Verner, J. and Stevens, E. (1980) Range expansion and diurnal changes in dispersion of the Brown-headed Cowbird in the Sierra Nevada. *Auk* **97:** 253–267.
Schafer, E. W., Bowles, W. A. and Hurlbut, J. (1983) The acute oral toxicity, repellency, and hazard potential of 998 chemicals to one or more species of wild and domestic birds. *Archives of Environmental Contamination and Toxicology* **12:** 355–382.
Schleidt, W. M. (1973) Tonic communication: Continual effects of discrete signs in animal communication systems. *Journal of Theoretical Biology* **42:** 359–386.
Searcy, W. A. (1977) The effect of sexual selection on male Red-winged Blackbirds (*Agelaius phoeniceus)*. Ph.D. thesis, University of Washington, Seattle, WA.
Searcy, W. A. (1979a) Size and mortality in male Yellow-headed Blackbirds. *Condor* **81:** 304–305.
Searcy, W. A. (1979b) Male characteristics and pairing success in Red-winged Blackbirds. *Auk* **96:** 353–363.
Searcy, W. A. (1979c) Sexual selection and body size in male Red-winged Blackbirds. *Evolution* **33:** 649–661.
Searcy, W. A. (1979d) Morphological correlates of dominance in captive male Red-winged Blackbirds. *Condor* **81:** 417–420.
Searcy, W. A. (1979e) Female choice of mates: A general model for birds and its application to Red-winged Blackbirds. *American Naturalist* **114:** 77–100.
Searcy, W. A. (1986a) Dual intersexual and intrasexual functions of song in Red-winged Blackbirds. *Acta XIX Congresus Internationalis Ornithologici*, **1:** 1373–1381.
Searcy, W. A. (1986b) Are female Red-winged Blackbirds territorial? *Animal Behaviour* **34:** 1381–1391.
Searcy, W. A. (1988) Do female Red-winged Blackbirds limit their own breeding densities? *Ecology* **69:** 85–95.
Searcy, W. A. (1989) Function of male courtship vocalizations in Red-winged Blackbirds. *Behavioral Ecology and Sociobiology* **24:** 325–331.
Searcy, W. A. and Brenowitz, E. A. (1988) Sexual difference in species recognition of avian song. *Nature* **332:** 152–154.
Searcy, W. A. and Yasukawa, K. (1981) Sexual size dimorphism and survival of male and female blackbirds (Icteridae). *Auk* **98:** 457–465.
Searcy, W. A. and Yasukawa, K. (1983) Sexual Selection and Red-winged Blackbirds. American Scientist **71:** 166–174.
Searcy, W. A. and Yasukawa, K. (1989) Alternative models of territorial polygyny in birds. *American Naturalist* **134:** 323–343.
Searcy, W. A. and Yasukawa, K. (1990) Use of the song repertoire in intersexual and intrasexual contexts by male Red-winged Blackbirds. *Behavioral Ecology and Sociobiology* **27:** 123–128.
Searcy, W. A. and Yasukawa, K. (1995) *Polygyny and Sexual Selection in Red-winged Blackbirds*. Princeton University Press, Princeton, NJ.

Selander, R. K. (1958) Age determination and molt in the Boat-tailed Grackle. *Condor* **60:** 353–376.
Selander, R. K. (1965) On mating systems and sexual selection. *American Naturalist* **69:** 129–141.
Selander, R. K. (1966) Sexual dimorphism and differential niche utilization in birds. *Condor* **68:** 113–151.
Selander, R. K. (1972) Sexual selection and dimorphism in birds. In Campbell, B. (ed.) *Sexual Selection and the Descent of Man*, pp. 180–230. Aldine, Chicago, IL.
Selander, R. K. and Giller, D. R. (1960) First year plumages of the Brown-headed Cowbird and Red-winged Blackbirds. *Condor* **62:** 202–214.
Shelley, L. O. (1930) Companionate feeding activities of a Spotted Sandpiper and a Red-winged Blackbird. *Auk* **47:** 78–79.
Shepard, P. E. K. (1962) A breeding record of the Red-winged Blackbird in Alaska. *Condor* **64:** 440.
Sherman, P. W. (1989) Mate-guarding as paternity insurance in Idaho ground squirrels. *Nature* **338:** 418–420.
Shutler, D. and Weatherhead, P. J. (1991a) Owner and floater Red-winged Blackbirds: Determinants of status. *Behavioral Ecology and Sociobiology* **28:** 235–241.
Shutler, D. and Weatherhead, P. J. (1991b) Basal song rate variation in male Red-winged Blackbirds: Sound and fury signifying nothing? *Behavioral Ecology* **2:** 123–132.
Shutler, D. and Weatherhead, P. J. (1992) Surplus territory contenders in male Red-winged Blackbirds: Where are the desperados? *Behavioral Ecology and Sociobiology* **31:** 97–106.
Shutler, D. and Weatherhead, P. J. (1994) Movement patterns and territory acquisition by male Red-winged Blackbirds. *Canadian Journal of Zoology* **72:** 712–720.
Sibley, C. G., Ahlquist, J. E. and Monroe, B. L. (1988) A classification of the living birds of the World based on DNA-DNA hybridization studies. *Auk* **105:** 409–423.
Sibley, C. G. and Ahlquist, J. E. (1990) *Phylogeny and Classification of Birds*. Yale University Press, New Haven, CT.
Simmers, R. W. (1975) Variation in the vocalizations of male Red-winged Blackbirds (*Agelaius phoenicieus*) in New England and New York. Ph.D. thesis, Cornell University, Ithaca, NY.
Smith, D. G. (1972) The role of the epaulets in the Red-winged Blackbird social system (*Agelaius phoenceus*). *Behaviour* **41:** 251–268.
Smith, D. G. (1976) An experimental analysis of the function of Red-winged Blackbird song. *Behaviour* **56:** 136–156.
Smith, D. G. (1979) Male singing ability and territory integrity in Red-winged Blackbirds. (*Agelaius phoenceus*). *Behaviour* **68:** 193–206.
Smith, D. G. and Norman, D. O. (1979) 'Leader-follower' singing in Red-winged Blackbirds. *Condor* **81:** 83–84.
Smith, D. G. and Reid, F. A. (1979) Roles of the song repertoire in Red-winged Blackbirds. *Behavioral Ecology and Sociobiology* **5:** 279–290.
Smith, H. G. and Montgomerie, R. D. (1991) Nestling American Robins compete with siblings by begging. *Behavioral Ecology and Sociobiology* **29:** 307–312.
Smith, H. M. (1943) Size of breeding populations in relation to egg-laying and reproductive success in the eastern Red-wing (*Agelaius p. phoeniceus*). *Ecology* **24:**183–207.
Smith, J. K. and Zimmerman, E. G. (1976) Biochemical genetics and evolution of North American blackbirds, Family Icteridae. *Comparative Biochemisty and Physiology* **53B:** 319–324.
Smith, S. M. (1978) The 'underworld' in a territorial sparrow: Adaptive strategy for floaters. *American Naturalist* **112:** 571–582.
Smith, S. M. (1987) Responses of floaters to removal experiments on wintering chicadees. *Behavioral Ecology and Sociobiology* **20:** 363–367.

Smith, W. J. (1991) Singing is based on two markedly different kinds of signaling. *Journal of Theoretical Biology* **152:** 241–253.
Snelling, J. C. (1968) Overlap in feeding habits of Red-winged Blackbirds and Common Grackles resting in a cattail marsh. *Auk* **85:** 560–585.
Stacey, P. B. and Ligon, J. D. (1987) Territory quality and dispersal options in the Acorn Woodpecker, and a challenge to the habitat saturation model of cooperative breeding. *American Naturalist* **130:** 654–676.
Steward, V. B., Smith, K. G. and Stephen, F. M. (1988) Red-winged Blackbird predation on periodical cicadas (Cicadidae: *Magicicada* spp.): Bird behavior and cicada responses. *Oecologia* **76:** 348–352.
Stewart, P. A. (1978) Survival tables for starlings, Red-winged Blackbirds, and Common Grackles. *North American Bird Bander* **3:** 93–94.
Stickel, W. H., Stickel, L. F, Dyrland, R. A. and Hughes, D. L. (1984a) Aroclor 1254 residues in birds: Lethal levels and loss rates. *Archives of Environmental Contamination and Toxicology* **13:** 7–13.
Stickel, W. H., Sticket, L. F., Dyrland, R. A. and Hughes, D. L. (1984b) DDE in birds: Lethal residues and loss rates. *Archives of Environmental Contamination and Toxicology* **13:** 1–6.
Stickley, A. R. and Ingram, C. R. (1977) Methiocarb as a bird repellent for mature sweet corn. *Proceedings Bird Control Seminar* **7:** 228–238.
Stone, W. B., Overmann, S. R. and Okoniewski, J. C. (1984) Intentional poisoning of birds with parathion. *Condor* **86:** 333–336.
Strehl, C. (1978) Asynchrony of hatching in Red-winged Blackbirds and survival of late and early hatching birds. *Wilson Bulletin* **90:** 653–655.
Strehl, C. E. and White, J. (1986) Effects of super-abundant food on breeding success and behavior of the Red-winged Blackbird. *Oecologia* **70:** 178–186.
Teather, K. L. (1992) An experimental study of competition for food between male and female nestlings of the Red-winged Blackbird. *Behavioral Ecology and Sociobiology* **31:** 81–87.
Teather, K. L, Muma, K. E. and Weatherhead, P. J. (1988) Estimating female settlement from nesting data. *Auk* **105:** 196–200.
Thorpe, W. H. (1958) The learning of song patterns of birds, with special reference to the song of the Chaffinch, *Fringilla Coelebs*. *Ibis* **110:** 535–570.
Thorpe, W. H. (1961) *Bird song: The Biology of Vocal Communication and Expression in Birds*. Cambridge University Press, Cambridge, UK.
Trivers, R. (1972) Parental investment and sexual selection. In Campbell, B. (ed.) *Sexual Selection of the Descent of Man*, pp. 136–179. Aldine, Chicago, IL.
Verner, J. (1964) Evolution of polygamy in the Long-billed Marsh Wren. *Evolution* **18:** 252–261.
Verner, J. and Ritter, L. V. (1983) Current status of the Brown-headed Cowbird in the Sierra National Forest. *Auk* **100:** 355–368.
Verner, J. and Willson, M. F. (1966) The influence of habitats on mating systems of North American passerine birds. *Ecology* **47:** 143–147.
Verner, J. and Willson, M. F. (1969) Mating systems, sexual dimorphism, and the role of male North American passerine birds in the nesting cycle. *Ornithological Monographs* **9:** 1–76.
Voights, D. K. (1973) Food overlap of two Iowa marsh icterids. *Condor* **75:** 392–399.
Walters, M. J. and Harding, C. F. (1988) The effects of an aromatization inhibitor in the reproductive behavior of male zebra finches. *Hormones and Behavior* **22:** 207–218.
Ward, P. and Zahavi, A. (1973) The importance of certain assemblages of birds as 'information centres' for finding food. *Ibis* **115:** 517–534.
Wasserman, F. E. (1977) Mate attraction function of song in the White-throated Sparrow. *Condor* **79:** 125–127.
Weatherhead, P. J. (1981) The dynamics of Red-winged Blackbird populations at four late summer roosts in Quebec. *Journal of Field Ornithology* **52:** 22–27.

Weatherhead, P. J. (1984) Mate choice in avian polygyny: Why do females prefer older males? *American Naturalist* **123:** 873–875.
Weatherhead, P. J. (1985) The birds' communal connection. *Natural History* **94:** 34–41.
Weatherhead, P. J. (1989) Sex ratios, host-specific reproductive success, and impact of Brown-headed Cowbirds. *Auk* **106:** 358–366.
Weatherhead, P. J. (1990a) Nest defence as shareable paternal care in Red-winged Blackbirds. *Animal Behaviour* **39:** 1173–1178.
Weatherhead, P. J. (1990b) Secondary sexual traits, parasites, and polygyny in Red-winged Blackbirds, *Agelaius phoeniceus*. *Behavioral Ecology* **1:** 125–130.
Weatherhead, P. J. (1995) Effects on female reproductive success of familiarity and experience among male Red-winged Blackbirds. *Animal Behaviour* **49:** 967–976.
Weatherhead, P. J. and Bider, J. R. (1979) Management options for blackbird problems in agriculture. *Phytoprotection* **60:** 145–155.
Weatherhead, P. J. and Boag, P. T. (1995) Pair and extra-pair mating success relative to make quality in Red-winged Blackbirds. *Behavioral Ecology and Sociobiology* **37:** 81–91.
Weatherhead, P. J and Hoysak, D. J. (1984) Dominance structuring of a Red-winged Blackbird roost. *Auk* **101:** 551–555.
Weatherhead, P. J. and Robertson, R. J. (1977a) Harem size, territory quality, and reproductive success in the Red-winged Blackbird (*Agelaius phoeniceus*). *Canadian Journal of Zoology* **55:** 1261–1267.
Weatherhead, P. J. and Robertson, R. J. (1977b) Male behavior and female recruitment in the Red-winged Blackbird. *Wilson Bulletin* **89:** 583–592.
Weatherhead, P. J. and Robertson, R. J. (1979) Offspring quality and the polygyny threshold: "The sexy son hypothesis". *American Naturalist* **113:** 201–208.
Weatherhead, P. J., Bendell, B. E. and Stewart, R. K. (1981) The impact of predation by Red-winged Blackbirds on European corn borer populations. *Canadian Journal of Zoology* **59:** 1535–1538.
Weatherhead, P. J., Greenwood, H. and Clark, R. G. (1987) Natural selection and sexual selection on body size in Red-winged Blackbirds. *Evolution* **41:** 1401–1403.
Weatherhead, P. J., Hoysak, D. J., Metz, K. J. and Eckert, C. G. (1991) A retrospective analysis of red-band effects on Red-winged Blackbirds. *Condor* **93:** 1013–1016.
Weatherhead, P. J., Metz, K. J., Bennett, G. F. and Irwin, R. E. (1993) Parasite faunas, testosterone, and secondary sexual traits in male Red-winged Blackbirds. *Behavioral Ecology and Sociobiology* **33:** 13–23.
Weatherhead, P. J., Montgomerie, R., Gibbs, H. L. and Boag, P. T. (1994) The cost of extra-pair fertilizations to female Red-winged Blackbirds. *Proceedings of the Royal Society of London Series* B **258:** 315–320.
Webster, M. S. (1991) The dynamics and consequences of intrasexual competition in the Montezuma Oropendola: Harem-polygyny in a Neotropical bird. Ph.D. thesis, Cornell University, Ithaca, NY.
Webster, M. S. (1994) Female defence polygyny in a Neotropical bird, the Montezuma Oropendola. *Animal Behaviour* **48:** 779–794.
Weeden, J. S. and Falls, J. B. (1959) Differential responses of male Ovenbirds to recorded songs of neighboring and more distant individuals. *Auk* **76:** 343–351.
Westneat, D. F. (1990) Genetic parentage in the Indigo Bunting: A study using DNA fingerprinting. *Behavioral Ecology and Sociobiology* **27:** 67–76.
Westneat, D. F. (1992a) Do female Red-winged Blackbirds engage in a mixed mating strategy? *Ethology* **92:** 7–28.
Westneat, D. F. (1992b) Nesting synchrony by female Red-winged Blackbirds: Effects on predation and breeding success. *Ecology* **73:** 2284–2294.
Westneat, D. F. (1993) Polygyny and extra-pair fertilizations in eastern Red-winged Blackbirds (*Agelaius phoeniceus*). *Behavioral Ecology* **4:** 49–60.

Westneat, D. F. (1994) To guard mates or go forage: Conflicting demands affect the paternity of male Red-winged Blackbirds. *American Naturalist* **144:** 343–354.

Westneat, D. F. (1995). Paternity and paternal behaviour in the Red-winged Blackbird, *Agelaius phoeniceus. Animal Behaviour* **49:** 21–35.

Westneat, D. F., Sherman, P. W. and Morton, M. L. (1990) The ecology and evolution of extra-pair copulations in birds. In Power, D. M. (ed.) *Current Ornithology*, vol. 7, pp. 331–369. Plenum Press, New York, NY.

White, S. B., Dolbeer, R. A. and Bookhout, T. A. (1985) Ecology, bioenergetics, and agricultural impacts of a winter-roosting population of blackbirds and starlings. *Wildlife Monographs* **93:** 1–42.

Whittingham, L. A. (1989) An experimental study of paternal behavior in Red-winged Blackbirds. *Behavioral Ecology and Sociobiology* **25:** 73–80.

Whittingham, L. A. and Robertson, R. J. (1993) Nestling hunger and parental care in Red-winged Blackbirds. *Auk* **110:** 240–246.

Whittingham, L. A., Kirkconnell, A. and Ratcliffe, L. M. (1992) Differences in song and sexual dimorphism betwen Cuban and North American Red-winged Blackbirds (*Agelaius phoeniceus*). *Auk* **109:** 928–933.

Wiens, J. A. (1963) Aspects of cowbird parasitism in southern Oklahoma. *Wilson Bulletin* **75:** 130–138.

Wiens, J. A. (1965) Behavioral interactions of Red-winged Blackbirds and Common Grackles on a common breeding ground. *Auk* **82:** 356–374.

Wiens, J. A. and Dyer, M. I. (1975) Simulation modelling of Red-winged Blackbird impact on grain crops. *Journal of Applied Ecology* **12:** 63–82.

Wiley, R. H. and Hartnett, S. A. (1976) Effects of interactions with older males on behavior and reproductive development in first-year male Red-winged Blackbirds *Agelaius phoeniceus. Journal of Experimental Zoology* **196:** 231–242.

Williams, J. F. (1940) The sex ratio of nestling Eastern Red-wings. *Wilson Bulletin* **52:** 267–277.

Willson, M. F. (1966) Breeding ecology of the Yellow-headed Blackbird. *Ecological Monographs* **36:** 51–77.

Willson, M. F. and Orians, G. H. (1963) Comparative ecology of Red-winged and Yellow-headed Blackbirds during the breeding season. *Proceedings of the XVI International Congress of Zoology* **3:** 342–346.

Wilson, E. O. (1975) *Sociobiology*. Harvard University Press, Cambridge, MA.

Wilson, S. W. (1978) Food size, food type and foraging sites of Red-winged Blackbirds. *Wilson Bulletin* **90:** 511–520.

Wimberger, P. H. (1988) Food supplement effects on breeding time and harem size in the Red-winged Blackbird (*Agelaius phoeniceus*). *Auk* **105:** 799–802.

Wingfield, J. C., Ball, G. F., Duffy, A. M., Hegner, R. E. and Ramenofsky, M. (1987) Testosterone and aggression in birds. *American Scientist* **75:** 602–608.

Wingfield, J. C., Hegner, R. E., Dufty, A. M. and Ball, G. F. (1990) The 'challenge hypothesis': Theoretical implications for patterns of testosterone secretion, mating systems, and breeding strategies. *American Naturalist* **136:** 829–846.

Wittenberger, J. F. (1976) The ecological factors selecting for polygyny in altricial birds. *American Naturalist* **110:** 779–799.

Wittenberger, J. F. and Hunt, G. L. (1985) The adaptive significance of coloniality in birds. In Farner, D. S., King, J. R. and Parkes, K. C. (eds) *Avian Biology*, vol. 8, pp. 1–78. Academic Press, Orlando, FL.

Wood, H. B. (1938) Nesting of Red-winged Blackbirds. *Wilson Bulletin* **50:** 143–144.

Woronecki, P. P., Dolbeer, R. A. and Stehn, R. A. (1981) Response of blackbirds to mesurol and sevin applications on sweet corn. *Journal of Wildlife Management* **45:** 693–701.

Wright, M. O. (1907) The Red-winged Blackbird. *Bird-Lore* **9:** 93–96.

Wright, P. L. and Wright, M. H. (1944) The reproductive cycle of the male Red-winged Blackbird. *Condor* **46:** 46–59.
Yasukawa, K. (1978) Aggressive tendencies and levels of a graded display: Factor analysis of response to song playback in the Red-winged Blackbird (*Agelaius phoeniceus*). *Behavioral Biology* **23:** 446–459.
Yasukawa, K. (1979) Territory establishment in Red-winged Blackbirds: Importance of aggressive behavior and experience. *Condor* **81:** 258–264.
Yasukawa, K. (1981a) Song and territory defense in the Red-winged Blackbird. *Auk* **98:** 185–187.
Yasukawa, K. (1981b) Song repertoires in the Red-winged Blackbird (*Agelaius phoeniceus*): A test of the Beau Geste hypothesis. *Animal Behaviour* **29:** 114–125.
Yasukawa, K. (1981c) Male quality and female choice of mate in the Red-winged Blackbird (*Agelaius phoeniceus*). *Ecology* **62:** 922–929.
Yasukawa, K. (1987) Breeding and nonbreeding season mortality of territorial male Red-winged Blackbirds. *Auk* **104:** 56–62.
Yasukawa, K. (1989) The costs and benefits of a vocal signal: The nest-associated "Chit" of the female Red-winged Blackbird, *Agelaius phoeniceus*. *Animal Behaviour* **38:** 866–874.
Yasukawa, K. (1990) Does the 'Teer' vocalization deter prospecting female Red-winged Blackbirds? *Behavioral Ecology and Sociobiology* **26:** 421–426.
Yasukawa, K. and Searcy, W. A. (1981) Nesting synchrony and dispersion in Red-winged Blackbirds: Is the harem competitive or cooperative? *Auk* **98:** 659–668.
Yasukawa, K. and Searcy, W. A. (1982) Aggression in female Red-winged Blackbirds: A strategy to ensure male parental investment. *Behavioral Ecology and Sociobiology* **11:** 13–17.
Yasukawa, K. and Searcy, W. A. (1985) Song repertoires and density assessment in Red-winged Blackbirds: Further tests of the Beau Geste hypothesis. *Behavioral Ecology and Sociobiology* **16:** 171–175.
Yasukawa, K., Blank, J. L. and Patterson, C. B. (1980) Song repertoires and sexual selection in the Red-winged Blackbird. *Behavioral Ecology and Sociobiology* **7:** 233–238.
Yasukawa, K., Bick, E. I., Wagman, D. W. and Marler, P. (1982) Playback and speaker-replacement experiments on song-based neighbor, stranger, and self discrimination in male Red-winged Blackbirds. *Behavioral Ecology and Sociobiology* **10:** 211–215.
Yasukawa, K., Boley, R. A. and Simon, S. E. (1987a) Seasonal change in the vocal behavior of female Red-winged Blackbirds, *Agelaius phoeniceus*. *Animal Behaviour* **35:** 1416–1423.
Yasukawa, K., Knight, R. L., and Skagen, S. K. (1987b) Is courtship intensity a signal of male parental care in Red-winged Blackbirds (*Agelaius phoeniceus*)? *Auk* **104:** 628–634.
Yasukawa, K., McClure, J. L., Boley, R. A. and Zanocco, J. (1990) Provisioning of nestlings by male and female Red-winged Blackbirds, *Agelaius phoeniceus*. *Animal Behaviour* **40:** 153–166.
Yasukawa, K., Whittenberger, L. K. and Nielsen, T. A. (1992a) Anti-predator vigilance in the Red-winged Blackbird (*Agelaius phoeniceus*): Do males act as sentinels? *Animal Behaviour* **43:** 961–969.
Yasukawa, K., Boley, R. A., McClure, J. L. and Zanocco, J. (1992b) Nest dispersion in the Red-winged Blackbird. *Condor* **94:** 775–777.
Yasukawa, K., Leanza, F. and King, C. D. (1993) An observational and brood-exchange study of paternal provisioning in the Red-winged Blackbird, *Agelaius phoeniceus*. *Behavioral Ecology* **4:** 78–82.
Young, H. (1963) Age-specific mortality in the eggs and nestlings of blackbirds. *Auk* **80:** 145–155.
Zahavi, A. (1975) Mate selection – a selection for a handicap. *Journal of Theoretical Biology* **53:** 205–214.

INDEX